Biological Systems

Papers from SCIENCE, *1988-1989*

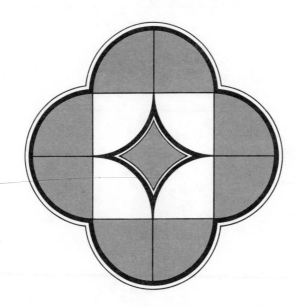

Edited by Barbara R. Jasny and Daniel E. Koshland, Jr.

AMERICAN ASSOCIATION FOR THE ADVANCEMENT OF SCIENCE

Library of Congress Cataloging-in-Publication Data

Jasny, Barbara and Koshland, Daniel E., Jr.
    Biological systems: papers from Science, 1988–1989 / Barbara Jasny and Daniel E. Koshland, Jr.
        p.   cm.
    Includes bibliographic references.
    ISBN 0-87168-351-2
    1. Biological systems.   2. Biology—Research—Methodology
I. Jasny, Barbara.   II. Koshland, Daniel.   III. Science (Weekly).

QH324.B474   1989
574'.01'1—dc20   89-18005
                                                                                  CIP

Publication No. 89-16S

© 1990 by the American Association for the Advancement of Science
1333 H Street, N.W., Washington, D.C. 20005

Biological Systems

Papers from Science, *1988-1989*

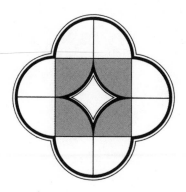

Contents

Preface vii
    Daniel E. Koshland, Jr.

Biological Systems: The Choice Is Yours ix
    Barbara R. Jasny

1. Retroviruses 1
    Harold Varmus

2. Research on Bacteria in the Mainstream of Biology 17
    Boris Magasanik

3. Genetic Engineering of Bacteria from Managed and Natural Habitats 25
    Steven E. Lindow, Nickolas J. Panopoulos, Beverly L. McFarland

4. Yeast: An Experimental Organism for Modern Biology 39
    David Botstein and Gerald R. Fink

5. *Dictyostelium discoideum*: A Model System for Cell-Cell Interactions
   in Development 47
    Peter Devreotes

6. *Xenopus laevis* in Developmental and Molecular Biology 57
    Igor B. Dawid and Thomas D. Sargent

7. The Nematode *Caenorhabditis elegans* 67
    Cynthia Kenyon

8. Parasitic Protozoans and Helminths: Biological and
   Immunological Challenges 77
    Adel A. F. Mahmoud

9. *Drosophila melanogaster* as an Experimental Organism 91
    Gerald M. Rubin

10. Plants: Novel Developmental Processes 105
    Robert B. Goldberg

11. Genetically Engineered Plants for Crop Improvement 121
    Charles S. Gasser and Robert T. Fraley

12. Fish as Model Systems 135
    Dennis A. Powers

13. Contributions of Bird Studies to Biology 147
    Masakazu Konishi, Stephen T. Emlin, Robert E. Ricklefs, John C. Wingfield

14. The Rat as an Experimental Animal 163
    Thomas J. Gill III, Garry J. Smith, Robert W. Wissler, Heinz W. Kunz

15. Transgenic Animals 179
    Rudolf Jaenisch

16. Cetaceans 193
    Bernd Würsig

17. Primates 209
    Frederick A. King, Cathy J. Yarbrough, Daniel C. Anderson,
    Thomas P. Gordon, Kenneth G. Gould

18. The Human as an Experimental System in Molecular Genetics 225
    Ray White and C. Thomas Caskey

19. Fetal Research 235
    John R. Hansen and John T. Sladek, Jr.

20. How Many Species Are There on Earth? 245
    Robert M. May

Index 261

Preface

It could be argued that modern biology acts as if the proper study of mankind is the fly, the frog, the bacterium, and the nematode. If those seem too simple, then the rat and the ape are elevated to stand in for the human. Historically, a variety of biological species have been studied for their own intrinsic interest and as a way of understanding humans, because research on humans themselves is illegal, costly, or far too complex.

Despite the fact that millions of species exist in the world today, modern research has concentrated on only a few. The reason for this concentration is quite clearcut: there is a certain amount of spadework that must be done on any system before productive research can begin. This includes understanding the basic metabolism, finding the appropriate nutritional conditions for the organism, and eliminating the special toxic elements that prevent good growth. The first scientists to approach such a system always have to work through these difficult problems, so switching from one type of rat that has been thoroughly studied to a second type of rat means that these housekeeping parameters will have to be determined again. The old world, in which it was a courtesy not to work on the same species as one's colleague, has largely disappeared. The result is a major advance in the efficiency of modern science.

Many species are still interesting per se, but others are studied because they cause illness, because they are needed to feed human populations, or because they are good models for general phenomena. The vast majority of these systems have been studied for their relevance to the health of humans. Therefore, findings such as the universality of the genetic code support an assumption of early biologists, that all these so-called lower species are appropriate organisms for understanding more about humans. This underlying similarity is fortunate, because simpler species such as the bacterium and the fly offer beautifully stripped down, Matisse-like analogs for the Constable-like organism known as *Homo sapiens*.

Of course, humans cannot be left out of the picture. The ultimate verification that other species can be models for humans is to show that the lessons learned from lower species are applicable to humans. Moreover, there are some features of humans, such as their ability to tell the investigator how they feel, which cannot be reproduced in lower species. The human genome project now underway will, together with more sophisticated and noninvasive techniques, make the human an ever more accessible experimental animal. Humans are the ultimate experimental resource—their hemoglobins have been isolated in hospitals all over the world and they have provided a wealth of treasured family histories.

Assembling these major experimental systems in one volume was designed partly to help graduate students, assistant professors, and others entering a new field to choose an experimental system that is appropriate to their research. This book can also be useful to scientists who have run up against roadblocks or limitations in the systems they are currently using. Additionally, it may be of value to investigators who do not plan to utilize a different system, but who would like to know its advantages and limitations so that they can properly evaluate the scientific literature. Finally, and most important of all, this collection includes systems that are fascinating in themselves, presenting scientific puzzles and challenges that make them stimulating reading.

Daniel E. Koshland, Jr.
Science
December 1989

Biological Systems: The Choice Is Yours

We have been studying the biological systems around us since the earliest humans walked the earth. The cave paintings found at Lasceaux in France are only one example of the attention that was being paid, more than 15,500 years ago, to animals such as wild ox, horses, bison, deer, and ibex (1). Observation and study of other species were essential for human survival. Human manipulation of plants can be traced back to the transition between the Paleolithic and Neolithic periods—the development of agriculture. The conceptual underpinnings for modern studies of plant development (Goldberg, Chapter 10) and the genetic engineering of agricultural plants (Gasser and Fraley, Chapter 11) can be found in the experiments of the early plant breeders. Thus, the acquisition and improvement of our food supply can be seen as one of the earliest driving forces behind biological research.

Another compelling reason for studying biological systems has been so that we can better understand the rules governing all living things, ourselves included. Aristotle's work "De Partibus Animalium" illustrates that even in ancient times there was an appreciation of the underlying similarities in developmental and physiological processes (2). He could not have imagined how much we have learned from some of the most primitive organisms. For example, as discussed by Varmus (Chapter 1), retroviral genes have their homologs in the human proto-oncogenes and studies of these genes have revolutionized our understanding of the etiology of cancer. In addition, retroviruses are now recognized as devastating animal and human pathogens. Basic research in bacteria (Magasanik, Chapter 2) and yeast (Botstein and Fink, Chapter 4) is shedding light on such ubiquitous processes as the regulation of gene expression and the control of cell division.

A major emphasis in this book is on use of various biological systems to investigate questions relating to development. Fundamental principles of early development and differentiation can be learned from slime mold (Devreotes, Chapter 5) and from frog oocytes (Dawid and Sargent, Chapter 6). Certain systems, such as the nematode, are particularly suited to the elaboration of "fate maps," in which the lineage of individual cells can be followed with exquisite precision (Kenyon, Chapter 7). *Drosophila,* which has for many years occupied a position as the quintessential "intermediate organism" for mutant studies and for elaborating genetic principles, is now being used for studies of homeotic genes and in studying the development of the nervous system (Rubin, Chapter 9). Humans and primates exhibit many of the same physical and behavioral manifestations of old age.

Another question that has absorbed scientists and the general public is, "How do biological systems interact with their surroundings?" Systems that operate in a three-dimensional environment such as birds (Konishi *et al.*, Chapter 13), fish (Powers, Chapter 12), and cetaceans (Würsig, Chapter 16) face unique problems of orientation and navigation. However, all species must maintain homeostasis in the face of variations in temperature or ionic conditions. For example, fish have developed antifreeze proteins to withstand extreme cold and cetaceans have adapted to withstand changes in pressure.

Living systems must cope with more than their physical surroundings; they must interact with their neighbors. Birds have been critical in the development and testing of models of predator-prey relationships as well as in studying the value of altruism and kinship to an individual and to the group. Animal communication is an exciting area that is being actively studied in several systems described in this volume such as birds, cetaceans, and primates. Experimental psychology could not have developed without the laboratory rat (Gill *et al.*, Chapter 14), and the complexity of primate interactions (King *et al.*, Chapter 17) has provided insights into our own behaviors.

An interaction between species that poses its own problems and challenges is parasitism, especially when, as for certain protozoans and helminths, humans are the final host (Mahmoud, Chapter 8). Understanding of the life cycles of these organisms is being used in the development of vaccines.

In an ideal world we would be caretakers of all living things. Biological research has frequently been used in promoting animal welfare *(3)*. Breeding programs have helped maintain primate species that might otherwise have become extinct. Vaccines that protect animals against such diseases as leukemia, rinderpest, and anthrax could not have been developed without research on cattle, cats, and sheep. The development of surgical and diagnostic techniques for future use in humans have had immediate applications in veterinary medicine.

Robert May's article (Chapter 20) indicates the enormous number of species with whom we share the Earth. Research promoting biological welfare will continue to have a mounting urgency when one considers estimates that by the year 2000 an average of 100 species will become extinct every day *(4)*.

A positive outcome of the animal rights furor is that it has forced scientists to reevaluate the necessity for animals in their experiments *(5)*. Improved understanding of statistical analysis has enabled scientists to design better experiments that use fewer animals. In vitro tests such as cytotoxicity testing with corneal epithelial cells have begun to provide an alternative to the use of rabbits in the Draize tests *(6)*. Defined media are replacing fetal serum for many cell culture systems *(7)*. However, according to the National Research Council's 1988 report, while alternatives to the use of animals should be considered, "there is no chance of replacing all animals in research and testing in the foreseeable future" *(8)*.

Higher animals have been and will continue to be used as models for understanding human pathology despite the pressure from animal rights groups. Many of the major advances in medicine have come through the use of animals—such as the development of insulin for the treatment of diabetes (which was primarily studied in dogs), the polio vaccine (which was tested in primates), cancer therapies, and organ transplantation *(8)*. Rats will continue to be extremely important in studies of organ transplantation, as models for cardiovascular disease and alcoholism, and in carcinogenicity assessment. The development of open heart surgery was based on research in cats and dogs.

Despite the usefulness of animals as models of human processes, our absorption with the medical histories, health, and quality of life of our own species has given the human some important advantages as a biological system in its own right. However, use of humans in research poses an entirely different set of ethical problems. How can experiments be properly controlled and evaluated if placebos are unethical? To what use will genetic information be put in our society? Will it be available for decisions on job suitability or insurability of individuals who may have inherited a predisposition towards a disease with a multifactorial etiology? Does knowledge of carrier status improve quality of life? Who should be tested? White and Caskey (Chapter 18) discuss the advantages and disadvantages of humans as an experimental system in molecular genetics.

There is a lot of truth to the statement that no life is more precious to us than that of our young. At the very forefront of human biological research, relying on some of the newest technological advances, is the study of the human fetus. This research, as described by Hansen and Sladek (Chapter 19), holds out the promise of curing cardiac arrhythmias, immune disorders, and hydrocephalus before infants are born. Here too there are considerable moral and ethical dilemmas to be met.

How does an investigator choose the right organism to answer a scientific question? While certain processes are so essential to life that they have been conserved throughout evolution, other phenomena can only be studied in particular organisms. Cost is also a sizeable factor. Experiments that require very large numbers of animals can not reasonably be done in primates, as they are an expensive and endangered species, but certain kinds of research that are directly applicable to human health require them. Bacteria are not a good model system for studying human psychology, but they are inexpensive, easy to maintain and use in hundreds of replicate experiments, and are relatively unlovable.

The advent of genetic engineering means that it is no longer necessary to tailor the question to fit the biological system; the potential exists to specifically design the organism that can best be used to answer a particular question. Retroviruses are excellent vectors for transferring genetic information into mammalian embryos to create transgenic animals (Jaenisch, Chapter 15). In the foreseeable future, transgenic animals will be critical as models for an assortment of diseases and to understand the consequences of turning on and turning off tissue-specific gene expression. The need to create systems that are better suited to our economic needs will also continue—such as higher-yield plants or bacteria that degrade pollutants or help plants resist cold (Lindow *et al.,* Chapter 3). However, the authors dealing with the genetic engineering of plants and bacteria for applied uses discuss issues this technology raises in terms of release of novel organisms into the environment.

The authors in this book were given a uniquely difficult chore. They were specifically told not to emphasize their own work or individual experiments but to evaluate why they had picked the particular system they did. Although this volume could not cover all of the systems used in research, we have tried to present a broad spectrum. We hope that this survey of the advantages and disadvantages of particular systems will help new investigators make better choices and understand the rationales behind the selection of these systems by other scientists.

Barbara R. Jasny
Science
December 1989

References

1. J. A. Moore, *Amer. Zool.* **28**, 449 (1988).
2. Aristotle, "De Partibus Animalium," in *The Basic Works of Aristotle*, R.M. McKeon, Trans. (Random House, New York, 1941), p. 643.
3. F. M. Loew, *ILAR News* **30**, 13 (1988).
4. N. Myers, *Bioscience* **39**, 39 (1989).
5. Office of Technology Assessment, Congress of the United States, *Alternatives to Animal Use in Research, Testing, and Education* (Marcel Dekker, Inc., New York, 1988).
6. A. M. Goldberg and J. M. Frazier, *Sci. Am.* **261**, 24 (1989).
7. J. Mather, Ed., *Mammalian Cell Culture: The Use of Serum-Free Hormone Supplemented Media* (Plenum, New York, 1984).
8. Committee on the Use of Laboratory Animals in Biomedical and Behavioral Research, National Research Council, *Use of Laboratory Animals in Biomedical and Behavioral Research* (National Academy Press, Washington, DC, 1988), p. 46.

Retroviruses

Harold Varmus

Retroviruses are special entries on the menu of biological systems surveyed in this volume. This is so because retroviruses necessarily intersect with many other "systems" (for example, humans, other primates, transgenic animals, *Drosophila*, and yeast) through infection of cells, through strong similarities with mobile genetic elements that reside in eukaryotic chromosomes, or through intimate associations with cellular genes that are instrumental in retrovirus-induced cancers. Furthermore, retroviruses are unusual parasites in that they insinuate themselves into the life-styles of their hosts in revealing and often unprecedented ways: by converting their genes from RNA to a DNA form; by incorporating viral DNA stably into chromosomes of somatic or germ cells to form a provirus; by mutating, and even capturing, cellular genes; by rarely impairing, and often potentiating, the growth of their host cells; and by entrusting gene expression to host mechanisms under the direction of viral signals.

The objective of this selective review is to introduce readers to several of these themes, to evoke the flavor of the discipline of retrovirology, and to raise some questions that might attract the next generation of disciples. Retroviruses will be viewed here from three perspectives potentially attractive to those seeking exciting experimental prospects: (i) as models for the study of fundamental biological problems, including transfer of genetic information, DNA recombination, regulated gene expression, growth control, and macromolecular assembly; (ii) as problems posed by their pathogenic potential in human and animal hosts, where they cause diseases such as AIDS and many forms of cancer; and (iii) as tools for genetic manipulations ranging from gene therapy to mutagenesis.

Growth and Development of the Retrovirus Community

Viruses of the type we now call retroviruses were among the earliest known viruses, first discovered about 80 years ago as filterable agents that cause cancers in chickens and anemia in horses (*1*). For many years, however, they had a small following in the scientific community, due, in part, to the lack of reliable cell culture and biochemical techniques and, in part, to skepticism in some quarters about the significance of tumor viruses that had no apparent counterparts in mammals. These attitudes began to shift with the discoveries of viruses, later proved to be retroviruses, that cause mammary carcinomas and leukemias in mice (*2*) and with the development of quantitative assays for chicken sarcoma and murine leukemia viruses in cultured cells (*3*). By the

H. Varmus is an American Cancer Society Professor of Molecular Virology in the Department of Microbiology and Immunology and the Department of Biochemistry and Biophysics at the School of Medicine of the University of California, San Francisco, CA 94143. This chapter is reprinted from *Science* **240**, 1427 (1988).

late 1960s and 1970s, the retrovirus community exhibited nearly logarithmic growth as major milestones were passed: the discovery of reverse transcriptase (*4*), the discovery of proviruses transmitted in the germ line (*5*), and the discovery of cellular progenitors of retroviral oncogenes (*6*). In the past few years, the growth of the retroviral community has been further accelerated by several connections between retroviruses and human diseases: discoveries of human retroviruses that cause adult T cell leukemia/lymphoma (*7*) and AIDS (*8*), and identification of human oncogenes, related to retroviral oncogenes, that are active in human cancers (*9*). These advances have galvanized widespread interest in retroviruses and their oncogenes in the medical community, among politicians and public interest groups, in biotechnology firms and their entrepreneurial supporters, and even in the public at large.

No longer a cottage industry, retrovirology has merged with several other disciplines as a consequence of some remarkable discoveries during this decade. (i) The structure of the provirus revealed that retroviruses belonged to a larger class of mobile genetic elements, called retrotransposons (or retroposons), important to investigators working on many eukaryotic organisms, especially yeast and *Drosophila* (*10–12*). (ii) Reverse transcription was assigned a central role in the replication of other viruses [hepatitis B (*13*) and cauliflower mosaic viruses (*14*)] and in the transposition and generation of other kinds of DNA, mostly in eukaryotes but recently in bacteria as well (*15*). (iii) Once recognized as naturally occurring vectors for host-derived oncogenes, retroviruses were studied as gene vectors by medical geneticists, developmental biologists, and practically anyone who wished to determine the phenotypic consequences of expressing a cloned gene in a cultured animal cell (*16*). (iv) With the attribution of biochemical functions to the products of retroviral oncogenes, investigation of viral tumorigenesis became closely linked to the study of growth factors, their receptors, signal transducers, protein kinases, and transcriptional regulators (*17*). Several of these disciplinary fusions will resurface in later discussions.

The Essential Facts About Retroviruses: The 3-Minute Course

Retroviruses resemble other animal viruses in several respects, but differ from all others in containing an RNA genome that replicates through a DNA intermediate (*4, 10–12*). The extracellular virus particle is composed of a genome (single-stranded RNA) wrapped in a core of viral protein that is, in turn, surrounded by an envelope studded with viral glycoproteins and derived from the membrane of the previous host cell (Figs. 1 and 2A). Although multiplication occurs only within cells and depends on cellular functions, an infecting retrovirus also brings along an organized collection of viral enzymes and RNA designed to direct the synthesis of a double-stranded DNA copy of the RNA genome (reverse transcription) and the precise joining of that DNA to the host chromosome (integration).

The life cycle

Retroviruses attach to cells with the help of normal cell surface proteins specifically recognized by viral envelope proteins, and they enter either through a mechanism, receptor-mediated endocytosis, that cells have evolved to ingest beneficial extracellular substances such as growth factors or through direct fusion with the plasma membrane (*18*). Entry initiates the conversion of a quiescent, enveloped particle into an enzymatically active nucleoprotein complex that performs its idiosyncratic functions, reverse transcription and integration, with little or no help from the host cell (Fig. 2A). Once its provirus is ensconced in a cell chromosome, however, a retrovirus becomes highly dependent on its host—for replication of the provirus as part of its chromosomal context, for transcription of the provirus by RNA polymerase II, for processing of RNA transcripts by mechanisms normally used to cap, polyadenylate, and splice host RNAs, and for translation of resulting messenger RNAs by host polyribosomes (Fig. 2, A and B). These routine cellular functions are regulated by viral

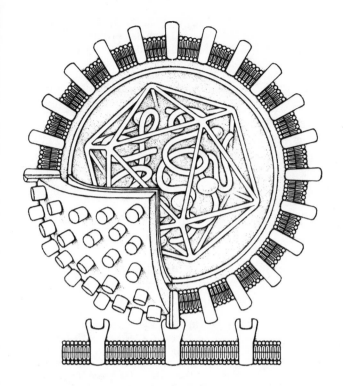

Fig. 1. Schematic view of a retrovirus particle. Two identical single strands of viral RNA and viral enzymes (reverse transcriptase, integrase, and protease) are drawn within an icosahedral viral core, and the core is surrounded by an envelope that is derived from host membranes enriched with viral glycoprotein. Interaction of envelope glycoprotein with a host-encoded cell surface receptor is shown at the bottom. [Adapted from (*93*); copyright 1987 *Advances in Oncology*]

signals that determine the efficiency of transcription and splicing and mediate the occasional bypass of termination codons during translation, as discussed later. Finally, new virus particles are assembled and released. Still poorly characterized recognition signals draw together RNA and core proteins from the cytoplasm to associate with envelope glycoproteins embedded in plasma membranes, and a virus-encoded protease later cleaves viral polyproteins into the smaller components found in mature virus particles.

The genome

The viral genome, as found in virus particles, is a complex of two identical chains of RNA, making retroviruses diploid, which is an oddity among viruses. Each viral RNA molecule is base-paired with a specific host transfer RNA that primes DNA synthesis, another curious feature of retroviruses. All retroviral genomes are organized in a standard format, best appreciated in comparison with a DNA version of

the viral genome, the provirus integrated within host DNA (Fig. 2B). Sequences that regulate the structural transformations of the viral genome during the life cycle are clustered near the ends of the RNA: signals for initiation and progression of DNA synthesis, for integration, for transcription of the provirus into RNA, for RNA processing, and for packaging RNA into progeny particles.

During the intricate maneuvers used to synthesize viral DNA, sequences present once near the ends of viral RNA (U3 and U5) are duplicated to generate long terminal repeats (LTRs), several hundred base pairs in length, at the ends of proviral DNA (Fig. 2B). The LTRs encompass many of the regulatory signals in the viral nucleotide sequence, and they are distinctive features that unite proviruses structurally with the broad collection of eukaryotic transposable elements known as retrotransposons (Fig. 3). Between these regulatory regions are coding sequences (open reading frames) for the major structural proteins of the virus particle (the *gag* frame encodes the core proteins, the *env* frame the envelope

A

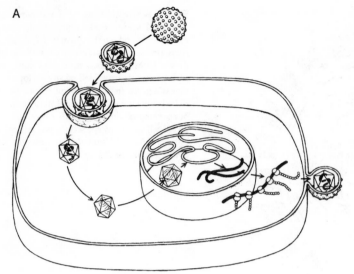

Fig. 2. Two views of the retrovirus life cycle. (**A**) A virus particle entering a cell at the upper left, uncoating to form a nucleoprotein complex in which viral RNA (black) is copied into DNA (white) by reverse transcriptase. After migration to the nucleus, the complex mediates integration into a host cell chromosome (long white ribbon). Synthesis of viral RNA and proteins (beaded chains) leads to assembly of particles that exit the cell at the right by budding through the plasma membrane.

B

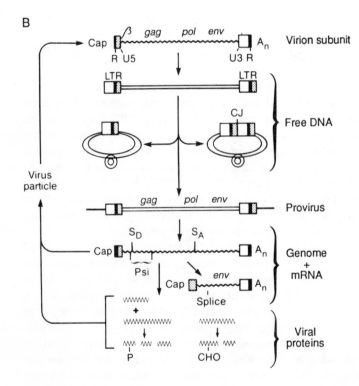

(**B**) The molecular transformations of the indicated species of the viral genome during the life cycle. Cap, capped nucleotide at 5' end of viral RNA; A_n, polyadenylic acid at 3' end of viral RNA; R, repeated sequence at ends of viral RNA; U3 and U5, unique sequences duplicated during DNA synthesis; LTR, long terminal repeat; SD and SA, splice donor and acceptor sites, respectively; Psi, signal for packaging of viral RNA; P and CHO, modifications of viral proteins by phosphorylation and glycosylation, respectively; RNA, wavy lines; DNA, straight lines; protein, jagged lines. The tRNA primer is positioned next to U5 on the virion subunit. [Adapted from (*11, 12, 31*)]

glycoproteins); for the enzymes found in particles (a protease, reverse transcriptase, and integration protein, at least two of which are encoded by the *pol* frame); and for proteins with specialized, intracellular functions, exhibited only by those retroviruses endowed with oncogenes or regulatory genes.

Genetic behavior

During the virus life cycle, several interesting genetic and quasi-genetic phenomena may occur, especially if cells are infected by more than one virus: production of heterozygotic dimeric genomes, formation of pseudotypes at

General structure

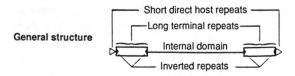

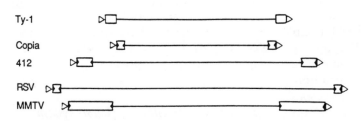

Fig. 3. Representative retrotransposons of yeast (Ty-1) and *Drosophila* (copia, 412) compared with proviruses of Rous sarcoma virus (RSV) and mouse mammary tumor virus (MMTV). Symbols are explained at top. [Adapted from (*11*); used with permission, copyright 1983 Academic Press]

high frequencies (particles with core proteins and genome provided by one virus and envelope proteins by another), frequent deletions and nucleotide substitutions, and recombination between related, coinfecting viruses. [Recombination between retroviruses is surprisingly efficient, but its mechanistic basis has not been resolved (*19*).] Another, more specialized genetic attribute of retroviruses, their ability to cause insertion mutations, is fundamental to several of the important interactions between these agents and their hosts. The mutations may cause either a recessive loss of function due to gene disruption or a dominant gain of function due to stimulation of expression of genes adjacent to an insertion site.

Biological behavior

Although united by their life cycle and the conserved features of their genomes, retroviruses are unusually diverse in their biological manifestations. Found in all vertebrates in which they have been sought, they can be transmitted horizontally in the form of infectious extracellular particles or genetically in the form of endogenous proviruses integrated within the germ line. The physiological consequences of infection range widely. In some instances, retroviruses produce no apparent effects, even when virus is produced in large amounts. In other cases, retroviruses potentiate growth of cultured cells or cause cancer in the intact host. A few retroviruses are deleterious to

cells and produce a variety of destructive diseases, most notably immunodeficiency syndromes such as AIDS.

Retroviruses as Model Systems for Studying Eukaryotic Biology

Retroviruses as exemplars of information flow from RNA to DNA

The very name "retrovirus" embodies the biochemical property for which this class of viruses is most famous: the capacity to copy its RNA into DNA (reverse transcription). Work with these viruses first convinced the scientific community that transfer of information in biological systems was not limited to conventional transcription (DNA to RNA), translation (RNA to protein), and replication (DNA to DNA and RNA to RNA).

Although reverse transcription was first encountered in the retrovirus life cycle, it is hardly unique to retroviruses (*20*); it is now recognized as a widespread phenomenon in eukaryotic cells, viruses, and in certain bacteria (*13–15*). Indeed, as much as 10% of the eukaryotic genome may be composed of products of reverse transcription (*21*). Moreover, reverse transcription seems to have been a logically necessary event at a crucial stage in evolution, the transition from the earlier RNA-dominated world to our DNA-dominated one (*22*).

Despite the many settings in which reverse

transcription is now believed to occur, retroviral reverse transcriptases are still the only ones that have been studied in a satisfactory manner. This is due to the ease with which the enzymes can be solubilized and purified from retroviral particles (*23*), the availability of retroviral reverse transcriptases made in *Escherichia coli* by recombinant DNA technology (*24*), and the existence of several mutants in the viral *pol* gene (*10*). (The reverse transcriptases of hepatitis B and cauliflower mosaic viruses have not been solubilized from particles, those encoded by retrotransposons have not been harvested in sufficient amounts for serious biochemistry, those responsible for synthesizing other components of vertebrate genomes have not been identified, and those found naturally in bacteria have been discovered too recently to be properly evaluated.)

Retroviral reverse transcriptases display many unexpected properties (*11, 12, 23*). They use RNA as natural primers, including the host transfer RNA base-paired near the 5' end of the viral genome. They are "jumping polymerases" that transfer nascent strands between templates at least twice during synthesis of retroviral DNA. They have a second enzymatic activity, located in a different domain of the protein, that digests RNA to oligonucleotides once it has been copied into DNA [ribonuclease (RNase) H]. Finally, they will copy virtually any RNA template, once provided with a suitable DNA or RNA primer, making them popular reagents for nucleotide sequencing and for cloning complementary DNA copies of mRNAs. Still, major issues about these unusual and important enzymes are unresolved: virtually no structural work has been done, the active site for polymerization has not been defined, and few useful inhibitors have been identified.

An appreciation of the workings of retroviral reverse transcriptases has been central to the discoveries of reverse transcription in the life cycles of other viruses and in the transposition cycle of eukaryotic transposons. Because a minus strand copy of virion RNA is always made before the plus strand, retroviral DNA synthesis appears asymmetrical, in contrast to traditional semiconservative DNA synthesis. A similar asymmetry of viral strands is apparent in livers infected by hepatitis B viruses: minus DNA strands are much more abundant than plus strands (*25*). On the basis of this provocative clue, stronger evidence for reverse transcription was sought and obtained (*13*). Retrovirus-based principles have also been helpful in the design of experiments showing that the yeast transposon Ty migrates by reverse transcription of an RNA copy of the element (*26*) and in the characterization of retrotransposition in *Drosophila* (*27*).

Retroviruses provide an efficient system for specialized eukaryotic recombination

Information about recombination in vertebrate organisms is scanty in general, although a specialized system for somatic recombination rearranges some genes in immune cells (*28*) and homologous recombination is known to occur at high frequency between extrachromosomal DNAs (*29*) and at low frequency with chromosomal DNA (*30*). Integration of viral DNA into host chromosomes by a specialized and precise recombinational mechanism provided by the virus is a cardinal feature of the retrovirus life cycle (*11, 12, 31*), and one that differentiates retroviruses from several other viruses (such as simian virus 40) whose DNAs are integrated infrequently and haphazardly (*32*). Indeed, integration of retroviral DNA is the recombination event in higher eukaryotes that occurs with greatest efficiency and precision, and it is the most amenable to biochemical and genetic analysis. Moreover, the analogous structures of endogenous proviruses and retrotransposons imply that the insertion of these elements uses the same mechanism.

A few facts about retroviral integration are now well established: (i) the structure of the product, proviral DNA, is invariant, with viral DNA always joined to host DNA 2 bp from the ends of the LTRs (*33*); (ii) specific sequences near the ends of the LTRs are necessary for the reaction (*34*) and can be considered analogs of bacteriophage attachment (*att*) sites; and (iii) the reaction requires a viral integration protein, IN, that is encoded near the 3' end of the *pol* gene but is not involved in reverse transcription (*35*). Still, many important aspects of retroviral integration have remained in dispute, including the nature of the immediate precursor to the provirus, the organization of viral ingredients, the preferential use of sites in host chromosomes, the function of IN, and the enzymological characteristics of the reactions.

These issues are now more easily approached because the integration reaction can be studied in a cell-free system, with murine leukemia virus (MLV) DNA as the integrating species and naked bacteriophage λ DNA as a target (*36*). The reaction is driven by a viral nucleoprotein machine derived from infected cells and composed of recently synthesized viral DNA associated with viral proteins from parental particles. Direct examination of the intermediates in the reaction indicates that the immediate precursor to the provirus is a linear duplex, with two nucleotides missing from the 3' end of each strand, a result of the action of IN protein (*37*). This system and refinements of it may ultimately provide a picture of retroviral integration that rivals the view of integration of temperate bacteriophage DNAs into the *E. coli* chromosome (*38*).

Retroviruses as guides to the molecular basis of cancer

The cancer-causing properties of retroviruses probably provide the most common motivations for choosing to work with these agents. Oncogenic retroviruses, isolated from such vertebrates as fish, chickens, rodents, cats, subhuman primates, and humans, induce sarcomas (tumors of mesenchymal origin), various kinds of leukemias, and, less often, epithelial malignancies (the most common human cancers),

including carcinomas of the breast, kidney, and liver (*10*). Given their small genomes and the regularity with which many induce characteristic forms of cancers in convenient laboratory animals, retroviruses represent seductively simple instruments with which to ask how something as complex as an animal cell can be converted into a cancer cell.

Almost all oncogenic retroviruses seem to fall into two camps [an important exception, human T cell leukemia virus (HTLV), is discussed below]. One group, typified by Rous sarcoma virus (RSV), carries a viral oncogene responsible for the swift induction of tumors in animals and the efficient transformation of cells in culture (*39*). The others, exemplified here by avian leukosis virus (ALV) and mouse mammary tumor virus (MMTV), lack a viral oncogene, do not transform cells in culture, but regularly cause tumors after long latency through what appears to be a multistep process (*40*).

Both groups of viruses cause tumors, at least in part, through the agency of cellular genes (Fig. 4) (*39–41*). Viruses like RSV have captured (transduced) host genes through a mechanism that probably begins with a proviral insertion mutation, ultimately placing the captured genes under the control of viral signals and mutating their coding sequences as well. Viruses without oncogenes stimulate expression of cellular genes through adjacent proviral insertions that override normal control elements and sometimes alter the structure of the gene products. Because both the transduced and insertionally activated genes contribute to cancerous change, they are called oncogenes, and their normal progenitors, proto-oncogenes. Evidence to date generally supports the premise that proto-oncogenes are important regulators of cell growth or development (*17*,

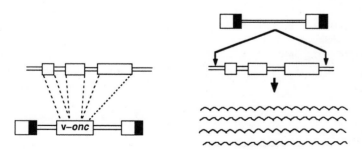

Fig. 4. Simplified representations of the two common mechanisms by which retroviruses harness cellular proto-oncogenes to cause cancers. The figure on the left indicates that a typical viral oncogene (v-*onc*) is derived from exons of a cellular gene by transduction; the figure on the right indicates that proviral insertion mutations on either side of the exons of a cellular gene may cause augmented expression of the gene.

41). Many of the proto-oncogenes discovered through the use of retroviruses are also sometimes targets for nonviral, somatic mutations believed to lead to human cancer (*9*).

Retroviral transduction and molecular cloning of provirally activated genes have together been responsible for isolation of the vast majority of proto-oncogenes, which now number about fifty (see Table 1 for examples). The profound influence of these genes upon the study of eukaryotic cells is apparent from the following historical synopses.

The cellular origin of v-src. Among retroviral oncogenes, the v-*src* gene of RSV was especially susceptible to genetic and biochemical maneuvers in the era before molecular cloning, because RSV is the only retrovirus with a transduced oncogene that can replicate without a helper virus. Temperature-sensitive and deletion mutants of v-*src* established that virus multiplication could be dissociated from neoplastic transformation and that transformation required continued expression of the viral oncogene (*42, 43*). With molecular probes defined through the use of

deletion mutants, v-*src* was shown to be closely related to, and presumably derived from, a normal and highly conserved cellular gene, c-*src* (*6*), thereby establishing a paradigm for more than 20 other retroviral oncogenes (*39, 41*).

A family of protein kinases. Subsequent discoveries that the protein products of v- and c-*src* are protein kinases (*39, 44*), which specifically phosphorylate tyrosine residues (*45*), drew attention to the idea that tyrosine phosphorylation might be central to growth control, even though tyrosine phosphate constitutes less than 1% of the amino acid–derived phosphate in cellular proteins (*46*). The simple assays developed to detect kinase activities in immune complexes containing *src* proteins and the amino acid sequence motifs emblematic of the active domains of protein-tyrosine kinases have helped identify well over a dozen such kinases, including several that are transmembrane receptors for polypeptide growth factors (*46*). Some of these receptors have proven to be themselves the products of proto-oncogenes: the epidermal growth factor receptor gene is the progenitor of the avian virus

Table 1. Categories of proto-oncogenes and modes of retroviral activation. The table provides a few examples of proto-oncogenes (most discussed in the text) whose protein products have been located within or outside the cell and assigned physiological and biochemical functions. The two columns at the right indicate whether the genes have been naturally transduced by retroviruses and whether they have been encountered as targets for proviral insertion mutations. Further details and references can be found in the text and in (*9*), (*17*), and (*36*). PDGF, platelet-derived growth factor; EGF, epidermal growth factor; CSF-1, colony stimulating factor–1.

Location of protein product	Proposed function	Example	Retroviral transduction	Insertional activation
Secretory vesicles and extracellular space	Growth factor, ligand for membrane receptor	*sis*/PDGF	+	−
		int-1	−	+
Plasma membrane				
Transmembrane	Growth factor receptor with protein-tyrosine kinase activity	*erb*B/EGF receptor	+	+
		fms/CSF-1 receptor	+	+
Inner face	Protein-tyrosine kinase	*src*	+	−
		abl	+	−
	Signal transduction, GTP binding and hydrolysis	*ras*	+	+
Cytoplasm	Protein-serine (threonine) kinase	*mos*	+	+
Nucleus	DNA binding, transcriptional regulation	*jun*	+	−
		fos	+	+
		myc	+	+

oncogene, v-*erb*B (*47*), and the receptor for the macrophage growth factor, CSF-1, is encoded by the progenitor of the feline virus oncogene, v-*fms* (*48*). The ligand for yet another receptor with protein-tyrosine kinase activity, the platelet-derived growth factor (PDGF) receptor, is partially encoded by the *sis* proto-oncogene (*49*).

An oncogene explains a mutant chromosome. An odd version of chromosome 22, called the Philadelphia chromosome (Ph^1), was one of the first visible signs of a reproducible genetic lesion to be noted in human leukemia cells (*50*). More than 20 years later, c-*abl*, the cellular homolog of the oncogene of Abelson-MLV, normally found on chromosome 9 where it encodes a relatively inactive protein-tyrosine kinase, was shown to be broken and fused to another gene on chromosome 22, forming both Ph^1 and a hybrid protein with augmented kinase activity (*51*).

ras oncogenes in human tumors. Nearly a decade ago, human tumors and cells transformed by chemical mutagens were found to contain active oncogenes through experiments in which DNA from such cells and tumors was used to induce oncogenic properties in a tissue culture cell line (*52*). Subsequently the genes were identified in most cases as mutant versions of *ras* genes (*53*), proto-oncogenes first discovered as the progenitors of the *ras* oncogenes of murine sarcoma viruses (*39*). Like *src* proteins, *ras* proteins have a measurable biochemical function, guanosine triphosphate (GTP) binding and hydrolysis, and they also belong to a large family of proteins that includes GTPases (G proteins), which convey extracellular signals to adenylate cyclases (*54*).

An oncogene implicated in transcriptional control. The *jun* oncogene, recently discovered in an avian sarcoma virus, is related to the yeast *GCN4* gene, a regulator of transcription, on the basis of both structural and functional tests (*55*). Because *GCN4* protein was known to bind the same DNA sequence as does a mammalian transcriptional activation complex, AP-1, the cellular homolog of v-*jun* seemed a good candidate to encode a component of AP-1; recent evidence argues strongly that it does (*56*). Yet another proto-oncogene (c-*fos*, the

progenitor of the oncogene of FBJ murine osteosarcoma virus) encodes a second component of AP-1, a protein that enhances the DNA binding properties of the *jun* protein by forming a heterodimer with it (*57*).

Genetic rearrangements of c-myc. ALV, a cause of B cell lymphomas in chickens, was the first retrovirus shown to cause insertion mutations during tumor induction (*58*). The significance of the mutations was established by the identity of the activated gene: c-*myc* (*59*), a gene already labeled a proto-oncogene because several retroviruses carry it in transduced form (*39*). Thus retroviruses are central to both arms of the argument that first implicated a specific cellular gene in an oncogenic process: the progenitor of a retroviral oncogene is also the target for mutation by a provirus without an oncogene. The c-*myc* insertion mutations proved to be paradigms for other kinds of genetic rearrangements—chromosomal translocations and gene amplification—that affect c-*myc* and its close relatives, N-*myc* and L-*myc*, with particular frequency (*9*).

Development and int-1. The first gene found activated by MMTV proviral insertions in mammary cancers (*60*), *int*-1, is not related to any oncogenes transduced by retroviruses. The most striking features of this proto-oncogene suggest a role in development: in mice, the gene is normally expressed in only two places, the neural tube in midgestational embryos and postmeiotic cells in testes (*61*); in *Drosophila*, the homolog of *int*-1 has been identified as *wingless*, a gene required for a normal gradient of cells in larval body segments (*62*). The *int*-1 protein is likely to act as an intercellular signal for both normal development and neoplastic growth (*63*).

These brief tales are meant to suggest the multiplicity of ways in which genes brought to light by retroviruses are shaping the study of oncogenesis, growth control, and development. These genes provide an unfulfilled opportunity to describe in molecular terms how cells behave when responding to signals for normal growth or when violating restraints during neoplastic growth. To do this, it will be necessary to learn more about the biochemical properties of the proteins involved and the identities of their relevant biochemical targets.

Could the genes we now call proto-oncogenes have been found without the assistance of retroviruses? It is possible that growth factor receptors would ultimately have led to genes like *src*, that transcription factors would have led to *jun*, and that *Drosophila* genes implicated in development would have led to genes like *int*-1. But progress by such routes, and perception of the relationship to cancer, would have been painfully slow.

Retroviruses as probes for host regulatory mechanisms

The utility of retroviruses for understanding properties of their host cells is particularly obvious during viral gene expression, when the provirus is dependent on host machinery for transcription, RNA processing, translation, and protein modification (Fig. 2). The simplicity of their genomes and their harmonious existence within their hosts made retroviruses especially attractive reagents for working on such topics in the years before molecular cloning. Although any cellular gene can now, in principle, be isolated for such purposes, retroviruses continue to be widely used, in part because of historic precedent, and in part because a retroviral provirus presents a remarkable opportunity to examine adaptation of a parasite to its host. Thus the host provides the machinery, and the provirus encodes the signals that regulate expression, sometimes in surprising ways. This formula has inspired study of many aspects of retroviral gene expression; a few examples illustrate its attractions.

Transcription. Signals that modulate retroviral transcription are lodged mainly in the U3 region of the LTR and were among the first eukaryotic promoters and enhancers to be carefully studied (*11, 12*). The regulatory elements are recognized by host transcription factors (*64*), determine how well a virus will grow in different cell types (*11, 12*), and influence the oncogenic spectrum of each retrovirus (*40, 65*). Sequences in the MMTV LTR are binding sites for the glucocorticoid receptor, and the MMTV promoter is stimulated by the hor-

mone-receptor complex (*66*). The MMTV LTR was the first transcriptional initiator shown to be subject to primary regulation by glucocorticoid hormones (*67*), it is still commonly used to examine the mechanism of regulation, and it is a popular promoter for achieving inducible expression of heterologous genes in eukaryotic cells (*68*).

Splicing. Even retroviruses that make only a single spliced subgenomic RNA must regulate the ratio of spliced RNA and its precursor, since the precursor is also used as both genomic RNA and mRNA (see Fig. 2B). The signals for maintaining appropriate levels are not known, but they may be differently interpreted in different hosts. Thus in avian cells less than half of RSV RNA is spliced to make two subgenomic RNAs, but in mammalian cells nearly all is spliced to form one subgenomic species (*69*). In addition, some retroviruses [human immunodeficiency virus (HIV), HTLV, and their close relatives] have more complex splicing patterns and encode viral proteins that influence the ratio of spliced to unspliced RNAs (*70, 71*).

Translation. The mechanisms used for synthesis of retroviral *pol* proteins illustrate the capacity of the eukaryotic translational apparatus to do unexpected things in response to retroviral signals. All retroviruses express their *pol* genes as *gag-pol* fusions, but *gag* and *pol* are separated by a stop codon in MLV; read in different, briefly overlapping frames in RSV and HIV; and separated by a third frame (*pro*, for the viral protease) in MMTV and HTLV (Fig. 5). Rather than commandeer the host's splicing apparatus to create mRNAs with a single open reading frame for *gag-pol* (or *gag-pro-pol*) proteins, retroviruses have instead exploited previously unrecognized potentials of vertebrate ribosomes to insert an amino acid occasionally in response to the nonsense codon at the end of MLV *gag* (*72*) or to shift reading frames at defined sites and frequencies when translating the other viral RNAs (*73*). Neither of these phenomena have been encountered during the translation of cellular mRNA and would, of course, be deleterious if allowed to occur frequently and without purpose. But they have clear benefits for retroviruses: structural (*gag*) proteins can be made in large

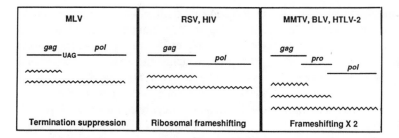

Fig. 5. Strategies for expression of retroviral *pol* genes. Arrangements of the *gag*, *pol*, and *pro* reading frames of some representative retroviral genomes (straight lines) are shown in relation to their protein products (wavy lines). Underlying reading frames are −1 with respect to the frames above. Suppression and frameshifting events occur with frequencies ranging from 5 to 25%, so that *gag-pol* or *gag-pro-pol* polyproteins are about 5% as abundant as *gag* proteins. MLV, murine leukemia virus; RSV, Rous sarcoma virus; HIV, human immunodeficiency virus, MMTV, mouse mammary tumor virus; BLV, bovine leukemia virus; HTLV, human T cell leukemia virus.

amounts and catalytic (*pro* and *pol*) proteins in relatively small amounts; and *pol* products can be incorporated into viral cores through attached *gag* components.

What is the basis of such translational control? The retroviral frameshifts move the ribosome into the −1 reading frame in response to at least two sets of instructions in the viral RNA: a short sequence at the frameshift site, and secondary structure downstream of the site (74). Although no cellular genes have yet been found to use the frameshifting potential of eukaryotic ribosomes, at least one other class of viruses, the coronaviruses, take advantage of it (75), and at least one retrotransposon, the Ty element of yeast, mediates frameshifting in the +1 direction (76). More recently, retrovirus-like frameshifting has been implicated in the expression of the transposase gene of the bacterial transposon, IS1 (77).

Retroviruses as structural models

Retroviruses are excellent models for thinking about complex interactions among macromolecules in eukaryotic cells. The retrovirus life cycle, like that of several other animal viruses, is rich with such interactions. At the outset, a retroviral glycoprotein must recognize a cell surface protein, bind to it, and mediate uptake of the virus particle into the cell. How do these events occur? Do they differ from other internalization processes? What are the structural transformations of the particle that accompany virus entry and activate reverse transcription of

viral RNA? What is the organization of the nucleoprotein complex that makes viral DNA and mediates its integration? Later in the life cycle, viral RNA, transfer RNA, and products of *gag*, *pol*, and *env* must correctly assemble into particles. What are the rules that govern virus assembly? In particular, how do *gag* proteins interact to form cores? How does RNA get into cores? How is the protease activated to process polyproteins into mature components? How does the core associate with a region of plasma membrane enriched with envelope glycoproteins? In this area, there are few accomplishments to recount: it is a true frontier. Recent determination of the crystal structures of RSV and HIV proteases, demonstrating that formation of the active site depends upon dimerization, is an exciting step towards solution of some of the problems of virus assembly and maturation (78).

Retroviruses as Pathogens in Humans and Animals

To this point, we have considered several ways in which retroviruses have been informative about general problems in eukaryotic biology. But retroviruses have an unusual property as a biological system: they are problems themselves, because they are causative agents of disease, including lethal diseases of humans.

HIV and AIDS. Discovery of a retrovirus as the cause of AIDS (8) has had many effects on this discipline. Precepts about retroviruses are now matters of public health, funding for retroviral research in both public and private sectors is growing rapidly, and "retrovirus" is in the lexicon of common speech. But there have also been important effects on the retrovirologist's view of his or her own science generated by the discovery of HIV: a renaissance of interest in the retrovirus life cycle, the pathogenesis of cytotoxic infections, and the immune response to retroviruses.

Several factors have contributed to the revitalized study of the life cycle. Most obvious is the incentive to identify steps in replication at which intervention might be successful, despite the generally dismal experience with antiviral drugs. (i) Identification of the cell surface protein, CD4, as the receptor (or a major component of the receptor) for HIV (*79*) permits a deliberate assault on the initial step in replication and offers a potent ligand for virus particles (*80*). (ii) Although the only drug to show clear clinical benefit thus far, azidothymidine, attacks the "obvious" step, reverse transcription (*81*), recent perceptions about retroviral integration, transcription, translation, proteolysis, and assembly encourage a search for new drugs to act at other steps as well. (iii) The HIV genome has been discovered to encode no less than five novel proteins in addition to those encoded by *gag*, *pol*, and *env*: one protein is a positive regulator of HIV gene expression, acting mainly to increase levels of viral RNA (*82*); another influences the relative abundance of various spliced and unspliced HIV mRNAs (*70*); a third augments the infectiousness of HIV (*83*); and yet another resembles host G proteins (*84*). The mechanisms by which such proteins regulate virus production are largely unknown, but of obvious importance.

Because most retroviruses are not cytopathic and because there has been little incentive to develop vaccines against retroviral diseases in most animals, studies of retroviral pathogenesis (other than oncogenesis) and of the immune response to retroviral infection have lagged behind work on the molecular biology of these viruses. AIDS has dramatically changed these attitudes. The immune response must be understood in order to develop clinical tests for HIV and for the consequences of infection; to devise better strategies for vaccination in the face of discouraging levels of neutralizing antibodies in infected patients (*85*) and disappointing vaccine trials in chimpanzees (*86*); and to evaluate hypotheses about pathogenesis, taking into account both antiviral and autoimmune components and the complications of infecting cells that are themselves central to an immune response (*87*).

HTLV. The first isolates of HTLVs from patients with an uncommon but virulent leukemia (*7*) vindicated a decade of frustrating and sometimes embarrassing efforts to find human oncogenic retroviruses to match their counterparts in animals (*88*). The HTLVs have proven to be curious agents, difficult to understand as pathogens and difficult to study as infectious viruses. Unlike common oncogenic retroviruses of animals, HTLVs neither carry host-derived oncogenes nor activate cellular proto-oncogenes by insertion mutation. Instead, their oncogenic action has been provisionally ascribed to an open reading frame that lies between the *env* gene and the 3'LTR and encodes a protein that acts as a positive effector of transcription from the HTLV LTR and from certain cellular promoters (*89*). However, models for tumorigenesis must also account for the prolonged latency and infrequent occurrence of disease in infected people, the lack of viral gene expression in primary tumor tissue, and the dearth of direct transformation assays for individual viral genes (*90*).

Although the number of cases of HTLV-associated leukemia and lymphoma is relatively small, many people are infected, especially in Japan and the Caribbean and among intravenous drug abusers in this country and Europe (*91*). Thus better understanding of the biological properties of the HTLVs and their relatives (bovine leukemia virus and simian T cell leukemia virus) is urgently needed.

Retroviruses as Tools for Studying Development, Delivering Genes, and Curing Diseases

Retroviruses can also be used as technical devices for genetically altering host cells and organisms. When put to such purposes, retroviruses themselves are not usually at the heart of the scientific question; instead they need to be understood only as far as necessary to achieve other objectives: marking cells with a recognizable provirus, causing mutations with interesting phenotypes, expressing a favorite gene in a desired cell type, or correcting a genetic deficiency.

The incentive to use retroviruses as genetic vectors originated with the perception that retroviruses with viral oncogenes are naturally occurring genetic vectors (Fig. 4). The ways in which oncogenes are incorporated into viral genomes and efficiently expressed have now been appropriated by investigators designing vectors of their own, to deliver to chosen cells any of the large collection of genes made available by molecular cloning. In general, this enterprise has been hugely successful: experiments in many fields of biology now depend on retrovirus vectors to deliver genes to cultured cells and occasionally to animals. The vectors can be grown to high titers, they often carry two genes in various arrangements, and they are available in models that do or do not initiate an infection that spreads to surrounding cells (16).

It is anticipated that retroviral vectors will ultimately be used to correct human genetic deficiencies (16). However, reliable expression of transmitted genes has yet to be achieved after infection of hematopoietic stem cells, the usual targets in current strategies, and it is not yet known whether infected cells will persist in the host in numbers adequate to ameliorate symptoms. The safety of retrovirus vectors has also not been fully evaluated. Improved design of vectors intended for human gene therapy thus remains a major technical challenge, but one that promises relief from any disease caused by a recessive mutation in a gene available for delivery.

In the meantime, retroviruses are becoming important tools in developmental biology because proviruses can stably and benignly mark cells for tracing lineages and can initiate insertion mutations with developmental consequences. (i) Cells have been marked with MLV proviruses by infecting preimplantation embryos to learn when cells become committed to a single lineage or organ (92); by infecting hematopoietic stem cells to show that a single cell can serve as the source of all blood cells (93); and by infecting retinas to identify the heterogeneous descendants of a single cell (94). (ii) Natural or experimental infection of the mouse germ line with MLV has occasionally produced insertion mutations with especially interesting effects: an endogenous provirus on chromosome 9 is responsible for the light hair pigmentation characteristic of *dilute* mice (95), and an insertion into an intron of a collagen gene after infection of the mouse germ line produced a recessive mutation lethal to midgestational embryos (96). (iii) Retroviral infection of embryonic stem cells in culture may ultimately permit production of mouse strains with a wide variety of important genetic lesions. Some success has already been achieved with a known X-linked gene, *hprt*, as a mutational target for MLV; metabolic selection of cells in which *hprt* was disrupted recently led to production of HPRT-deficient mice from the mutant stem cells (97).

A Final Perspective on Retrovirology

This review has stressed the several facets of retroviruses that have attracted people to them in increasing numbers over the past two decades: retroviruses as potent models for understanding many aspects of both normal and cancerous eukaryotic cells; as important pathogens in humans and animals; as technical devices for gene delivery, mutagenesis, and lineage marking; and, most simply, as inherently fascinating microbes. Although the success of retrovirology is apparent from the sheer numbers of people who now work with these viruses, a more persuasive measure is the major impact the discipline has had upon the way we now think about many important topics in contemporary biology: the pathways for transfer of genetic information in eukaryotic cells, the causes of cancer and other major diseases, the genes and proteins that contribute to normal growth and development, and the regulation of gene expression.

Despite the extraordinary productivity of retrovirology in recent years, questions in all branches of the discipline seem larger and more numerous than ever. The major outlines of the replicative cycle are firmly drawn, but mechanisms of central events—virus entry, integration, regulated expression, and assembly—are just now coming into view and have assumed a greater urgency because of AIDS. A monumental list of oncogenes and proto-oncogenes has been assembled, but the biochemical activities crucial to growth control and neoplasia await discovery. Important human pathogens have been identified among retroviruses, but strategies for prevention and cure are still desperately needed. Strong evidence for the utility of retroviruses as genetic reagents is in hand, but insertional mutagenesis is not yet a simple experimental device, and gene therapy with retroviral vectors is not yet ready for clinical trials. Such deficiencies are the legacy of progress, and an invitation to the future.

References and Notes

1. H. Vallee and H. Carre, *C. R. Acad. Sci. Ser. D.* **139**, 331 (1904); P. Rous, *J. Exp. Med.* **13**, 397 (1911); V. Ellermann and O. Bang, *Fizentralbl. Bakteriol.* **46**, 595 (1908).
2. J. J. Bittner, *Science* **84**, 162 (1936); L. Gross, *Proc. Soc. Exp. Biol. Med.* **76**, 27 (1951).
3. H. M. Temin and H. Rubin, *Virology* **6**, 669 (1958).
4. H. M. Temin and S. Mizutani, *Nature* **226**, 1211 (1970); D. Baltimore, *ibid.*, p. 1209.
5. R. J. Huebner and G. J. Todaro, *Proc. Natl. Acad. Sci. U.S.A.* **64**, 1087 (1969); D. R. Lowy, W. P. Rowe, N. Teich, J. W. Hartley, *Science* **174**, 155 (1971); W. P. Rowe, *J. Exp. Med.* **136**, 1272 (1972).
6. D. Stehelin, H. E. Varmus, J. M. Bishop. P. K. Vogt, *Nature* **260**, 170 (1976).
7. B. J. Poiesz *et al.*, *Proc. Natl. Acad. Sci. U.S.A.* **77**, 7415 (1980).
8. F. Barré-Sinoussi *et al.*, *Science* **220**, 868 (1983); R. C. Gallo *et al.*, *ibid.* **224**, 500 (1984); J. A. Levy *et al.*, *ibid.* **225**, 840 (1984).
9. J. M. Bishop, *ibid.* **235**, 305 (1987); H. E. Varmus, *Annu. Rev. Genet.* **18**, 553 (1984).
10. R. A. Weiss, N. Teich, H. Varmus, J. Coffin, Eds., *Molecular Biology of Tumor Viruses: RNA Tumor Viruses* (Cold Spring Harbor Laboratory, Cold Spring Harbor, NY, 1982, 1985), vol. 1 and vol. 2, respectively.
11. H. E. Varmus, in *Mobile Genetic Elements*, J. Shapiro, Ed. (Academic Press, New York, 1983), pp. 411–503.
12. _____ and R. Swanstrom, in *(10)*, vol. 1, pp. 369–512.
13. J. Summers and W. S. Mason, *Cell* **29**, 403 (1982).
14. R. Hull and S. N. Covey, *Trends Biol. Sci.* **8**, 119 (1982).
15. For eukaryotes: A. M. Weiner, P. L. Deininger, A. Efstratiadis, *Annu. Rev. Biochem.* **55**, 631 (1987); for prokaryotes: M. Beljanski and M. Beljanski, *Biochem. Genet.* **12**, 163 (1974); S. Inouye, M.-Y. Hsu, S. Eagle, and M. Inouye, *Cell* **56**, 709 (1989); D. Lim and W. K. Maas, *ibid.*, p. 891; H. E. Varmus, *ibid.*, p.721.
16. E. Nichols, *Human Gene Therapy* (Harvard Univ. Press, Cambridge, MA, 1988); J. Coffin, in *(10)*, vol. 2, pp. 17–73.
17. H. E. Varmus and J. M. Bishop, Eds., *Cancer Surv.* **5** (1986); H. E. Varmus, in *Molecular Basis of Blood Diseases*, G. Stamatoyannopoulos, A. W. Nienhuis, P. Leder, P. W. Majerus, Eds. (Saunders, Philadelphia, 1987), pp. 271–346.
18. C. A. Mims, *J. Infect. Dis.* **12**, 199 (1986).
19. J. M. Coffin, *J. Gen. Virol.* **42**, 1 (1979); E. Hunter, *Curr. Top. Microbiol. Immunol.* **79**, 295 (1978).
20. H. E. Varmus, *Sci. Am.* **257**, 56 (September 1987).
21. H. M. Temin, *Mol. Biol. Evol.* **2**, 455 (1986).
22. J. D. Watson, V. Hopkins, T. Roberts, J. Steitz, A. Weiner, *Molecular Biology of the Gene* (Benjamin-Cummings, Menlo Park, CA, ed. 4, 1987), vol. 2, chap. 28, pp. 1098–1163.
23. I. M. Verma, *Biochim. Biophys. Acta* **473**, 1 (1977).
24. M. J. Roth, N. Tanese, S. Goff, *J. Biol. Chem.* **260**, 9326 (1985); N. Tanese, J. Sodroski, W. Haseltine, S. Goff, *J. Virol.* **59**, 743 (1986); A. Hizi, C. McGill, S. H. Hughes, *Proc. Natl. Acad. Sci. U.S.A.* **85**, 1218 (1988).
25. W. S. Mason, C. Aldrich, J. Summers, J. M. Taylor, *Proc. Natl. Acad. Sci. U.S.A.* **79**, 3997 (1982).
26. D. J. Garfinkel, J. D. Boeke, G. R. Fink, *Cell* **42**, 507 (1985); J. D. Boeke, D. J. Garfinkel, C. A. Styles, G. R. Fink, *ibid.* **40**, 491 (1985).
27. A. J. Flavell and D. Ish-Horowicz, *ibid.* **34**, 415 (1983); I. R. Arkhipova *et al.*, *ibid.* **44**, 555 (1986).
28. S. Tonegawa, *Sci. Am.* **253**, 122 (October 1985); L. Hood, M. Kronenberg, T. Hunkapiller, *Cell* **40**, 225 (1985); S. Akira, K. Okazaki, H. Sakano, *Science* **238**, 1134 (1987).
29. R. S. Kucherlapati, E. M. Eves, K.-Y. Song, B.

S. Morse, O. Smithies, *Proc. Natl. Acad. Sci. U.S.A.* **81**, 3153 (1984); C. T. Wake, F. Vernaleone, J. H. Wilson, *Mol. Cell. Biol.* **5**, 2080 (1985); P. K. Bandyopadhyay, S. Watanabe, H. M. Temin, *Proc. Natl. Acad. Sci. U.S.A.* **81**, 3476 (1984).

30. K. R. Thomas, K. R. Folger, M. R. Capecchi, *Cell* **44**, 419 (1986); F. L. Lin, K. Sperle, N. Sternberg, *Proc. Natl. Acad. Sci. U.S.A.* **82**, 1391 (1985); R. M. Liskay and J. L. Stachelek, *Cell* **35**, 157 (1983).

31. H. E. Varmus and P. Brown, in *Mobile DNA Elements*, M. Howe and D. Berg, Eds. (American Society of Microbiology, Washington, DC, 1989) pp. 53-108.

32. M. Botchan, W. Topp, J. Sambrook, *Cell* **9**, 269 (1976); G. Ketner and T. J. Kelly, Jr., *Proc. Natl. Acad. Sci. U.S.A.* **73**, 1102 (1976).

33. S. H. Hughes *et al.*, *Cell* **15**, 1397 (1978); T. W. Hsu, J. L. Sabran, G. E. Mark, R. V. Guntaka, J. M. Taylor, *J. Virol.* **28**, 810 (1979).

34. A. T. Panganiban and H. M. Temin, *Nature* **306**, 155 (1983); J. Colicelli and S. P. Goff, *J. Mol. Biol.* **199**, 47 (1988).

35. L. A. Donehower and H. E. Varmus, *Proc. Natl. Acad. Sci. U.S.A.* **81**, 6461 (1984); P. Schwartzberg, J. Colicelli, S. P. Goff, *Cell* **37**, 1043 (1984); A. T. Panganiban and H. M. Temin, *Proc. Natl. Acad. Sci. U.S.A.* **81**, 7885 (1984).

36. P. O. Brown, B. Bowerman, H. E. Varmus, J. M. Bishop, *Cell* **49**, 347 (1987).

37. A. P. Brown, B. Bowerman, H. E. Varmus, J. M. Bishop, *Proc. Natl. Acad. Sci. U.S.A.* **86**, 2525 (1989).

38. T. Fujiwara and R. Craigie, *ibid.*, p. 3065.

39. J. M. Bishop and H. E. Varmus, in (*10*), vol. 1, pp. 999–1108.

40. N. Teich, J. Wyke, T. Mak, A Bernstein, W. Hardy, *ibid.*, pp. 785–998.

41. J. M. Bishop, *Annu. Rev. Biochem.* **52**, 301 (1983).

42. G. S. Martin, *Nature* **227**, 1021 (1970); P. K. Vogt, *Virology* **46**, 939 (1971).

43. M. Linial, in (*10*), vol. 1, pp. 649–783.

44. M. S. Collett and R. L. Erikson, *Proc. Natl. Acad. Sci. U.S.A.* **75**, 2021 (1978); A. D. Levinson, H. Oppermann, L. Levintow, H. E. Varmus, J. M. Bishop, *Cell* **15**, 561 (1978).

45. T. Hunter and B. M. Sefton, *Proc. Natl. Acad. Sci. U.S.A.* **77**, 1311 (1980).

46. T. Hunter and J. A. Cooper, *Annu. Rev. Biochem.* **54**, 897 (1985).

47. J. Downward *et al.*, *Nature* **307**, 521 (1984).

48. C. J. Sherr *et al.*, *Cell* **41**, 665 (1985).

49. R. F. Doolittle *et al.*, *Science* **221**, 275 (1983); M. D. Waterfield *et al.*, *Nature* **304** 35 (1983).

50. P. C. Nowell and D. A. Hungerford, *Science* **132**, 1497 (1960).

51. J. Groffen *et al.*, *Cell* **36**, 93 (1984); J. B. Konopka, S. M. Watanabe, O. N. Witte, *ibid.* **37**, 1035 (1984).

52. C. Shih, B.-Z. Shilo, M. P. Goldfarb, A. Dannenburg, R. A. Weinberg, *Proc. Natl. Acad. Sci. U.S.A.* **76**, 5714 (1979).

53. C. J. Tabin *et al.*, *Nature* **300**, 143 (1982); C. J. Der, T. G. Krontiris, G. M. Cooper, *Proc. Natl. Acad. Sci. U.S.A.* **79**, 3637 (1982).

54. M. Barbacid, *Annu. Rev. Biochem.* **56**, 779 (1987).

55. P. K. Vogt, T. J. Bos, R. F. Doolittle, *Proc. Natl. Acad. Sci. U.S.A.* **84**, 3316 (1987); K. Struhl, *Cell* **50**, 841 (1987).

56. D. Bohmann *et al.*, *Science* **238**, 1386 (1987).

57. R. Gentz, F. J. Rauscher III, C. Abate, T. Curran, *Science* **243**, 1695 (1989); E. Ziff and T. Kouzarides, *Nature* **336**, 646 (1988); P. Sassone-Corsi, L. J. Ransone, W. W. Lamph, I. M. Verma, *ibid.*, p. 692; R. Turner and R. Tjian, *Science* **234**, 1689 (1989).

58. B. G. Neel, W. S. Hayward, H. L. Robinson, J. Fang, S. M. Astrin, *Cell* **23**, 323 (1981); G. S. Payne *et al.*, *ibid.*, p. 311.

59. W. S. Hayward, B. G. Neel, S. M. Astrin, *Nature* **290**, 475 (1981).

60. R. Nusse and H. E. Varmus, *Cell* **31**, 99 (1982).

61. A. Jakobovits, G. M. Shackleford, H. E. Varmus, G. R. Martin, *Proc. Natl. Acad. Sci. U.S.A.* **83**, 7806 (1986); G. M. Shackleford and H. E. Varmus, *Cell* **50**, 89 (1987); D. G. Wilkinson, J. A. Bailes, A. P. McMahon, *ibid.*, p. 79.

62. F. Rijsewijk *et al.*, *Cell* **50**, 649 (1987); C. V. Cabrera, M. C. Alonso, P. Johnston, R. G. Phillips, P. A. Lawrence, *ibid.*, p. 659; N. Baker, *EMBO J.* **6**, 1765 (1987).

63. J. Papkoff, A. M. C. Brown, H. E. Varmus, *Mol. Cell. Biol.* **7**, 3978 (1987); G. Morata and P. Lawrence, *Dev. Biol.* **56**, 227 (1977).

64. G. Nabel and D. Baltimore, *Nature* **326**, 711 (1987); N. A. Speck and D. Baltimore, *Mol. Cell. Biol.* **7**, 1101 (1987); K. A. Jones, J. T. Kadonaga, P. A. Luciw, R. Tjian, *Science* **232**, 755 (1986).

65. P. A. Chatis, C. A. Holland, J. W. Hartley, W. P. Rowe, N. Hopkins, *Proc. Natl. Acad. Sci. U.S.A.* **80**, 4408 (1983); D. Celander and W. A. Haseltine, *Nature* **312**, 159 (1984).

66. K. Yamamoto, *Annu. Rev. Genet.* **199**, 209 (1985).

67. G. Ringold, K. R. Yamamoto, G. M. Tomkins, J. M. Bishop, H. E. Varmus, *Cell* **6**, 299 (1975).

68. E. B. Jakobovits, J. E. Majors, H. E. Varmus, *ibid.* **38**, 757 (1984).

69. W. S. Hayward, *J. Virol.* **24**, 47 (1977); S. R.

Weiss, H. E. Varmus, J. M. Bishop, *Cell* 12, 983 (1977); N. Quintrell, S. H. Hughes, H. E. Varmus, J. M. Bishop, *J. Mol. Biol.* 143, 363 (1980).

70. M. B. Feinberg, R. F. Jarrett, A. Aldovini, R. C. Gallo, F. Wong-Staal, *Cell* 46, 807 (1986); J. Sodroski *et al.*, *Nature* 321, 412 (1986).

71. J. Inoue, M. Yoshida, M. Seiki, *Proc. Natl. Acad. Sci. U.S.A.* 84, 3653 (1987); M. Hidaki, J. Inoue, M. Yoshida, M. Seiki, *EMBO J.* 7, 519 (1988).

72. Y. Yoshinaka, I. Katoh, T. D. Copeland, S. Oroszlan, *Proc. Natl. Acad. Sci. U.S.A.* 82, 1618 (1985).

73. T. Jacks and H. E. Varmus, *Science* 230, 1237 (1985); T. Jacks, K. Townsley, H. E. Varmus, J. Majors, *Proc. Natl. Acad. Sci. U.S.A.* 84, 4298 (1987); R. Moore, M. Dixon, R. Smith, G. Peters, C. Dickson, *J. Virol.* 61, 480 (1987).

74. T. Jacks, H. D. Madhani, F. R. Masiarz, H. E. Varmus, *Cell* 55, 447 (1988).

75. I. Brierley *et al.*, *EMBO J.* 6, 3779 (1987).

76. J. Clare and P. Farabaugh, *Proc. Natl. Acad. Sci. U.S.A.* 82, 2829 (1985); W. Wilson, M. H. Malim, J. Mellor, A. J. Kingsman, S. M. Kingsman, *Nucleic Acids Res.* 14, 7001 (1986).

77. R. Y. Sekine and E. Ohtsubo, *Proc. Natl. Acad. Sci. U.S.A.* 86, 4609 (1989).

78. M. Miller *et al.*, *Nature* 337, 576 (1989); M. A. Naviar *et al.*, *ibid.*, p. 615; A. M. Skalka, *Cell* 56, 911 (1989).

79. A. G. Dalgleish *et al.*, *Nature* 312, 763 (1984); D. Klatzmann *et al.*, *ibid.* p. 767; P. J. Maddon *et al.*, *Cell* 47, 333 (1986); J. S. McDougal *et al.*, *Science* 231, 382 (1986).

80. Q. J. Sattentau and R. A. Weiss, *Cell* 52, 631 (1988).

81. H. Mitsuya and S. Broder, *Nature* 325 773 (1987).

82. C. A. Rosen *et al.*, *ibid.* 319, 555 (1986); B. R. Cullen, *Cell* 46, 973 (1986); A. G. Fisher *et al.*, *Nature* 320, 367 (1986).

83. A. G. Fisher *et al.*, *Science* 237, 888 (1987); K.

Strebel *et al.*, *Nature* 328, 728 (1987).

84. B. Guy *et al.*, *Nature* 330, 266 (1987).

85. R. A. Weiss *et al.*, *Nature* 316, 69 (1985).

86. P. W. Berman *et al.*, *Proc. Natl. Acad. Sci. U.S.A.* 85, 5200 (1988).

87. A. S. Fauci, *Science* 239, 617 (1988).

88. R. Weiss, in (*10*), vol. 1, pp. 1205–1281.

89. J. G. Sodroski, C. A. Rosen, W. A. Haseltine, *Science* 225, 381 (1984); M. Seiki, J. I. Inoue, T. Takeda, M. Yoshida, *EMBO J.* 5, 561 (1986); W. C. Greene *et al. Science* 232, 877 (1986).

90. I. S. Y. Chen, W. Wachsman, J. D. Rosenblatt, A. J. Cann, *Cancer Surv.* 5, 329 (1986).

91. Y. Hinuma *et al.*, *Proc. Natl. Acad. Sci. U.S.A.* 78, 6476 (1981); R. S. Tedder *et al.*, *Lancet* ii, 125 (1984); R. C. Gallo, *Sci. Am.* 255, 88 (December 1986).

92. P. Soriano and R. Jaenisch, *Cell* 46, 19 (1986).

93. J. E. Dick, M. C. Magli, D. Huszar, R. A. Phillips, A. Bernstein, *ibid.* 42, 71 (1985); I. R. Lemischka, D. H. Raulet, R. C. Mulligan, *ibid.* 45, 917 (1986).

94. J. Price, D. Turner, C. Cepko, *Proc. Natl. Acad. Sci. U.S.A.* 84, 156 (1987).

95. N. A. Jenkins, N. G. Copeland, B. A. Taylor, B. K. Lee, *Nature* 293, 370 (1981); N. G. Copeland, K. W. Hutchison, N. A. Jenkins, *Cell* 33, 379 (1983).

96. R. Jaenisch *et al.*, *Cell* 32, 209 (1983).

97. M. R. Kuehn, A. Bradley, E. J. Robertson, M. J. Evans, *Nature* 326, 295 (1987).

98. H. E. Varmus, *Adv. Oncol.* 3, 3 (1987).

99. I thank members of my laboratory for helpful discussions of topics reviewed here, P. Brown for pertinent collaboration on (*31*), M. Bishop for long-standing shared interests in retroviruses, L. Levintow for textual criticism, N. Hardy for the drawings in Figs. 1 and 2A, J. Marinos for Fig. 4 and help with the manuscript, and T. Jacks for Fig. 5. Supported by grants from NIH.

2

Research on Bacteria in the Mainstream of Biology

Boris Magasanik

In their introductory chapter to the recently published treatise *Escherichia coli and Salmonella typhimurium, Cellular and Molecular Biology* (*1*), Schaechter and Neidhardt conclude with the statement: "Not everyone is mindful of it, but all cell biologists have two cells of interest: the one they are studying and *Escherichia coli*" (*2*). This view correctly reflects the great contribution the study of this prokaryotic organism has made to the current concepts of the biology of eukaryotic microbial, plant, and animal cells. Yet, less than 50 years ago, in 1954, Kluyver and Van Niel, two eminent microbiologists, found it necessary to devote five lectures at Harvard University to convince their audience that the study of microbes could make a major contribution to biology (*3*). As late as 1942, J. S. Huxley expressed the view that bacteria may lack a genetic system analogous to that of higher organisms (*4*). It was only in 1943, when Luria and Delbrück reported the results of their experiments on the statistics of mutation in *E. coli,* that it was clearly shown that changes in the phenotype that had been observed in bacteria were not due to a direct effect of the environment, but arose from spontaneous genetic alteration followed by Darwinian selection (*5*). In 1944, the identification by Avery and his collaborators of the

material responsible for the transformation of cells of *Streptococcus pneumoniae* as DNA, whose presence in the nuclei of higher cells was well established, confirmed the concept of the unity of all living forms (*6*). Finally, the report by Lederberg and Tatum in 1946 concerning genetic recombination in *E. coli* made it clear that bacteria were proper objects for genetic study (*7*).

The discovery of a method that greatly facilitated the isolation of auxotrophic mutants independently by Davis and by Lederberg in 1948 was the next important step in the recognition of the utility of bacteria for genetic studies (*8*). These mutants allowed the isolation of rare recombinants; the great advantage of using bacteria with generation times much shorter than those of other organisms could thus be readily exploited. In addition, these mutants were used for the elucidation of the biosynthetic pathways leading to amino acids and nucleotides (*9*), and for the elucidation of the organization of the genetic material (*10*). Some of the great advances in our understanding of the general principles of metabolic regulation, the synthesis of macromolecules, and the regulation of gene expression that followed these initial discoveries have already been well described by Judson (*11*).

B. Magasanik is at the Department of Biology, Massachusetts Institute of Technology, Cambridge, MA 02139. This chapter is reprinted from *Science* 240, 1435 (1988).

Recent Results with *Escherichia coli*

It is clear, in retrospect, that a young biologist embarking on a career in the 1950s could look forward to a lifetime of research in the mainstream of biology by choosing to work on bacteria. It is my purpose here to discuss whether a similar prognosis can be made for those who choose bacteria as their object of study at the present time. Obviously, the scene is a different one. The concepts derived from the study of the bacterial cell have greatly enhanced our ability to study the properties of eukaryotic cells. The results of some of these studies have been interpreted as indicating that, with the exception of their most basic properties, eukaryotic cells are so different from prokaryotic cells that further study of bacteria is not likely to produce results of general importance for our understanding of the properties of all types of cells. It has also been suggested that essentially everything of importance for understanding the biology of the bacterial cell has been discovered and that further exploration will only reveal additional facts that can be derived from previously established principles.

That neither of these views is correct was vividly brought home to me when I began to write this article in the first days of March 1988. The issue of *Science* of 26 February 1988 and of *Nature* of 25 February 1988 had arrived on the same day. Each contained a report of an unexpected finding in cells of *E. coli* with considerable significance for the molecular biology of the cells of higher organisms. The report in *Science* (*12*) provided evidence for the presence of a 50-nucleotide untranslated region in gene 60 of the *E. coli* phage T4. It had already been shown that some other genes of this bacteriophage code for messenger RNA subject to processing by a self-splicing mechanism originally discovered in, and thought to be characteristic of, eukaryotic cells (*13*). The remarkable discovery concerning gene 60 was that in this case the untranslated region is not removed prior to translation. Apparently, it is possible for the translation apparatus to bypass a region of the message by an as yet unknown mechanism. Elucidation of this mechanism will result in a better understanding of protein synthesis in all cells.

The report in *Nature* dealt with the discovery of an *E. coli* gene for a new species of transfer RNA that accepts serine and cotranslationally inserts selenocystein into a polypeptide (*14*). Selenopeptides are subunits of formate dehydrogenase produced by *E. coli* under anaerobic conditions. The codon for this tRNA, responsible for the incorporation of an amino acid beyond the canonical 20 into a protein, is UGA, whose other role is chain termination. Selenocystein is also an essential component of mammalian glutathione peroxidase and may be incorporated into this polypeptide by a corresponding mechanism. The manner of discovery of the gene for this unusual tRNA in *E. coli* is a good example of the advantage of using a simple microorganism for the precise analysis of a biological phenomenon. It was found that mutation in any one of three genes results in the inability of the cell to incorporate selenium into formate dehydrogenase, a step required for the formation of the active enzyme. When these genes were cloned by complementation it was found that two of the genes code for polypeptides and the third gene for a previously unknown tRNA. The fact that alteration of the tRNA by mutation results in the inability of the cell to produce the selenium-containing enzyme provides clear evidence for the role of the tRNA in this process.

What these examples show is that the very fact that so much is known about *E. coli* makes it possible to recognize the unusual in new observations and to devise experimental approaches that will reveal the significance of new findings.

Regulation of Gene Expression

Our knowledge of *E. coli* and its relatives makes it possible to delineate the areas whose exploration in the near future, let us say in the next 10 or 15 years, will significantly contribute to our understanding of cell biology. One such area is the regulation of gene expression. It was the brilliant exploration of the regulation of β-

galactosidase synthesis by Monod and Jacob that led to the discovery of mRNA as the unstable intermediate between the DNA and the polypeptide, to the recognition of the promoter as the site on the DNA where RNA polymerase initiates the transcription of a gene or of a group of genes, and to the concept that a specific regulatory protein can control gene expression by binding to a specific site on the DNA (15). On the basis of their study of the induction of β-galactosidase synthesis by the addition of β-galactosides to the growth medium, Jacob and Monod proposed a simple mechanism for the regulation of the expression of all genes and even suggested the circuitry that would allow this mechanism to be used to organize the differentiation of the cells of higher organisms (15). They postulated that the expression of a gene or of a group of linked genes is negatively regulated by a specific macromolecule, the repressor, which can be reversibly modified by interaction with a specific small molecule, the effector. Either the unmodified or the modified repressor would block the transcription of the regulated gene by binding to the site on the DNA used by RNA polymerase to initiate the transcription. In the former case the effector would induce and in the latter case repress the expression of the gene.

The clear exposition of this hypothesis indicated the experiments that had to be done to test its validity for other systems and encouraged the development of new methods to test its specific predictions. The results of these investigations eventually revealed the inadequacy of this simple model. The expression of genes is not exclusively regulated at the initiation of transcription and does not necessarily involve specific regulatory macromolecules. Rather, the regulation of gene expression is characterized by a diversity of the mechanisms and by the initially unsuspected complexity of the individual systems (16).

It is now evident that regulatory proteins may not only block, but may also activate the initiation of transcription by RNA polymerase and that they may in either case bind to a site on the DNA far from the binding site for RNA polymerase (17). These observations suggest complex interactions between the regulatory protein and the RNA polymerase involving bending of the intervening DNA. The fact that these phenomena have been observed not only in bacterial cells, but also in animal cells, emphasizes their generality (18). One of the important goals of future research will be to elucidate the nature of the interactions between regulatory proteins and RNA polymerase. The advantage of the bacterial system for such investigations is the fact that it is much easier here than in other systems to correlate the expression of the gene in the intact cell with its expression in a system composed of highly purified components. It is therefore possible to identify without ambiguity the exact role of the individual components of complex regulatory systems.

This approach has led to a new understanding of the role of RNA polymerase in the expression of different genes. Bacteria contain several different σ subunits, the product of specific genes (19). These σ subunits combine with core RNA polymerase to enable it to distinguish different classes of promoters. Apparently, the bacterial cell uses this ability in a variety of ways. In the case of regulation of gene expression in response to nitrogen availability, the specific σ factor appears to reserve a portion of the RNA polymerase for the transcription of the regulated genes, but is not directly responsible for the regulatory response (20). In the case of regulation of gene expression in response to a rise in temperature, that is, "heat shock," it is the increased accumulation of specific σ factor resulting from the heat shock that causes increased expression of the regulated genes (21). Finally, in the case of sporulation in *Bacillus subtilis*, a number of different σ subunits are involved in the orderly execution of a program (22).

In addition to the σ subunit, other regulatory proteins have been shown to associate temporarily with RNA polymerase to regulate proper elongation and termination (23). Thus, there is no principal difference between the RNA polymerase of bacteria and the RNA polymerase of eukaryotic cells. It is clear that further study of the regulation of the initiation and progress of transcription in bacteria will make valuable contributions to our understanding of the corresponding mechanisms in eukaryotic cells.

The complexity of individual regulatory systems is particularly apparent in the global response of the bacterial cell to profound changes in its environment (24). Thus, a sudden change in the temperature, deprivation of a source of carbon, nitrogen, or phosphorous, the lack of oxygen, damage to DNA by irradiation, each triggers the activation of the expression of a set of different genes. This activation requires the sensing by the cell of the environmental change and the transmission of this information to the protein responsible for the activation of the expression of the multigene system. The simplicity of the bacterial cell which lacks the compartments such as the nuclei, mitochrondria, and vacuoles of eukaryotic cells makes it possible to study precisely the interactions of the proteins and small molecules that determine the response of such complex cascade systems.

The study of the global systems has barely begun and each of the global control systems of *E. coli* that has been thus far investigated has provided new and unexpected information. The fact that heat shock elicits the production of a set of similar proteins in bacteria, yeast, and the cells of higher organisms suggests that the study of this system in *E. coli* will be of importance for our understanding of this response in eukaryotic cells (21).

Another area of global control of gene expression is the response of the cell to changes in the availability of oxygen and of other electron acceptors. There is some evidence that the regulation of gene expression in these cases involves control of the supercoiling of the DNA (25). It is likely that further study of this regulatory mechanism will lead to the elucidation of complex interactions essential for the proper function of all cells.

The study of regulation in response to nitrogen utilization has revealed that a cascade system composed of three proteins is involved in the activation of the effector (NR_I) that is, in turn, responsible for the activation of the expression of the regulated genes (26). Two of these proteins are also part of a cascade system

responsible for the regulation of glutamine synthetase activity in response to nitrogen availability.

NR_I shares considerable homology in its NH_2-terminal domain with a large number of bacterial effectors, some of which have been identified as activators of transcription. One of these effectors, the product of the *cheY* gene of *Salmonella typhimurium*, is an activator of the flagellar motor (27). Effectors are modulated, in response to an environmental stimulus, by a second class of proteins (modulators) that are homologous in their COOH-terminal domains (27). For example, NR_I is converted to the form capable of activating the transcription of the regulated genes by phosphorylation (26). This conversion is brought about by the attachment of the γ-phosphate of ATP to a histidine residue, presumably located in the COOH-terminal domain of a modulator (NR_{II}), followed by its transfer to an aspartate residue in the NH_2-terminal domain of NR_I (26). In several additional effector-modulator systems, the modulator phosphorylates the effector in this manner (28).

Two component systems of this type govern the cellular responses to chemoattractants, phosphate availability, nutritional status, osmolarity, and anaerobiosis (28). Moreover, they have recently been identified as key regulators of the virulence of bacteria for plant and animal cells. They control the respective abilities of *Agrobacterium tumefaciens* to induce tumors in susceptible plants, of *Vibrio cholerae* and *Bordetella pertussis* to produce toxins and of *Shigella flexneri* and *S. typhimurium* to invade mammalian cells (29). These recent discoveries open exciting prospects for scientists interested in the study of disease and for those interested in the evolution of cellular regulatory mechanisms.

DNA and Membrane Activities

I have tried in the preceding paragraphs to give specific examples of problems of general biological significance that can most readily be studied in a bacterial system. All the examples dealt with the regulation of the synthesis and function of cytoplasmic proteins. It is this area where most is known, and yet, where, as my examples attempt to show, a great deal remains to be discovered. In addition, there are other less well explored areas of cell biology that can be more readily studied in prokaryotic rather than in eukaryotic cells. In these cases the advantages are again the simple noncompartmented structure of the bacterial cell, the ready accessibility of the genetic material, and the ease with which an alteration in the macromolecular composition of the cell can be related to a change in its physiology.

Of particular importance are also the transactions of DNA, such as replication, repair, and recombination (*30*). Much remains to be learned with regard to the physical and biochemical nature of these processes and the responsible macromolecules.

Another important question is how can an environmental stimulus be transmitted to the interior of the cell. In a number of instances this signal transduction is accomplished by proteins located in the cytoplasmic membrane that have domains protruding on both sides of the membrane. Some modulators, described in the preceding section (but not including NR_{II}), have this property (*28*). Apparently, the interaction of the extracellular, periplasmic domain of these membrane proteins with a molecule outside the cell affects their intracellular domain. The responsible mechanism can be most readily elucidated by the isolation of mutants defective in signal transduction, followed by the localization of the mutation to a region of the gene encoding a specific domain of the membrane protein (*31*).

The importance of the proteins located within the cell membrane or in the periplasmic space for proper cell function is evident. It is therefore of interest to discover the mechanisms responsible for placing these proteins in their compartments. The recent demonstration that the protein SecA, which exists in a soluble and in a membrane-bound form, is essential for this process is an important advance, as the process can now be studied outside the cell (*32*).

The cytoplasmic membrane is not only responsible for the transport of molecules in and out of the cytoplasm, but also plays an important role in energy metabolism. Thus, it combines the function of the cell membrane and the mitochondria of eukaryotic cells. In this case the different techniques that have been developed for the study of the function of cytoplasmic macromolecules can be used to identify the specific steps in the construction of this complex cell constituent.

This knowledge is in turn required for still another area of research that can be most readily investigated in bacteria, the coordination of the synthesis of DNA and cell division. The cell cycle constitutes an orderly process comprising a continuous increase in cell length accompanied by DNA replication and nucleoid segregation, which is followed by septation and cell division (*33*). The simplicity of *E. coli*, in which the DNA is organized as a single chromosome that is not physically separated from the cytoplasm but is in contact with the cell membrane, makes this system very attractive for the study of this mechanism. It is likely that the study of this bacterial mechanism will provide important clues for our understanding of the more complex regulation of the cell cycle in eukaryotic cells.

In the case of rod-shaped bacteria such as *E. coli,* one of the last stages of cell division is the formation of a division septum at the midpoint of the long axis of the cell. The exact placement of the septum is important because an unequal division would not allow the appropriate partition of the genetic material. Mutations that cause the placement of the septum at the poles result in minicells lacking chromosomal DNA. A recent study, identifying two gene products (MinC and MinD) as inhibitors of septation and a third gene product (MinE) as an antagonist of these inhibitors at a specific membrane site opens this complex problem to analysis at the molecular level (*34*).

Differentiation

It is also likely that the study of the differentiation of bacterial cells will reveal mechanisms generally used in cellular differentiation. There are, in particular, four organisms whose development has been studied in recent years (*35*): *Caulobacter crescentus*, whose multiplication depends on transition between swarmer cells equipped with a flagellum and stalked sessile cells (*36*); *B. subtilis*, whose cells can give rise to metabolically inactive spores which can in turn give rise to vegetative cells (*37*); *Myxobacteria*, which occur as multicellular masses, but can aggregate to form characteristic fruiting bodies and then segregate into myxospores (*38*); and *Cyanobacteria*, which form heterocysts when grown with N_2 as sole nitrogen source (*39*). The technology that has been developed for the study of *E. coli*, the selection of mutants, and the cloning of individual genes can be applied to the study of these organisms.

An unexpected observation in the case of the cyanobacterium *Anabaena* has an important parallel in the differentiation of cells of vertebrate organisms. When the filamentous *Anabaena* is deprived of a source of nitrogen, approximately every tenth cell in the chain develops into a heterocyst with specific features that provide an anaerobic environment for nitrogen fixation. This differentiation also results in the activation of the expression of the *nif* genes whose products are the enzymes responsible for the reduction of N_2 to ammonia. The activation involves genetic rearrangements providing the operon structure apparently required for the expression of the *nif* genes (*40*). Although such recombinational switching has been previously observed in prokaryotes (*41*), this is the first example of recombinational switching as a specific response to an environmental change. The best example of a genetic rearrangement in a developmental program of eukaryotic cells is that of genes coding for immunoglobulin (*42*). Thus, the same molecular mechanism is used in a bacterium and in vertebrates.

Applications

The study of the complex genetics, biochemistry, and molecular biology of nitrogen fixation, an area not only of theoretical but also practical interest, has led to the investigation of the interaction of the prokaryotic *Rhizobia* and the eukaryotic legume root cells (*43*). This interaction results in the formation of the root nodule where bacteria contained within plant cells have differentiated into organelles whose only function is to supply the plant with the ammonia they obtain by the reduction of dinitrogen. The elucidation of the mechanisms responsible for this interaction between a eukaryotic and a prokaryotic cell that results in symbiosis will be of importance for our understanding of corresponding interactions that result in parasitism, such as the invasion of animal cells by pathogenic bacteria. The investigation of such medical problems should prove attractive to young scientists whose primary interest is the application of science to problems confronting human society. We must remember that the identification of DNA as genetic material resulted from the study of the ability of a bacterium to cause a disease. The alert investigator attracted to a field of study because of the practical application of the results of the investigation may well make a fundamental discovery that enlarges our understanding of cell biology. The new technology makes it possible to study and to alter important properties of bacteria such as *Streptomyces* used for the production of antibiotics, of *Clostridia* that could again become important agents for the production of organic compounds (*44*), and of *Thiobacilli* used for the mining of copper and other metals (*45*).

Evolutionary Considerations

The aim of scientific research is to generate hypotheses whose value is judged by the range of their applicability. Yet, although considerable progress has been made in the elucidation of the mechanisms responsible for the regulation of gene expression in bacteria, no hypothesis has emerged that would explain why a particular molecular mechanisms is used in any given case. For example, there are three instances when addition of a compound to the growth medium of *E. coli* results in the formation of a specific enzyme that allows that compound to be used as a source of carbon and energy: but in each instance a different molecular mechanism is responsible for the induction. Lactose brings about the inactivation of a protein that blocks the initiation of transcription of the gene for β-galactosidase (*46*); maltose causes the activation of a protein required for the initiation of transcription of the gene for amylomaltase (*47*); and tryptophan causes suppression of the termination of transcription initiated at a site upstream from the gene for tryptophanase (*16*). We do not know whether the specific molecular mechanism in each case was selected as the one best suited for the purpose or whether the mechanisms are essentially equivalent and were selected purely by chance in the course of their evolution (*48*).

We also lack any knowledge of the selective forces responsible for the appearance of eukaryotic cells (*49*). An attractive hypothesis, which is based on the recent comparison of the nucleotide sequences of ribosomal RNA from different cells and from mitochrondria and plastids, suggests that more than 3.5 billion years ago the progenitor of all cells, a progenote, evolved into the separate kingdoms of the archaebacteria, the eubacteria, and the prokaryotic progenitors of the eukaryotes (*50*). The last gave rise perhaps 2 billion years later to the eukaryotic cell by endosymbiosis with the eubacterial ancestors of mitochrondria and plastids (*51*). Yet it does not appear that the more elaborate structure of the yeast cell either significantly improves or diminishes its ability to grow in the same environment as a prokaryotic cell. Thus, more than 100 years ago, Pasteur found that the reason for a failure in the industrial production of ethanol from sugar beets was the growth of prokaryotic *Lactobacilli* rather than eukaryotic yeast in the fermentation mixture that obviously provided a good environment for either organism. Moreover, we know now that *E. coli* and *S. cerevisiae* can respond equally rapidly to changes in their environment by using corresponding molecular mechanisms to change their enzymatic composition.

We also know, as discussed earlier, that not only eukaryotic, but also prokaryotic cells can differentiate and that both in the prokaryotic *Anabaena* and the eukaryotic lymphocyte this process involves genetic rearrangement. We must conclude that the common progenitor of all cell types had already, more than 3.5 billion years ago, acquired many of the complex molecular mechanisms we find in all cells, but that further evolution has resulted in the diversity we observe. We can appreciate the possibilities of diversification during such a time interval by considering that *E. coli* and *S. typhimurium*, two closely related but clearly distinct organisms, are thought to have diverged about 150 million years ago (*52*).

At present we know the changes in phenotype that can result from mutations in individual genes, from gene fusions that may cause a reassortment in protein domains, and from more drastic alterations of the genome by plasmids acquired from other organisms. All these genetic mechanisms must have had a role in evolution. However, we know almost nothing about the environmental conditions that were responsible for the selection of genetically altered cells. In fact, we know very little about the natural environments of microbial cells and therefore can assess only to a limited degree the normal physiological role of their complex molecular mechanisms. It is apparent that comparative study of different microorganisms with the use of as yet unformulated concepts and new methods is required for the discovery of the intricate relations between environment and organism in which the organism was changed by environment and in turn changed the environment. An understanding of these relationships may lead to a comprehensive theory of cell biology. A young biologist embarking on a career at the present time could perhaps look forward to a lifetime of research in the mainstream of biology by choosing to study the evolution of the microbial cell.

References and Notes

1. F. C. Neidhardt, Ed., Escherichia coli and Salmonella typhimurium. *Cellular and Molecular Biology* (American Society for Microbiology, Washington, DC, 1987).
2. M. Schaechter and F. C. Neidhardt in (*1*), p. 1.

3. A. J. Kluyver and C. B. Van Niel, *The Microbe's Contribution to Biology* (Harvard Univ. Press, Cambridge, MA, 1956).

4. J. S. Huxley, *Evolution: The Modern Synthesis* (Harper, New York, 1942), p. 645.

5. S. E. Luria and M. Delbrück, *Genetics* **28**, 491 (1943).

6. O. T. Avery *et al.*, *J. Exp. Med.* **79**, 137 (1944).

7. J. Lederberg and E. L. Tatum, *Nature* **158**, 558 (1946); J. Lederberg, *Annu. Rev. Genet.* **21**, 23 (1987).

8. J. Lederberg and N. Zinder, *J. Am. Chem. Soc.* **70**, 4267 (1948); B. D. Davis, *ibid.*, p. 4267.

9. E. Umbarger and B. D. Davis, in *The Bacteria*, I. C. Gunsalus and R. Y. Stanier, Eds. (Academic Press, New York, 1962), p. 167; B. Magasanik, in *ibid.*, p. 295.

10. F. Jacob and E. L. Wollman, *Sexuality and the Genetics of Bacteria* (Academic Press, New York, 1961), p. 155.

11. H. F. Judson, *The Eighth Day of Creation: Makers of the Revolution in Biology* (Simon and Schuster, New York, 1979).

12. W. M. Huang *et al.*, *Science* **239**, 1005 (1988).

13. F. K. Chu *et al.*, *Proc. Natl. Acad. Sci. U.S.A.* **81**, 3049 (1984); F. K. Chu *et al.*, *Cell* **45**, 157 (1986).

14. W. Leinfelder *et al.*, *Nature* **331**, 723 (1988).

15. F. Jacob and J. Monod, *Cold Spring Harbor Symp. Quant. Biol.* **26**, 193 (1961); J. Monod and F. Jacob, *ibid.*, p. 389.

16. J. Beckwith in (*1*), p. 1439; R. Landrich and C. Yanofsky in (*1*), p. 1276; S. Gottesman in (*1*), p. 1308.

17. B. Magasanik and F. C. Neidhardt in (*1*), p. 1318; R. Schleif in (*1*), p. 1473.

18. Y. Gluzman and T. Shenk, Eds., *Enhancers and Eukaryotic Gene Expression* (Cold Spring Harbor Laboratory, Cold Spring Harbor, NY, 1983).

19. B. C. Hoopes and W. R. McClure, in (*1*), p. 1231.

20. L. J. Reitzer *et al.*, *J. Bacteriol.* **169**, 4279 (1987).

21. F. C. Neidhardt and R. A. Van Bogelen in (*1*), p. 1334; J. W. Erickson *et al.*, *Genes Develop.* **1**, 419 (1987).

22. R. Losick and J. Pero, *Cell* **25**, 582 (1981).

23. T. D. Yager and P. H. von Hippel in (*1*), p. 1241.

24. F. C. Neidhardt in (*1*), p. 1313.

25. N. Yamamoto and M. L. Droffner, *Proc. Natl. Acad. Sci. U.S.A.* **82**, 2077 (1985).

26. B. Magasanik, *Trend. Biochem. Sci.* **13**, 475 (1988).

27. A. M. Stock *et al.*, *Cold Spring Harbor Symp. Quant. Biol.* **53**, 49 (1988).

28. L. M. Albright *et al.*, *Annu. Rev. Genet.*, in press.

29. J. F. Miller *et al.*, *Science* **243**, 916 (1989).

30. R. McMacken *et al.*, in (*1*), p. 564; G. M. Weinstock, in (*1*), p. 1034; S. R. Kushner, in (*1*), p. 1044.

31. K. Oosawa and M. Simon, *Proc. Natl. Acad. Sci. U.S.A.* **83**, 6930 (1986).

32. R. Y. Cabelli *et al.*, *Cell* **55**, 683 (1988).

33. K. von Meyenburg and F. G. Hansen in (*1*), p. 1555; W. D. Donachie and A. C. Robinson, in (*1*), p. 1594.

34. P. A. J. Boer *et al.*, *Cell* **56**, 641 (1989).

35. R. Losick and L. Shapiro, Eds., *Microbial Development* (Cold Spring Harbor Laboratory, Cold Spring Harbor, NY, 1984).

36. B. Ely and L. Shapiro, in (*35*), p. 1.

37. R. Losick and P. Youngman, in (*35*), p. 63.

38. D. Kaiser, in (*35*), p. 197.

39. R. Haselkorn, *Annu. Rev. Plant. Physiol.* **29**, 319 (1978).

40. J. W. Golden *et al.*, *Nature* **314**, 419 (1985).

41. M. Silverman and M. Simon, in *Mobile Genetic Elements*, J. A. Shapiro, Ed. (Academic Press, New York, 1983), p. 223.

42. T. Honjo and S. Habu, *Annu. Rev. Biochem.* **54**, 803 (1985).

43. F. M. Ausubel, in (*35*), p. 275.

44. R. Y. Stanier, J. L. Ingraham, M. L. Wheelis, P. R. Painter, *The Microbial World* (Prentice-Hall, Englewood Cliffs, NJ, ed. 5, 1986), p. 657.

45. R. M. Atlas and R. Bartha, *Microbial Ecology: Fundamentals and Applications* (Addison-Wesley, Menlo Park, CA, 1981), p. 443.

46. J. Beckwith, in (*1*), p. 1444.

47. M. Schwartz, in (*1*), p. 1482.

48. M. A. Savageau, in *Biological Regulation and Development*, vol. 1, *Gene Expression*, R. F. Goldberger, Ed. (Plenum, New York, 1979), p. 57.

49. K. H. Schleifer and E. Stackebrandt, Eds., *Evolution of Prokaryotes* (Academic Press, New York, 1985).

50. C. R. Woese, in (*49*), p. 1; W. F. Doolittle and C. J. Daniels, in (*49*), p. 31.

51. M. W. Gray and W. F. Doolittle, *Microb. Rev.* **46**, 1 (1982).

52. H. Ochman and A. C. Wilson, in (*1*), p. 1649.

53. I thank J. Beckwith, M. S. Fox, D. E. Koshland, Jr., S. E. Luria, F. C. Neidhart, and G. C. Walker for their suggestions. Supported by PHS grants AM-13894 and GM-07446 from the National Institute of Arthritis, Diabetes, and Digestive and Kidney Diseases and the National Institute of General Medical Sciences, respectively, and by grant PCM84-00291 from the National Science Foundation.

3

Genetic Engineering of Bacteria from Managed and Natural Habitats

Steven E. Lindow, Nickolas J. Panopoulos, Beverly L. McFarland

Many distinct forms of bacteria exist in nature, each with potentially useful or detrimental attributes. Several strategies can be used to modify bacteria for useful purposes. In some instances, one or more genes for undesirable traits have been targeted for removal. There are also circumstances when the survival of useful microorganisms may be improved by single gene transfer or by genetic selection for tolerance to toxic substances. Sometimes a trait or process that is restricted to a given strain may prove useful in a habitat that is not readily exploited by that species. Although the adaptations that enable bacteria to colonize or survive in specific habitats are generally unknown, it is likely that many characteristics collectively determine survival. Thus, it is presently difficult or impossible to transfer all the genetic determinants enabling a bacterium to survive in a habitat to which it is not already adapted. However, certain traits that may be desirable to have expressed in a given environment are conferred by single genes or gene clusters, which can be transferred to and expressed in a bacterial strain indigenous to that environment. Most studies have emphasized the introduction of genes for novel traits into bacteria indigenous to the habitat to be exploited. Because of the wide scope of genetic engineering targets, this review will focus on the modification of some bacteria that affect important natural and industrial processes.

Plant-Microbe-Pest Interactions

Most bacterial species that reside on plant surfaces are not harmful to the plant and may even protect it from pathogens, other deleterious microorganisms, and insects. Many such species, particularly strains of *Pseudomonas*, are well adapted for growth and survival on leaves or roots of plants, with population sizes of 10^5 to 10^7 cells per square centimeter of plant surface being common. Attempts have been made to modify these bacteria by the addition of single genes so that they might protect crops against insect pests. For example, many lepidopteran insects are susceptible to the delta endotoxin produced by various strains of *Bacillus thuringiensis* (1). This bacterial species is found in diseased insects or in soil and plant debris and can cause low levels of mortality in susceptible insects in natural settings (1). The genes conferring production of several different *B. thuringiensis* delta endotoxins with different insect host specificities have been cloned and partially characterized (2), and effort has been

S. E. Lindow and N. J. Panopoulos are in the Department of Plant Pathology, 147 Hilgard Hall, University of California, Berkeley, CA 94720. B. L. McFarland is at the Chevron Research Company, Process Research Department, Richmond, CA 94802. This chapter is reprinted from *Science* 244, 1300 (1989).

directed toward determining the functional domains within the toxin so that hybrid toxins with altered host ranges or enhanced potency can be made (2). Commercially produced cultures of *B. thuringiensis* are effective insecticides, which are used on several agricultural and forest plant species. However, effective insect control requires repeated applications of *B. thuringiensis* since this species does not multiply on plants. Attempts have been made to overcome the spatial and economic limitations of foliar applications of commercially produced *B. thuringiensis* cells for insect control. For example, the δ endotoxin gene was introduced into the chromosome of *Pseudomonas fluorescens* (an effective colonizer of corn roots) in order to ensure stability of the gene and to minimize the risk of its transfer to other bacteria indigenous to corn roots (3). This recombinant strain attained population sizes similar to those of the parental strain on corn roots and did not differ from the parental strain in survival and dispersal characteristics as measured in laboratory studies. The modified *P. fluorescens* strain showed some toxicity to root cutworm but not to corn rootworm, which is a more important pest. There has not been a field test of the efficacy of this bacterium because environmental issues raised during an Environmental Protection Agency (EPA) review have necessitated additional research by the Monsanto Chemical Company, the initiator of this project. Whereas transfer of the *B. thuringiensis* endotoxin gene to root-colonizing bacteria may be potentially useful for increasing the number of habitats to which the toxin might be applied, its incorporation into an internal colonist of plants also has promise. *Clavibacter xyli* subspecies *cynodontis* is generally found inside Bermuda grass plants but can reach population sizes of $> 10^8$ cells per gram of stem tissue when inoculated into other plant species, including corn (4). Such a bacterium could be an efficient vector for the expression of cloned genes inside plants, and the *B. thuringiensis* δ endotoxin gene has been incorporated into this species (5). Field studies have been initiated with recombinant *C. xyli* strains for the control of leaf- or stem-feeding lepidopteran insects (6).

Chitin, a polymer of *N*-acetylglucosamine, is a structural component of many plant pests, including fungi and insects. Many bacteria, notably species within the genera *Serratia, Streptomyces*, and *Vibrio*, produce extracellular chitinases. Biological control of some soil-borne fungal diseases by soil-borne bacteria has been correlated with the production of chitinases (7). Inactivation of a chitinase gene in a soil bacterium reduced the ability of the bacterium to lower the incidence of fungal disease (8). Chitinases have been cloned from several strains of *Serratia marcescens* and from other bacteria (8, 9) and have been transferred into efficient plant colonizing bacteria such as *P. fluorescens*. However, the effectiveness of the recombinant strains in controlling fungal disease has not yet been reported.

Some bacteria possess traits that make them harmful to plants. When the genetic determinants for such traits are cloned, it may be possible to replace the native gene with a homologous gene that has been inactivated in vitro. This approach has been successfully applied to the control of frost injury to plants. Ice nucleation (Ice$^+$) strains of *Pseudomonas syringae* are common on the leaves of many plants that cannot tolerate ice formation and are therefore an important cause of frost damage to these plants (10). The gene conferring ice nucleation in *P. syringae* was cloned, and internal deletions within the structural gene were produced in vitro (11). Reciprocal exchange of the modified *ice* gene for the native chromosomal gene was accomplished by homologous recombination. The resultant Ice$^-$ mutants of *P. syringae* showed no difference in colonizing ability or survival on plants, or in other habitats, relative to the parental Ice$^+$ strains (12, 13). In both laboratory and field studies, the population size reached by Ice$^+$ *P. syringae* strains on leaves that had been previously colonized by Ice$^-$ mutant strains was much lower than that reached on leaves without such competitors (13, 14). Preliminary results indicate that significant control of plant disease by avirulent mutants of pathogens is also possible (15). Preemptive competitive exclusion of deleterious bacteria by bacteria of similar genotype (and thus similar habitat resource requirements) may be a useful general method for biological control.

Molecular genetic studies should enable researchers to analyze the relevance of antibiotic production by plant-associated bacteria in the biological control of deleterious bacteria and fungi. Inoculation of plant parts with certain bacterial strains can disrupt the plant-associated microbial communities and subsequently enhance plant growth or reduce the incidence or severity of plant diseases (*16*). Many bacteria used as inoculants produce antibiotic-like substances that are inhibitory to plant pathogens in vitro. Antibiotic-nonproducing mutants (generated by insertion of the transposon Tn5 or by chemical mutagenesis) of several bacterial strains have a reduced ability to antagonize deleterious fungi or bacteria on plants (*17, 18*). Similar genetic evidence for the inhibition of deleterious microorganisms by the production of efficient iron-sequestering agents (siderophores) has been obtained (*19*). In several cases, physical "tagging" of antibiotic biosynthesis genes by insertion of Tn5 has permitted their cloning (*17, 20*); in other cases, the genes have been identified in cosmid clones that complemented chemically induced mutants (*18*).

The regulation and the temporal and spatial patterns of antibiotic biosynthesis in natural environments, such as on leaves or roots, can be investigated by fusing antibiotic genes with "reporter genes" such as *lacZ*, *lux*, *cat*, and *gus*, whose products can be measured in vitro (*21*). A *lacZYA* reporter gene has been used to determine the transcriptional activity of an antifungal antibiotic operon of a *P. fluorescens* strain in response to the nutritional status in culture, and *lux* fusions with this operon have been used for the same purpose on seeds (*22*). A promoterless *ice* gene may prove to be a sensitive indicator of transcriptional activity of bacterial genes in complex natural environments such as plant tissue or soil (*23*). Reporter genes can also be used in cloning studies to identify those gene promoters of indigenous plant-associated bacteria that are induced by a particular set of environmental conditions, such as in response to root exudates that may precede fungal infection of roots.

Most bacteria are not pathogens of higher organisms. Plants possess defense mechanisms that are rapidly activated in response to attempted infection and only a few microbes have the complex sets of genes encoding attributes that enable them (i) to establish a successful parasitic existence with the host (basic compatibility), (ii) to produce pathogenicity and virulence factors, and (iii) to avoid, or overcome, defense responses of the host (*24, 25*). Among the factors required for bacterial pathogenesis on plants are (i) enzymes, such as pectate lyases, proteases, phospholipases, and glycosidases produced by soft rot *Erwinia* species, some *Xanthomonas* species, and a few other pathogens; (ii) toxins such as those produced by some pseudomonads; and (iii) plant growth hormones, such as indoleacetic acid and cytokinins, produced in large quantity by pathogens that cause plant hyperplasias (*24*). Genes for several of the above enzymes, for the phytohormones, and for several toxins have been cloned and their specific role in pathogenesis had been established (*24*). A number of "pathogenicity genes" exist whose functions are not yet known. In *Pseudomonas*, *Xanthomonas*, and *Erwinia amylovora*, large contiguous clusters of genes (*hrp*) and other unlinked loci are required for pathogenicity (*24, 25*). Many of these genes also are required in conjunction with avirulence genes (*avr*) for triggering the hypersensitive reaction, which is not a pathogenic response but is connected with the expression of resistance to heterologous pathogens or to avirulent (incompatible) pathogen races that normally cause disease on other cultivars of the same host species (*25*). Other genes are responsible for symptom production but not for the elicitation of the hypersensitive response (*24, 25*). Many of these genes seem to be conserved within taxa of phytopathogens (*24*).

Defensive reactions of plants include the production of the antimicrobial agents termed phytoalexins and the rapid necrosis of plant cells (the hypersensitive response), which is closely associated with the accumulation of phytoalexins (*24*). The nature of bacterial substances that can elicit defense-related processes is unknown. Such substances do not generally have the same biological specificity as their producers, some of which have a broad host range, although most are highly specialized, infecting only a limited number of host plants or only one or a few cultivars of a given species. This specificity is much better understood at the genetic level. Although both negative and positive factors in bacteria and plants may collectively define host range, *avr* genes appear to be the main host range determinants in various pathovars of *P. syringae* and *Xanthomonas campestris* (*25*). These genes are genetically dominant in merodiploids and act in conjunction with functionally corresponding plant resistance genes (*R*), which in most cases are also genetically dominant. Such *R-avr* gene pairs control the activation of host resistance. Several *avr* genes have been cloned, and four have been sequenced (*25*). An extension of this concept concerns the genetic basis of nonhost resistance to pathogens (*26*). Interspecies transfer of genomic libraries of bacterial pathogens has revealed cryptic *avr* genes that restrict bacterial pathogenesis on nonhost species. The presence of these genes could not have been inferred from classical genetic studies because different species of pathogens or plants cannot be easily crossed. These findings suggest a common basis for resistance in host cultivars and nonhost species.

Understanding the molecular basis of microbial pathogenesis, elucidation of resistance mechanisms, and cloning of native plant resistance genes may have an impact on crop protection strategies in the long term. However, some short-term applications have been considered: nonpathogenic mutants of *P. syringae* for frost control; similar mutants of *Pseudomonas solanacearum* that may also degrade fusaric acid, a putative toxin produced by vascular wilt Fusaria, as biological control agents of vascular wilts; and the construction of transgenic plants that express phytotoxin immunity genes derived from toxin-producing pathogens (*15, 27*).

Leguminous crops form symbiotic associations with *Rhizobium*, *Bradyrhizobium*, and *Frankia* species that fix atmospheric nitrogen in a form that can be used by the plant. The genes from these bacteria on which attention has focused include (i) the *nif* genes, which encode nitrogenase components; (ii) genes designated by various acronyms (*nod, hsn, fix, syr*) that collectively determine *Rhizobium* host range and nodule development and function; (iii) the *dct* genes, which are responsible for the energy-yielding metabolism of dicarboxylic acids in the nodule; (iv) the *hup* genes, which mediate the capture of hydrogen released as a consequence of nitrogenase function; and (v) genes for the biosynthesis of the phytotoxin rhizobitoxin by *Rhizobium japonicum*.

In the free-living nitrogen-fixing bacterium *Klebsiella pneumoniae*, 17 *nif* genes are organized into eight transcriptional units and occur as a cluster in a 24-kb region of the chromosome (*28*). Transfer of this cluster to *Escherichia coli*, a close taxonomic relative, confers nitrogen-fixing ability (*28*). Similarly, transfer of either Sym plasmids, which carry *nod* and host-range genes as well as *nif*, or certain cloned *nod* or *hsn* genes between rhizobia can extend, restrict, or have no effect on host range (*29*). Such a transfer also enables *Agrobacterium* to initiate nodulation on nonlegumes (*29*), which suggests that host range may be extended to nonleguminous hosts.

Attempts have been made to improve legume yields by modifying the expression of two specific genes in N-fixing symbionts. Strains of *B. japonicum* and *R. meliloti* that expressed the dicarboxylate transporter protein (DctA) and the *nif* activator protein (NifA), respectively, under nitrogen-regulated promoter control gave greater than 10% increase in biomass production on their respective hosts under greenhouse conditions (*30*).

A subset of *nod* genes form the *nodABC* operon, common to all *Rhizobia*, which is positively regulated by the *nodD* gene in the presence of plant-derived phenolic substances (flavones or isoflavones) (*31*). Some phenolic components in the root exudate antagonize *nodD*-mediated activation of *nodABC* and some rhizobia have several copies of *nodD* with apparently different specificities for phenolic inducers (*31*). These findings can be exploited in useful ways; for example, the nodulation of legume varieties, or progeny obtained in breeding programs, will be able to be more rapidly screened (*32*). Interstrain exchange of *nodD* alleles with different inducer or anti-inducer specificities, modification of other host specificity genes, and engineering of the *Nod* protein are also envisioned.

In other experiments, a plasmid carrying the *R. meliloti dct* genes enhanced dicarboxylate uptake and nitrogenase activity under microanaerobic conditions in vitro (*33*). About 25% of natural isolates of *R. japonicum* are Hup$^+$ and can grow autotrophically on H$_2$ and CO$_2$, whereas other strains are phenotypically Hup$^-$. When isogenic Hup$^+$ and Hup$^-$ strains were allowed to nodulate soybean roots in a contained N-free system, both the amount of total fixed N$_2$ and plant biomass produced were 10% greater in the Hup$^+$ strain (*34*). Finally, inactivation of phytotoxin biosynthesis by site-directed mutagenesis significantly improved strain performance (*35*).

The difficulty of introducing rhizobial strains into environments in which resident rhizobia are already present is a problem. Newly introduced strains encounter strong competition from resident strains (*36*). To establish effective nodulation in the rhizosphere, the inoculant must outnumber the indigenous population by at least 1000-fold. When indigenous *Rhizobium* populations in soil are high, this requirement leads to high inoculant costs. Little is known of the biochemical determinants of *Rhizobium* competitiveness, although a toxin (trifoliin) produced by the *Rhizobium leguminosarum* pathovar *trifolii* seems to be required for this strain to compete efficiently (*37*). Ultimately it may be possible to manipulate both host and rhizobial genes to obtain maximum efficiency of nodule formation and function and to tailor strains for unusual soil environments (*36, 37*).

Biodegradation of Xenobiotics and Toxic Waste Transformations

Genetic tools can be used to develop specific catabolic pathways for the degradation of xenobiotics in bacteria that can function under a wide range of environmental conditions. For example, the ability to degrade toluene was transferred from a mesophilic bacterium into the psychrophilic *Pseudomonas putida*, which could degrade toluate at temperatures as low as 0°C (*38*). Currently, the biodegradation effectiveness of recombinant bacterial strains at sites contaminated with toxic and hazardous waste has not been demonstrated; however, both successful (*39*) and unsuccessful (*40*) experiments with nonrecombinant bacteria have been reported. Some toxic chemicals may have structures that are resistant to microbial attack or may be present in mixtures that are incompatible for effective degradation, or in too low or too high a concentration. Successful biological treatment in situ will depend on (i) the introduction and establishment of microbes in the environment, (ii) an improvement in the rate and extent of xenobiotic degradation, and (iii) the resolution of problems inherent in heterogenous spatial distributions of pollutants, nutrients (including oxygen), and microorganisms (*39, 41*). The need to develop environmental processes to treat hazardous wastes is underscored by the fact that there are more than 900 designated hazardous waste sites in the United States, although the actual number is estimated to be closer to 10,000 (*40*).

Many bacterial genes for xenobiotic degradation have originated from strains isolated from contaminated waste sites and are often found on plasmids (*41, 42*). For plasmids encoding incomplete catabolic pathways, the degradation of recalcitrant chemicals as sole carbon and energy sources may require complementation of plasmid genes by host chromosomal genes to link the plasmid pathway with energy-yielding metabolism. Gene clusters subject to operon control are characteristic of most catabolic plasmids, and two or more regulons have been identified in some cases. Almost all of the plasmids characterized to date that have genes for xenobiotic catabolism are from gram-negative bacteria, predominantly *Pseudomomas* species (*43*).

Natural selection processes can be extremely slow in yielding improved xenobiotic degraders, especially when the acquisition of multiple catalytic activities is necessary, such as for the metabolism of compounds having structural elements or substituents rarely found in nature (*41*). Critical to the evolution of new metabolic activities is the relaxation of substrate specificity of enzymes and regulators without a concomitant loss in function (*41, 44*). Laboratory selection can speed up this process and provide control not available under natural conditions (*42*). Metabolic pathways can be engineered in the laboratory by (i) long-term batch incubations, (ii) soil perfusions, (iii) chemostat selection, (iv) in vivo genetic transfers, or (v) selection for the evolution of new catabolic or regulatory functions or the assembly of a new pathway in vitro (*41, 42, 44*). The first four techniques have been used to derive bacteria able to degrade a variety of xenobiotic and toxic wastes (Table 1). The first multiplasmid-containing strain constructed in the laboratory (*45*) was capable of oxidizing aliphatic, aromatic, terpenic, and lower molecular weight polynuclear aromatic hydrocarbons. This strain grew faster on crude oil than any of the parental isolates, and although it was patented in a landmark case, which recognized for the first time that man-made microbes were patentable (*45*), it has not been used commercially. The metabolic diversity of this strain was fairly limited compared to the actual number of compounds present in crude oil (> 3000), and it did not degrade higher molecular weight, condensed, or substituted hydrocarbon compounds, which tend to persist in the environment.

In vitro strain constructions require detailed genetic and biochemical information on the degradative pathways, which is nonexistent for many xenobiotics (*42*). When suffi-

cient biochemical and genetic information has been available, genetic engineering of metabolic pathways for recalcitrant compounds was successful (*42, 44*). For example, broad-specificity enzymes were recruited to extend the chlorocatechol pathway in several bacteria so that the degradation of chlorinated compounds through the tricarboxylic acid cycle was enhanced. *Pseudomonas* B13 has a pathway for the complete degradation of chlorocatechols, but the first enzyme of its chlorobenzoate pathway, which is encoded by chromosomal genes, has narrow substrate specificity (*44*). Recruitment of a broad-specificity dioxygenase from a toluene catabolic plasmid enabled this strain to degrade a wider range of chlorobenzoates and additional chloroaromatics after deleting an enzyme that misrouted intermediates to the meta-ring cleavage pathway rather than to the normal ortho-ring cleavage pathway (*44*).

Biodegradation processes at contaminated waste sites may be limited if complex mixtures of xenobiotics are present, if the inducer of the degradative pathway is not present, or if the pathway is blocked by inhibitors. Although individual chemicals can be completely degraded in soil and wastewater treatment plants, mixtures sometimes cannot (*41*). For example, dead-end metabolites result when chlorocatechols are cleaved by a meta-fission pathway and when methylcatechols are cleaved by an ortho-fission pathway. This nonproductive routing of degradation products during simultaneous degradation of chloro- and methyl-substituted aromatics can actually destroy the functioning of the biodegrading community (*41*). Enzyme recruitment overcame this prob-

Table 1. In vivo engineering of bacteria for the degradation of xenobiotic and toxic wastes.

Bacterium	Substrate
Pseudomonas cepacia	2,4,5-trichlorophenoxyacetic acid (*67*)
P. putida and *Pseudomonas* spp.	2,2-dichloropropionate (*68*)
P. putida and *Pseudomonas alcaligenes*	Chlorobenzenes (*69*)
Pseudomonas sp.	Chloroaniline, chlorosalicylate, chlorobenzoate, dichlorobenzene, amino-naphthalene sulfonates, hydroxy-naphthalene sulfonates, and other chlorophenols (*70*)
Alcaligenes sp.	Dichlorophenoxyacetic acid, mixed chlorophenols, 1,4-dichlorobenzene (*71*)
Acinetobacter sp.	4-chlorobenzoate (*41*)

lem by constructing catabolic routes with only one ring-fission mechanism (ortho-pathway) for chloro- and methylaromatics (*41*). The control of catabolic pathways can also be modified by placing key biodegradative enzymes that require inducers, some of which are pollutants themselves, under the control of new regulatory regions. For example, genetic engineering has been used to uncouple the *Pseudomonas mendocina* toluene monooxygenase genes from toluene induction, to derive *Pseudomonas* transconjugants that constitutively express the 2,4-dichlorophenoxyacetic acid degradation pathway, and to derive *E. coli* recombinant strains with enhanced polychlorobiphenyl degradative activity in the presence of exogenous catabolite repressor substances compared to the wild-type *Pseudomonas* donor strain (*38, 46*). The cloning of genes for modified enzymes that have useful catabolic properties (such as relaxed substrate-specificities or enhanced induction capabilities) provides an important repository of genetic diversity for future research (Table 2).

Production of Chemicals and Fuels

Recombinant DNA technology is having an impact on the microbial production of industrial chemicals and fuels (*47*). Plans have been announced for the production of amino acids on a commercial scale with recombinant bacteria including *Bacillus amyloliquefaciens* and *Lactobacillus casei* (*47*). Genetic engineering has been applied to fermentation processes to enable bacteria to use a wider variety of feedstocks, to biosynthesize new products, to accumulate intermediate metabolites via blocked pathways, or to increase product yields by enhanced synthesis of special enzymes.

In many commercial fermentation processes the cost of raw materials is the most expensive component (*47*). Industrially useful bacteria can be modified to use cheaper feedstocks such as D-xylose, lignocellulose, or cellulose. *Zymomonas mobilis* transconjugants were able to use lactose when they contained the *E. coli* lactose operon under the control of a

Z. mobilis promoter and the genes for galactose utilization from *E. coli* (*48*). The activity of cellulase synthesized in *Z. mobilis* was increased sixfold by placing a cellulase gene from *Cellulomonas uda* under the control of a strong *Z. mobilis* chromosomal promoter (*48*). An alternate approach for improving the economics of ethanol production has been to transfer *Z. mobilis* genes encoding the appropriate enzymes into other organisms, such as *E. coli* or *Klebsiella* (*49*). The genes encoding essential enzymes of the fermentative pathway for ethanol production in *Z. mobilis* were expressed at high levels in *E. coli*. The recombinant strain metabolized glucose and xylose to give almost the maximum theoretical yield of ethanol with significant decreases in the yields of formate, acetate, lactate, and butanediol (*49*). Similar results in strains of *Klebsiella planticola* have been obtained (*49*).

There are many bioprocessing applications for thermotolerant microbes and enzymes (*47, 50*), but genetic approaches with thermophiles have only recently been initiated (Table 2). The isolation of a highly transformable thermophile, *Bacillus stearothermophilus*, may facilitate advances in cloning thermophilic traits (*50*). Thermostable enzymes may also be produced by cloning their genes from thermophiles, provided selection for enzyme function can be achieved (*50*).

Mineral Processing

Most major copper mining companies use microbial extraction technology to obtain 10 to 20% of the world copper supply; microbial leaching of uranium is also used in Canada (*51*). Leaching rates are generally slow and metal recovery from the leachate is expensive. Commercial bioleaching operations in the mining industry are typically conducted outdoors with ores in heaps or pits, so many of the problems regarding introduction and establishment of improved bacterial strains apply to mining applications of biotechnology. Microbes responsible for the solubilization of metals tend to be acidophilic (*51*). Many are chemolithotrophs and obtain energy from the

Table 2. Examples of genetically engineered bacteria.

Host bacteria	Altered trait	Genes transferred (source organism)
Extension of substrate range		
Pseudomonas aeruginosa and *X. campestris*	Growth on whey for biopolymer production	Lactose metabolizing enzymes (*E. coli*) (72)
E. coli	Cellulose used as new feedstock	Various cellulose degrading enzymes (73)
Bacillus subtilis	Starch used as new feedstock	α-amylase (*B. subtilis*) (74)
Thermophilic enzymes added		
E. coli	Cellulose degradation	Various cellulose degrading enzymes (75)
E. coli	Starch degradation	α-amylase (*Dictyglomus thermophilum* and *Bacillus licheniformis*) (50)
E. coli	Raffinose removal	α-galactosidase (*B. stearothermophilus*) (76)
Biodegradation and waste treatment applications		
Pseudomonas	Altered pathway regulation	TOL plasmid catabolic enzymes (44)
E. coli	Enhanced degradation pathway for mono- and disubstituted chloroaromatics	Various genes (67) *Pseudomonas* spp. *Alcaligenes eutrophus*
E. coli	Metal removal from wastewater	Metallothionein (human) (56)
E. coli	Polychlorinated biphenyl metabolism	Entire pathway (*Pseudomonas* spp.) (38)
New biochemistry added		
Methylophilus methylotrophus	Increased efficiency of methanol conversion to single cell protein	Alcohol dehydrogenase (*B. stearothermophilus*) and glutamate dehydrogenase (*E. coli*) (77)
E. coli	Indigo production	Naphthalene oxidation enzymes (*P. putida*) (46)
E. coli	Improved oxygen metabolism	Hemoglobin structural gene (Vitreoscilla) (55)
E. coli	Control of intra- and extracellular biopolymers	Exopolysaccharide and polyhydroxybutyrate biosynthetic genes (*Z. ramigera* and *A. eutrophus*, respectively) (54)

oxidation of iron or sulfur found in ores such as pyrite, arsenopyrite, or chalcopyrite. *Thiobacillus ferrooxidans* is thought to play the main role in most metal-leaching operations (51). Genetic technology as applied to this group of microbes is not as advanced as for other species largely because of the unusual conditions required to grow them (51). Recently, plasmids have been constructed that may enhance the recovery of gold from arsenopyritic-pyritic ores by *T. ferrooxidans* and increase the resistance of *T. ferrooxidans* and *B. subtilis* to arsenite and arsenate (52). A *recA*-like gene from *T. ferrooxidans* has been cloned, and selectable shuttle cloning vectors have also been constructed for this bacterium (51). At least two companies plan to test genetically engineered organisms with enhanced bioleaching capabilities (52).

Wastewater Treatment Applications

Few aspects of municipal and industrial wastewater treatment processes are understood at the genetic and biochemical level. Some bacterial functions, such as the biodegradation of particular chemicals, may reside in one gene or gene cluster and be relatively easy to identify. Although it would be desirable to obtain improvements in other bacterial functions, such as better flocculation of heterotrophic bacteria in activated sludge, stronger attachment of bacteria to surfaces, increased growth rates of nitrifying bacteria, and decreased sensitivity to inhibitors, these are often complex and require multiple genes (53).

Many wastewater treatment processes require removal of biomass for disposal or recycling purposes. The most common example is the activated sludge process, which depends on the formation of aggregates of microbes for effective treatment (53). Although several microbes produce exopolysaccharides, which contribute to flocculation in aerobic wastewater treatment facilities, *Zoogloea* species have been implicated in particular (53). Recombinant DNA technology has been used to control biopolymer synthesis in *Zoogloea ramigera*. Several genes that are responsible for the production of different exopolysaccharides required for flocculation have been cloned from *Z. ramigera* strains (54). The ability to link the genes for biopolymer and surfactant production together with catabolic genes may improve wastewater treatment processes (40, 43). This coupling would also provide a way of keeping plasmid-containing cells within the bioreactor during continuous-flow operations, as the non-flocculated, nonplasmid bearing cells would leave the system in the effluent through the overflow (54).

The cloning of the *Vitreoscilla* structural gene for the oxygen-carrying compound hemoglobin and its expression in *E. coli* may further improve bioprocessing and biomining operations that are oxygen-limited (55). Evidence indicates that floc-forming bacteria have a lower affinity for oxygen than do certain filamentous bacteria (53). The linking of genes for biopolymer synthesis in floc-forming bacteria with an enhanced ability to scavenge oxygen from the environment could result in significant improvement in wastewater treatment.

Genetic technology could also improve the sequestration of metals in bacteria by enhancing the absorption of metals to the microbial cell surface or by increasing intracellular uptake. This would be useful for the removal of metals from aqueous solutions for both pollution control and the recovery of precious metals (51). Metallothioneins are low molecular weight peptides that are induced in response to increased metal concentrations (56). Human metallothionein has been cloned into *E. coli* and it has been proposed that these engineered bacteria be used as an immobilized cell system for removing metals from wastewaters (56). When the metallothionein fusion protein is induced in the presence of Cd^{2+} or Cu^{2+}, a direct correlation is found between the expression of the fusion protein and the bioaccumulation of Cd^{2+} and Cu^{2+}. The use of immobilized cells with high metallothionein responsiveness could provide a waste treatment system naturally responsive to variations in heavy metal concentrations.

Bacterial metal-binding proteins can be used either to remove phosphorus in wastewater treatment systems, so as to prevent further eutrophication of waters that receive industrial and municipal wastewater discharges, or as regenerable adsorbants or biosensors (51, 53, 57). For example, the phosphate-binding protein of *E. coli* has been cloned and expressed in strains that secrete proteins directly into the medium, so as to minimize product recovery costs (57). The recovered phosphate-binding protein was immobilized on a support matrix as a thermally regenerable adsorbant and onto a derivatized silicon chip to convert the binding of phosphate by the protein into an electronic signal for use as an on-line monitor of phosphate concentration (57).

Ecological Considerations for the Use of Genetically Engineered Bacteria

A large number of bacterial strains that are to be engineered for specific practical processes will need to be released into the environment (Table 3), although some bioprocessing activities will likely be conducted in a confined and well-controlled setting, such as a bioreactor. Many genetically engineered bacteria to be released in the open environment will be descendants of indigenous environmentally competent bacteria. The potential environmental impact of released recombinant bacteria will need to be examined. This subject has been the focus of much attention and several international conferences since 1975 (58), and this review cannot adequately address the many concerns that have already been stated in other proceedings (59).

Table 3. Authorized release of live recombinant bacteria into the open environment as of 1989.

Bacterium	Altered trait	Site of test	Purpose of test	Year of first test
P. syringae	Deletion of *ice* g (*13, 14*)	California	Control of frost damage to plants	1987
P. fluorescens	Deletion of *ice* g (*14*)	California	Control of frost damage to plants	1987
P. fluorescens	Addition of *lac*ZY to chromosome (*78*)	North Carolina	Assessment of spread of released bacteria	1987
R. meliloti	Additional copies of *nif* g	Wisconsin	To increase efficiency of N₂ fixation	1988
C. xyli	Introduction of *B. thuringiensis* delta endotoxin gene (*6*)	Maryland and France	Control of corn ear worm	1988
A. radiobacter	Deletion of *tra* g of Agrocin 84 plasmid (*62*)	Australia	Biological control of crown gall	1988
P. fluorescens	Addition of *lac*ZY to chromosome (*78*)	Washington	To assess movement and survival of biological control agent of take-all disease of wheat	1988

Regulations affecting the use of genetically engineered bacteria differ greatly between different countries (*60, 61*). One or more agencies in the United States have jurisdiction over a research activity or commercial biotechnology product, and excellent summaries of current agency jurisdiction in the United States have been published (*61*). The data required to support initial requests for field releases of recombinant bacteria have been formidable, and despite extensive documentation, not all requests for release of bacteria have been approved. Each experiment is currently addressed on a case-by-case basis.

Assessment of the environmental safety of released recombinant bacteria is rapidly taking advantage of advances in the methodology to sensitively and accurately detect specific bacterial strains or their genes. For example, bacterial DNA can now be efficiently extracted directly from environmental samples (such as soil) and identified and quantified by hybridization procedures (*62*). The polymerase chain reaction method for amplifying specific DNA sequences coupled with direct extraction of DNA from environmental samples increases the sensitivity of this method to as few as one cell per gram of soil (*63*). Information is also being generated on the frequency of gene exchange among bacteria in natural and managed environments (*64*). Several approaches to reduce or eliminate the potential persistence of modified bacteria also have been evaluated (*65*). For example, the *hok* gene, which encodes a protein that causes lethal collapse of the transmembrane potential of cells, has been placed under the control of the inducible *lac* promoter (*66*). The *hok* gene product can be induced to kill cells, if necessary, by application of inducer, thereby eliminating the recombinant cells from that environment. However, mutations that result in insensitivity to the *hok* gene product are strongly selected. Combinations of several conditional lethal blocks have yet to be tested for effectiveness and possible interference with intended biological performance.

Whereas tools now exist to genetically modify bacteria and to detect, disable, or measure cell activity in natural environments, a consensus has yet to be reached on what constitutes a safe release. There is need for better integration of research both on the ecology and molecular biology of bacteria and better focus on relevant questions that can be addressed by scientific methods. Modified bacteria, when properly applied, can become an important component of our environmental protection strategies in the future.

References and Notes

1. H. T. Dulmage and K. Aizawa, in *Microbial and Viral Pesticides*, E. Kurstack, Ed. (Dekker, New York, 1982), pp. 209–237; L. K. Miller, A. J. Lingg, L. A. Bulla, Jr., *Science* **219**, 715 (1983).
2. G. A. Held, L. A. Bulla, E. Ferrari, A. I. Aronson, S. A. Minnich, *Proc. Natl. Acad. Sci. U.S.A.* **79**, 6065 (1982); A. Klier, F. Fargette, J. Ribier, G. Rapport, *EMBO J.* **1**, 791 (1982); D. Lereclus *et al.*, *Biochimie* **67**, 91 (1985); A. I. Aronson, W. Beckman, P. Dunn, *Microbiol. Rev.* **50**, 1 (1986); B. B. Spear, in *Biotechnology in Agricultural Chemistry*, H. M. LeBaron, R. O. Mumma, R. C. Honeycutt, J. H. Duesing, Eds. (American Chemical Society, Washington, DC, 1987), pp. 204–214.
3. L. S. Watrud *et al.*, in *Engineered Organisms in the Environment: Scientific Issues*, H. O. Halverson, D. Pramer, M. Rogul, Eds. (American Society for Microbiology, Washington, DC, 1985), pp. 40–46; G. F. Barry, *Bio/Technology* **4**, 446 (1986).
4. G. Johnson, S. F. Tomasino, J. Flynn, *Phytopathology* **78**, 1540 (abstr.) (1988); P. W. Reeser and S. J. Kostka, *ibid.*, p. 1540 (abstr.).
5. S. J. Kostka, P. W. Reeser, D. P. Miller, *ibid.*, p. 1540 (abstr.).
6. S. J. Kostka, S. F. Tomasino, J. T. Turner, P. W. Reeser, *ibid.*, p. 1540.
7. E. W. Buxton, O. Khalifa, V. Ward, *Ann. Appl. Biol.* **55**, 83 (1965); R. Mitchell and M. Alexander, *Soil Sci. Soc. Am. Proc.* **26**, 556 (1962); I. Chet, Y. Henis, R. Mitchell, *Can. J. Microbiol.* **13**, 137 (1967); A. Ordentlich, Y. Elad, I. Chet, *Soil Biol. Biochem.* **19**, 747 (1987).
8. J. D. G. Jones, K. L. Grady, T. V. Suslow, J. R. Bedbrook, *EMBO J.* **5**, 467 (1986).
9. E. C. Wynne and J. M. Pemberton, *Appl. Environ. Microbiol.* **52**, 1362 (1986); R. L. Fuchs, S. A. McPherson, D. J. Drahos, *ibid.*, p. 504; A. T. Wortman, C. C. Somerville, R. R. Colwell, *ibid.*, p. 142; P. W. Robbins, C. Albright, B. Benfield, *J. Biol. Chem.* **263**, 443 (1988); L. Sundheim, *J. Agric. Sci. Finl.* **59**, 209 (1987).
10. S. E. Lindow, *Plant Dis. Rep.* **67**, 327 (1983); _____, D. C. Arny, C. D. Upper, *Appl. Environ. Microbiol.* **36**, 831 (1978); *Plant Physiol.* **70**, 1084 (1982); S. E. Lindow, *Annu. Rev. Phytopathol.* **21**, 363 (1983).
11. C. S. Orser, B. J. Staskawicz, N. J. Panopoulos, D. Dahlbeck, S. E. Lindow, *J. Bacteriol.* **164**, 359 (1985); R. L. Green and G. J. Warren, *Nature* **317**, 645 (1985).
12. J. Lindemann and T. V. Suslow, in *Proceedings of the 6th International Conference on Plant Pathogenic Bacteria*, E. L. Civerolo, Ed. (U.S. Department of Agriculture, College Park, MD, 1985), pp. 1005–1012.
13. S. E. Lindow, in *Engineered Organisms in the Environment: Scientific Issues*, H. O. Halvorson, D. Pramer, M. Rogul, Eds. (American Society for Microbiology, Washington, DC, 1985), pp. 23–25.
14. _____ and N. J. Panopoulos, in *The Release of Genetically Engineered Microorganisms*, M. Sussman, C. H. Collins, F. A. Skinner, Eds. (Academic Press, London, 1988), pp. 121-138; S. E. Lindow, in *Microbial Ecology*, F. Megusar and M. Gantar, Eds. (Slovene Society for Microbiology, Ljubljana, Yugoslavia, 1988), pp. 509–515; J. Lindemann and T. V. Suslow, *Phytopathology* **77**, 882 (1987).
15. S. E. Lindow, D. K. Willis, N. J. Panopoulos, *Phytopathology* **77**, 1768 (1987); D. A. Cooksey, *ibid.* **78**, 601 (1988).
16. T. J. Burr and A. J. Caesar, *CRC Crit. Rev. Plant Sci.* **2**, 1 (1984); M. N. Schroth and J. G. Hancock, *Annu. Rev. Microbiol.* **35**, 453 (1981); T. V. Suslow, in *Phytopathogenic Prokaryotes*, M. S. Mount and G. H. Lacy, Eds. (Academic Press, London, 1982), pp. 187–223.
17. L. S. Thomashow and D. M. Weller, *J. Bacteriol.* **170**, 3499 (1988).
18. P. R. Gill and G. J. Warren, *ibid.* **170**, 163 (1988).
19. J. E. Loper, *Phytopathology* **78**, 166 (1988); P. A. H. M. Bakker, A. W. Bakker, J. D. Marugg, P. J. Weisbeek, B. Schippers, *Soil Biol. Biochem.* **19**, 443 (1987); P. A. H. M. Bakker, P. J. Weisbeek, B. Schippers, *J. Plant Nutr.* **11**, 925 (1988).
20. N. L. Gutterson, T. J. Layton, J. S. Ziegle, G. J. Warren, *J. Bacteriol.* **165**, 696 (1986); A. R. Poplawsky, Y. F. Peng, A. H. Ellingboe, *Phytopathology* **78**, 426 (1988).
21. N. C. Franklin, *Annu. Rev. Genet.* **12**, 193 (1978); T. J. Silhavy and J. R. Beckwith, *Microbiol. Rev.* **49**, 398 (1985); S. E. Stachel, A. An, C. Flores, E. Nester, *EMBO J.* **4**, 891 (1985); T. J. Close and R. L. Rodriguez, *Gene* **20**, 305 (1982); J. J. Shaw and C. I. Kado, *Bio/Technology* **4**, 560 (1986).
22. N. G. Gutterson, J. S. Ziegle, G. J. Warren, T. J. Layton, *J. Bacteriol.* **170**, 380 (1988); G. J. Warren, H. R. Whitely, N. Gutterson, K. M. Timmis, M. J. Gasson, in *The Release of Genetically Engineered Microorganisms*, M. Sussman, C. H. Collins, F. A. Skinner, D. E. Stewart-Tull, Eds. (Academic Press, London, 1988), pp. 194–206.
23. P. B. Lindgren *et al.*, *EMBO J.*, **8**, 1291 (1989).
24. R. E. Mitchell, *Annu. Rev. Phytopathol.* **22**, 215 (1984); N. J. Panopoulos and R. C. Peet, *ibid.* **23**, 381 (1985); A. Kotoujanski, *ibid.* **25**, 405

(1987); R. O. Morris, *Annu. Rev. Plant Physiol.* 37, 509 (1986); M. J. Daniels, J. M. Dow, A. E. Osburn, *Annu. Rev. Phytopathol.* 26, 287 (1988); R. A. Dixon, *Biological Rev. Camb. Philos. Soc.* 61, 239 (1986).

25. N. T. Keen and B. J. Staskawicz, *Annu. Rev. Microbiol.* 42, 421 (1988).

26. M. C. Whalen, R. E. Stahl, B. J. Staskawicz, *Proc. Natl. Acad. Sci. U.S.A.* 85, 6743 (1988); D. K. Kabayashi, S. J. Tamaki, N. T. Keen, *ibid.* 86, 157 (1989).

27. A. Trigalet and D. Demery, *Physiol. Mol. Plant Pathol.* 28, 423 (1986); H. Toyoda *et al.*, *Proceedings of the 5th International Congress of Plant Pathology*, Kyoto, Japan, 22 to 27 August 1988, *Abstracts of Papers*, p. 237 (abstr. P-VI-I-33); R. D. Durbin and P. L. Langston-Unkefer, *Annu. Rev Phytopathol.* 26, 313 (1988).

28. G. N. Gussin, C. W. Ronson, F. M. Ausubel, *Annu. Rev. Microbiol.* 40, 567 (1986); R. Haselcorn, *ibid.*, p. 525; E. W. Triplet, G. P. Roberts, P. W. Ludden, J. Handelsman, *Am. Soc. Microbiol. News* 50, 15 (1989); L. Rossen, E. O. Davis, A. W. B. Johnston, *Trends Biochem. Sci.* 12, 430 (1987); C. W. Ronson and P. M. Astwood, in *Nitrogen Fixation Research Progress*, H. J. Evans, P. J. Botomley, W. E. Newton, Eds. (Nijhoff, Dordrecht, The Netherlands, 1985), pp. 201–209; G. Eisbrener and H. J. Evans, *Annu. Rev. Plant Physiol.* 34, 105 (1983); N. J. Brewin, in *Genes Involved in Plant-Microbe Interactions*, D. P. S. Verma and T. Hohn, Eds. (Springer-Verlag, Vienna, Austria, 1984), pp. 179–203.

29. M. A. Djordjevic and D. W. Gabriel, *Annu. Rev. Phytopathol.* 25, 145 (1987); J. L. Beynon and J. E. Beringer, *J. Gen. Microbiol.* 291, 351 (1981); F. Debelle *et al.*, *J. Bacteriol.* 170, 5718 (1988); P. J. J. Hooykaas, A. A. N. van Brussel, H. den Dulk-Ras, G. M. van Slogteren, R. A. Schilperoort, *Nature* 291, 351 (1981); G. Truchet *et al.*, *J. Bacteriol.* 157, 134 (1984); A. M. Hirsch *et al.*, *ibid.* 158, 1133 (1984).

30. F. Cannon, personal communication.

31. H. P. Spaink, C. A. Wijffelman, B. J. H. Okker, A. A. N. van Brussel, B. J. J. Lugtenberg, in *Cell to Cell Signals in Rhizobium-Legume Symbiosis*, S. Scannerini, Ed. (Springer-Verlag, Berlin; in press); M. A. Gottfert *et al.*, *J. Bacteriol.* 167, 881 (1986); M. A. Honma and F. M. Ausubel, *Proc. Natl. Acad. Sci. U.S.A.* 84, 8558 (1987).

32. Y. Kapulnick, C. M. Joseph, D. A. Phillips, *Plant Physiol.* 84, 1193 (1987).

33. K. Birkenhead, S. S. Manian, F. O'Gara, *J. Bacteriol.* 170, 184 (1988).

34. H. J. Evans *et al.*, in *Nitrogen Fixation and CO_2 Metabolism: A Steenbock Symposium in Honor of*

Robert H. Burris, P. W. Ludden and J. E. Burris, Eds. (Elsevier, New York, 1985); pp. 3–11.

35. K. Minamisawa, in *Nitrogen Fixation: 100 Years After*, H. Bothe, F. J. de Bruijin, W. E. Newton, Eds. (Fischer, Stuttgart, Federal Republic of Germany, 1988), p. 586.

36. J. E. Beringer and M. J. Bale, in *The Release of Genetically Engineered Microorganisms*, M. Sussman, C. H. Collins, F. A. Skinner, D. E. Stewart-Tull, Eds. (Academic Press, San Diego, CA, 1988), pp. 29–46; D. N. Dowling and W. J. Broughton, *Annu. Rev. Microbiol.* 40, 131 (1986); A. L. M. Hodgson and G. Stacey, *Crit. Rev. Biotechnol.* 4, 1 (1986).

37. E. W. Triplet and T. M. Barta, *Plant Physiol.* 85, 335 (1987).

38. F. J. Mondello, *J. Bacteriology* 17, 1725 (1989); R. Unterman *et al.*, in *Environmental Biotechnology: Reducing Risks from Environmental Chemicals Through Biotechnology*, (Plenum, New York, 1988), pp. 253–269; L. Wender, Ed., *Bioprocess Technol.* 11, 3 (1989); A. R. Harker, R. H. Olsen, R. J. Seidler, *J. Bacteriol.* 171, 314 (1989); R. J. Kolenc, W. E. Inniss, B. R. Glick, C. W. Robinson, C. I. Marfield, *App. Environ. Microbiol.* 54, 638 (1988).

39. M. D. Lee *et al.*, *CRC Crit. Rev. Environ. Control* 18, 29 (1988); R. M. Atlas, *Petroleum Microbiology* (Macmillian, New York, 1984).

40. J. E. McCarthy and M. E. A. Reisch, *Hazardous Waste Fact Book* (U.S. Congress, Congressional Research Service, Washington, DC, 1987); U.S. Congress, Office of Technology Assessment, *Are We Cleaning Up? Ten Superfund Case Studies*, a special report on OTA assessment on superfund (U.S. Congress Printing Office, Washington, DC, 1988).

41. D. D. Focht, in *Environmental Biotechnology Reducing Risks from Environmental Chemicals*, G. S. Omenn, Ed. (Plenum Press, New York, 1988), pp. 15–29; K. N. Timmis, F. Rojo, J. L. Ramos, *ibid.*, pp. 61–79; P. J. Chapman, *ibid.*, pp. 81–95; P. Adriaens and D. D. Focht, *ibid.*, p. 442.

42. Office of Technology Assessment, *New Developments in Biotechnology*, vol. 4, *U.S. Investment in Biotechnology* (U.S. Congress Printing Office, Washington, DC, 1988).

43. J. L. Ramos and K. N. Timmis, *Microbiol. Sci.* 4, 228 (1987); A. J. Weightman and J. H. Slater, in *Microorganisms in Action: Concepts and Applications in Microbial Ecology*, J. M. Lynch and J. E. Hobbie, Eds. (Blackwell Scientific, Oxford, 1988), pp. 322–347; J. F. Quensen and F. Matsumura, in *Treatment of Pesticide Wastes* (American Chemical Society, Washington, DC, 1984), pp. 327–341.

44. J. L. Ramos, A. Wasserfallen, K. Rose, K. N.

Timmis, *Science* 235, 593 (1987); J. L. Ramos, A. Stolz, W. Reineke, K. N. Timmis, *Proc. Natl. Acad. Sci. U.S.A.* 83, 8467 (1986); P. R. Lehrbach *et al. J. Bacteriol.* 158, 1025 (1984); I. Bartels, H.-J. Knackmuss, W. Reineke, *App. Environ. Microbiol.* 47, 500 (1984).

45. A. M. Chakrabarty, D. A. Friello, L. N. Bopp, *Proc. Natl. Acad. Sci. U.S.A.* 15, 3109 (1978); General Electric Company, U.S. Patents 72-260563, 72-0607,260,488 (1974).

46. B. D. Ensley *et al., Science* 222, 167 (1983); K. Furukawa and H. Suzuki, *Appl. Microbiol. Biotechnol.* 29, 363 (1988); N. Mermod, S. Harayama, K. N. Timmis, *Bio/Technology* 4, 321 (1986).

47. J. L. Glick, *Energy Technology*, 9th Energy Technology Conference, R. F. Hill, Ed. (Government Institutes, Washington, DC, 1982), pp. 1470–1474; H. W. Stokes, S. K. Picataggio, D. E. Eveleigh, *Adv. Sol. Energy,* 1, 113 (1983); *Biotechnology Newswatch* (McGraw-Hill, New York, August 1988), p. 2; P. F. Stanburg and A. Whitaker, Eds., *Principles of Fermentation Technology* (Pergamon Press, New York, 1984), pp. 1–9.

48. S. E. Buchloz, M. M. Dooley, D. E. Eveleigh, *J. Ind. Microbiol.* 4, 19 (1989); N. Misawa, T. Okamoto, K. Nakamura, *J. Biotechnol.* 7, 167 (1988).

49. A. D. Neale, R. K. Scopes, J. M. Keey, *App. Microbiol. Biotechnol.* 29, 162 (1988); L. O. Ingram, T. Conway, D. P. Clark, G. W. Sewell, J. F. Preston, *Appl. Environ. Microbiol.* 53, 2420 (1987); J. S. Tolan and R. K. Finn, *ibid.*, p. 2039.

50. S. Fukusumi, A. Kamizono, S. Horinouchi, T. Beppu, *Eur. J. Biochem.* 174, 15 (1988); I. C. Kim *et al., Korean J. Appl. Microbiol. Bioeng.* 16, 369 (1988); P. Oriel and A. Schwacha, *Enzyme Microb. Technol.* 10, 42 (1988); L. Wender, Ed., *Bioprocess. Technol.* 11, 2 (1989); H. Liao, T. McKenzie, R. Hageman, *Proc. Natl. Acad. Sci. U.S.A.* 83, 576 (1986).

51. A. A. Nicolaidis, *J. Chem. Technol. Biotechnol.* 38, 167 (1987); D. E. Rawlings, in *Biotechnology in Minerals and Metals Processing*, B. J. Scheiner, F. M. Doyle, S. K. Kawatra, Eds. (Cushing-Mallow, Ann Arbor, MI, 1989), pp. 3–8.

52. D. E. Rawlings and D. R. Woods, U.S. Patent 4 748 118 (1988); L. Wender, Ed., *Bioprocess. Technol.* 11, 1 (1989); R. S. Ramesar, D. R. Woods, D. E. Rawlings, *J. Gen. Microbiol.* 134, 1141 (1988).

53. B. E. Rittman, *Basic Life Sci.* 28, 215 (1987); M. Sezgin, D. Jenkins, D. Parker, *J. Water Polut. Control Fed.* 50, 362 (1978).

54. A. J. Sinskey, D. D. Easson, Jr., C. Rha, European Product Number 287576 (1988), and U.S. Patent 891136 (1986); D. D. Easson, Jr., O. P. Peoples, A. J. Sinskey, *Industrial Polysaccharides: Genetic Engineering Structure/Property Relations and Applications*, M. Yalpani, Ed. (Elsevier, Amsterdam, 1987), pp. 57–65.

55. C. Khosla and J. E. Bailey, *Mol. Gen. Genet.* 214, 158 (1988); *Nature* 331, 633 (1988).

56. F. M. Romeyer *et al., J. Biotechnol.* 8, 207 (1988).

57. C. E. Furlong *et al., in Environmental Biotechnology: Reducing Risks from Environmental Chemicals Through Biotechnology* (Plenum, New York, 1988), pp. 271–280.

58. J. M. Tiedje *et al., Ecology*, in press; M. Sussman, C. H. Collins, F. A. Skinner, D. E. Stewart-Tull, Eds., *The Release of Genetically Engineered Microorganisms* (Academic Press, London, 1988); H. O. Halvorson, D. Pramer, M. Rogul, Eds., *Engineered Organisms in the Environment: Scientific Issues* (American Society for Microbiology, Washington, DC, 1985).

59. L. Simonsen and B. R. Levin, *Trends Biotechnol.* 6, 527 (1988); P. J. Regal, *ibid.*, p. 536; S. A. Levin, *ibid.*, p. 547; Committee on the Introduction of Genetically Engineered Organisms into the Environment, *Introduction of Recombinant DNA–Engineered Organisms into the Environment: Key Issues* (National Academy Press, Washington, DC, 1987); U.S. Congress, Office of Technology Assessment, New Developments in Biotechnology—Field-Testing Engineered Organisms: Genetic and Ecological Issues (Government Printing Office, Washington, DC, 1988).

60. B. P. Ager, *Trends Biotechnol.* 6, S42 (1988).

61. D. T. Kingsbury, *ibid.*, p. S39; F. Betz, A. Rispin, W. Schneider, in *Biotechnology in Agricultural Chemistry*, H. M. LeBaron, R. O. Mumma, R. C. Honeycutt, J. H. Duesing, Eds. (American Chemical Society, Washington, DC, 1987), pp. 316–327.

62. D. A. Jones, M. H. Ryder, B. G. Clare, S. K. Farrand, A. Kerr, *Mol. Gen. Genet.* 212, 207 (1988); W. E. Holben, J. K. Jansson, B. K. Chelm, J. M. Tiedje, *Appl. Envrion. Microbiol.* 54, 703 (1988); J. S. Shim, S. K. Farrand, A. Kerr, *Phytopathology* 77, 463 (1987).

63. R. J. Steffen and R. M. Atlas, *App. Environ. Microbiol.* 54, 2185 (1988).

64. S. B. Levy and B. M. Marshall, in *The Release of Genetically Engineered Micro-Organisms*, M. Sussman, C. H. Collins, F. A. Skinner, D. E. Stewart-Tull, Eds. (Academic Press, London, 1988), pp. 61–76; J. L. Fox, *Am. Soc. Microbiol. News* 65, 259 (1989).

65. R. Curtiss III, *ibid.*, pp. 7–20.

66. A. K. Bej, M. H. Perlin, R. M. Atlas, *Appl. Environ. Microbiol.* **54**, 2472 (1988).

67. S. T. Kellogg, D. K. Chatterjee, A. M. Chakrabarty, *Science* **214**, 1133 (1981); J. J. Kilbane, D. K. Chatterjee, J. S. Karns, S. T. Kellogg, A. M. Chakrabarty, *App. Environ. Microbiol.* **44**, 72 (1982); D. Ghosal, I.-S. You, D. K. Chatterjee, A. M. Chakrabarty, *Science* **228**, 135 (1985); J. S. Karnes, J. J. Kilbane, S. Duttagupta, A. M. Chakrabarty, *App. Environ. Microbiol.* **46**, 1176 (1983).

68. J. H. Slater and D. Lovatt, in *Microbial Degradation of Organic Compounds*, D. T. Gibson, Ed. (Dekker, New York, 1984), pp. 439–485.

69. L. Krockel and D. D. Focht, *Appl. Environ. Microbiol.* **53**, 2470 (1987).

70. B. Nortemann, J. Baumgarten, H. G. Rast, H.-J. Knackmuss, *Appl. Environ. Microbiol.* **52**, 1195 (1986); J. A. M. DeBont, M. J. A. W. Vorage, S. Hartmans, W. J. J. van den Tweel, *ibid.*, p. 677; E. Dorn, M. Hellwing, W. Reineke, H.-J. Knackmuss, *Arch. Microbiol.* **99**, 61 (1974); J. Latorre, W. Reineke, H.-J. Knackmuss, *ibid.* **140**, 159 (1984); M. A. Rubio, K.-H. Engesser, H.-J. Knackmuss, *ibid.* **145**, 116 (1986); *ibid.*, p. 123.

71. S. W. Pirages, L. M. Curran, J. S. Hirschhorn, in *Impact of Applied Genetics in Pollution Control*, C. F. Kulpa, Jr., R. L. Irvine, S. A. Sojka, Eds., Symposium, University of Notre Dame, 24 to 26 May 1982 (Univ. of Notre Dame Press, Notre Dame, IN); G. Schraa *et al.*, *Appl. Environ. Microbiol.* **52**, 1374 (1986); J. C. Spain and S. F. Nishino, *ibid.* **53**, 1919 (1987); E. Schmidt, G. Remberg, H.-J. Knackmuss, *ibid.* **46**, 1038 (1983).

72. A. K. Koch, J. Reiser, O. Kappeli, A. Fiechter, *Bio/Technology* **6**, 1335 (1988); L. Wender, Ed., *Bioprocess. Technol.* **10**, 3 (1988).

73. H. J. Gilbert *et al.*, *J. Gen. Microbiol.* **134**, 3239 (1988).

74. M. Emori, T. Tojo, B. Maruo, *Agric. Biol. Chem.* **52**, 399 (1988).

75. V. L. Seligy *et al.*, *Biotechnol. Adv.* **2**, 201 (1984).

76. C. Ganter, A. Boeck, P. Buckel, R. Mattes, *J. Biotechnol.* **8**, 301 (1988).

77. J. D. Windass *et al.*, *Nature* **287**, 396 (1980); Imperial Chemical Industries, Ltd., European Product Number 35831 (1981); European Product Number 66994 (1982).

78. D. J. Drahos, B. C. Hennig, S. McPherson, *Bio/Technology* **4**, 439 (1986); D. J. Drahos *et al.*, in *The Release of Genetically-Engineered Micro-Organisms*, M. Sussman, *et al.*, Eds. (Academic Press, London, 1988), pp. 181–191.

79. We wish to thank our colleagues M. N. Schroth, B. J. Staskawicz, J. Loper, and E. Clark for comments, suggestions, and criticism during manuscript preparation. We also thank H. J. Evans and F. Cannon for useful comments and personal communications. Cited work in the authors' laboratories has been supported by grants DBM-8706129, DBM-8409723 PCM-8313052 from the NSF and grant DE-FG03-86ER13518 from the Department of Energy.

4

Yeast: An Experimental Organism for Modern Biology

David Botstein and Gerald R. Fink

The idea that a revolution is occurring in biological research has already achieved the status of cliché. Nonetheless, it is true that much of what can now be done experimentally could only be dreamed of as recently as 15 years ago. The agencies of this revolution are a set of new experimental tools. Foremost among these tools is, of course, the basic "recombinant DNA technology" itself: the ability to isolate individual genes from any organism and to determine their nucleotide sequences, thereby providing the amino acid sequence of any protein product. This prime tool has spawned a large number of generally useful technologies including the use of the cloned gene analytically to study the pattern of normal expression or to follow inheritance of the gene or its neighbors on the chromosome, the use of the cloned gene to produce essentially unlimited quantities of protein for study and for use as reagents, and, not least, the use of cloned genes to produce useful therapeutic agents.

Recombinant DNA technology grew directly out of classical molecular genetics, a field that concentrated on studies of bacteria (especially *Escherichia coli*) and their bacteriophages. The bacterial systems provided not only the materials for recombinant DNA technology (such as plasmid and phage vectors,

suitable hosts, and expression systems) but also, and more important, the intellectual basis (including the very idea of plasmid and replication origin, the concept of promoter, and the notion as well as the identity of the signals for beginning and ending transcription and translation). Bacteria paved the way because they had many advantages: they grow rapidly; they are easy to manipulate both biochemically and genetically; and they share fundamental properties with all other organisms (including DNA genes, messenger RNA and ribosome-based protein synthesis, and metabolic economy based on adenosine triphosphate and nicotinamide adenine dinucleotide). Moreover, they were excellent model systems for the invention and maturation of experimental designs and methods that proved also to be applicable to other systems.

There are, however, limitations to the application of both the facts and the methods of classical bacterial molecular genetics to eukaryotes. The biology of eukaryotic cells is significantly different from that of bacteria in fundamental ways, including the rules for transmission of genetic material, the type and function of subcellular organelles such as the mitochondria, and in basic aspects of metabolism and regulation. Even more obvious is the

D. Botstein is in the Department of Biology, Massachusetts Institute of Technology, Cambridge, MA 02139. G.R. Fink is at the Whitehead Institute, Massachusetts Institute of Technology, Cambridge, MA 02139. This chapter is reprinted from *Science* **240**, 1439 (1988).

fact that multicellular eukaryotic organisms consist of many types of specialized cells. The connection between the DNA and the phenotype is therefore more complicated than it is in bacteria. The methods of bacterial molecular genetics, even in their recombinant DNA incarnation, are not easily applied to eukaryotic cells derived from multicellular organisms. Profound problems inherent in the differentiation process, slow growth, large DNA contents, and poorly tractable genetics all make it difficult to apply the paradigms of bacterial genetics to higher eukaryotes directly.

The solution to this problem, of course, has been the development of experimental systems based on eukaryotic microorganisms. The most highly developed of these is *Saccharomyces cerevisiae* (*1*), a free-living yeast with excellent classical genetics and a fast growth rate (roughly half as fast as that of *E. coli*). It is this yeast that is used to make bread and a variety of alcoholic beverages (including beer and wine) throughout the world. Another evolutionarily unrelated yeast, *Schizosaccharomyces pombe*, shares most of the experimental advantages (*2*) but has developed less rapidly as a model system. The two yeasts sometimes have different advantages: one might choose *S. pombe* for its relatively large chromosomes or *S. cerevisiae* for the fact that its mitochondrial genome is dispensable and thus genetically more tractable. Comparisons between the two are very useful, and many laboratories have taken to using both systems.

These yeasts have proved to be good model systems; they are experimentally tractable yet at the same time typical enough so that lessons learned in the model have a good likelihood of still being true in many other organisms. Being microorganisms, they share with bacteria the simplicity and rapidity of growth and the suitability for biochemical and genetic methods that allows application of the full range of molecular genetic technology. Being eukaryotes, they share with their multicellular cousins many fundamental properties of cell biology (such as cytoskeletal organization, subcellular organelles, secretion systems, receptor and second messenger arrangements, metabolic regulation, and chromosome

mechanics). In the discussion that follows, we give some examples, often from our own work; these are meant as illustrations only. No attempt has been made to review the literature or to ascertain priority for any observations or ideas.

We think that the power of yeast molecular biology as a model for all eukaryotic biology derives from the facility with which the relation between gene structure and protein function can be established. As shown in Fig. 1, the application of the recombinant DNA technology allows one to associate a protein with its function in a number of ways, provided one can both clone a gene having only a mutation (that is, find a gene knowing only the mutant defect) and produce a mutation starting with the cloned gene (that is, determine the function by examining the consequence of the loss of a gene's activity). Two types of manipulation are involved: (i) the insertion of mutations made in vitro into the genome in their correct chromosomal context and (ii) the cloning, on a routine basis, of both the wild-type and mutant alleles of genes identified through mutations in vivo. Both of these kinds of manipulations are easily done in yeast because DNA introduced into yeast by transformation (*3*) behaves in a well-understood way.

Homologous Recombination

Yeast vector systems are of three generic types (*4*). When introduced into *S. cerevisiae* they

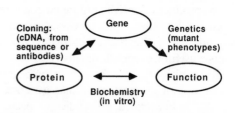

Fig. 1. Recombinant DNA technology allows yeast biologists to readily associate genes, the proteins they specify, and the biological functions the proteins perform. For given mutations, the gene can be cloned by complementation and the protein discovered by using the gene's DNA sequence. For a given protein, the gene can be cloned and the function of the protein analyzed by the production of mutations in vitro and replacement of the normal gene with these mutants.

allow the propagation of the cloned DNA in three different forms: as low-copy, autonomously replicating, stable, properly segregated plasmids [such vectors carry a yeast centromere (5) and are called YCp, for yeast centromere plasmid]; as high-copy, autonomously replicating, unstable, irregularly segregated plasmids [such vectors carry a replication origin from the yeast 2-μm plasmid (6) and are called YEp, for yeast episomal plasmid]; and finally as segments of DNA integrated by homologous recombination into the yeast genome (such vectors are called YIp, for yeast integrating plasmid). All the yeast vectors are also "shuttle vectors" that allow propagation and large-scale preparation of their DNA in *E. coli*.

The variety of options made possible by this array of vectors is surely a big advantage. It is now common for yeast workers to test the behavior of each new gene or mutation at both low copy and high copy. Informal standardization of vector design has made interchange of vector type (in vitro or in vivo) very easy. Quite frequently it turns out that overproduction of a normal gene or mutant allele has interesting consequences, even lethality (7, 8); in other cases many genes can suppress a mutation at high copy while only the correct gene does so at low copy (9).

It is the option of integration into the genome by homologous recombination that provides the yeast systems with their greatest advantage, for this is the feature that allows the movement of genes and mutations into and out of the yeast genome. The uses of integration begin with simple integration (3) of a YIp plasmid by homology in the cloned segment (Fig. 2a). Integration can be directed by the cleavage of the plasmid (10) and results in a duplication of the cloned gene and the concomitant addition of the vector sequences, including one or more selectable markers. The integration thus marks the locus of the cloned gene and is often used to map the locus. However, if the cloned copy contains a mutation, subsequent excision of the plasmid at a point other than the integration point will result in the placement of the mutation onto the genome (sometimes called "transplacement") in its correct position (11), lacking any vector sequences, just as if the

mutation had been made in vivo. This kind of "perfect construction" is routinely possible only in yeast and bacteria. Yet this is the only way that one can truly test the consequences of mutations made in vitro.

Many generally useful variations of this use of integrating plasmids have been devised. Two of these allow the recovery of mutations made in vivo onto vector plasmids that can then be propagated in *E. coli*. By integrating a YIp plasmid carrying the wild-type gene into a strain bearing the mutant allele, one can construct a heterogenotic duplication (Fig. 2a). If one then uses a suitable restriction enzyme (shown by the arrows) that cuts only outside the duplication, one can recover the mutation by circularizing the cut DNA and selecting in *E. coli* for the vector marker (12, 13). An alternative method (Fig. 2b) is to introduce a linear plasmid bearing a deletion of the region of the mutation. Yeast repair systems use the homology to repair the gap, resulting in a plasmid that has a copy of the mutant allele and that again can be selected for in *E. coli* (14). This normal repair reaction is the basis for using cloned genes to carry out fine-structure mapping of mutations directly in yeast (15).

Finally, homologous recombination can be used to prepare null mutations (including insertions, deletions, and "integrative disruptions") in yeast. Several ways of doing this have been devised. One is simply introduction of a deletion allele by the method (11) of Fig. 2a. Another is a replacement strategy by which an insertion (with or without concomitant deletion) in the gene containing a selectable marker is constructed; this construction is then used to replace the normal gene by transformation with linear DNA (16) (Fig. 2c). The third method, integrative disruption (17), consists of the integration of a YIp vector carrying an internal fragment of the gene (15); this results, after homologous recombination, in a partial duplication that splits the gene into two inactive parts (Fig. 2d). All of these systems have variations useful in different contexts. All share the enormous advantage that one can examine a null phenotype for any cloned gene by a straightforward procedure.

The biology of yeast is particularly helpful

here, allowing the recovery and detection of recessive lethal mutations. Introduction of the null mutation in a diploid strain allows subsequent meiosis to yield two normal progeny and two mutants (associated with vector markers). If these fail to grow, then the mutation is lethal. If they are viable they can be studied for any phenotype they might display. If null mutations are lethal, the function can be inferred from conditional-lethal mutations. Several procedures have been developed to screen for such mutations after in vitro mutagenesis of the cloned gene. Thus virtually any gene (*15, 18*), vital or not, can be tested for function.

Some confusion has occasionally arisen about the interpretation of mutant studies, especially when the results conflict with prior assumptions or biochemical evidence (*19–21*). Like any genetic result, apparent lethality or lack thereof must be interpreted with caution because it may depend on the particular circumstances used, or it might indicate the presence of a complicating factor. Failure to find lethality in a gene assumed (or even known) to carry out a vital function may simply indicate the presence of a second gene specifying the same protein or a different one with overlapping function. Likewise, finding of unexpected lethality may result from an unanticipated second function of a gene (for example, in spore germination) or the absence of a normally compensating function. Nevertheless, it sometimes happens that apparently essential proteins are dispensable for growth and vice versa; such results, even if unexpected, are still essentially the only way to test what is

really required by the living cell.

The cloning of a gene in yeast thus allows not only the standard recombinant analysis (sequence of the protein, analysis of regulation, production of the protein product, raising of antibodies) but also the immediate opportunity to study both point mutations and known null mutations. We argue, in fact, that proteins discovered elsewhere but present in yeast may best be studied first in yeast, for the access to genetic analysis of function in yeast is so much better. Indeed, many laboratories have taken this road, and major efforts to understand cytoskeletal proteins (such as actin and tubulin) (*22, 23*), proteins involved in secretion (such as clathrin) (*23, 19–21*) and cell cycle regulation (such as the RAS proteins) (*24*) are only the earliest beginnings of the use of yeast to make progress in understanding proteins common to all eukaryotes.

Saccharomyces cerevisiae: A Surprisingly Typical Eukaryote

One of the most surprising generalities to emerge from the widespread application of recombinant technology to the entire gamut of organisms is that most eukaryotic proteins are extremely well conserved in amino acid sequence. This conservation extends to *S. cerevisiae* (Table 1). Conservation is most extreme in ubiquitin and the cytoskeletal elements, but is still substantial (about 60% identity) for a great variety of enzymes and regulatory functions. Thus, it is not surprising that receptors in yeast strongly resemble

Table 1. Degree of identity in amino acid sequence between corresponding proteins of yeast (*Saccharomyces cerevisiae*) and humans (*Homo sapiens*). The data were generated from sequences in Genbank and other published sources.

Yeast or human protein	Identity (%) in amino acid sequence	Reference
Ubiquitin	96	(*27*)
Actin	89	(*28*)
β-tubulin	75	(*29*)
HMGCoA reductase	66	(*30*)
Cytochrome c	63	(*31*)
Citrate synthetase	62	(*32*)
RAS1/N-ras; RAS2/K-ras	60	(*33, 34*)
Glucose transporter	25	(*35*)

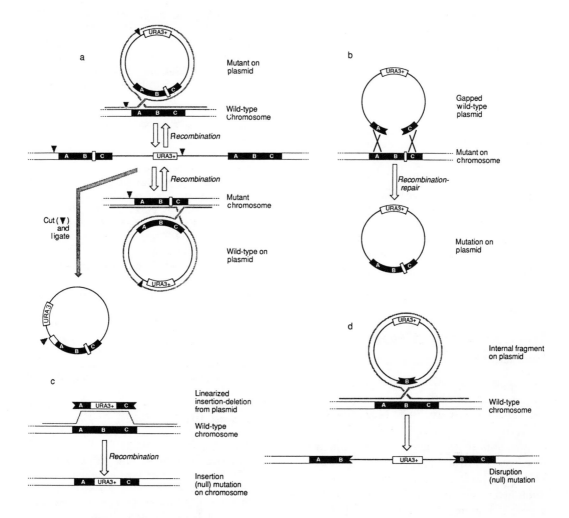

Fig. 2. The use of homologous recombination to transfer mutations into and out of the normal locus on a yeast chromosome. (a) Integration of a cloned gene (A-B-C) by homologous recombination into the mutant locus results in a heterogenotic duplication. The same duplication can be produced by homologous recombination of a mutant plasmid into the normal locus, as shown below. Depending on the position of the crossover event, excision of the plasmid by homologous recombination from the duplication can result in either a mutant or a wild-type gene at the locus. If one digests the DNA containing the duplication with a suitable restriction endonuclease (shown to the left) and ligates the fragments, one can obtain the mutation by selection for vector markers in *Escherichia coli* (*12*). (b) A mutation on a yeast chromosome can be recovered by recombination-repair after transformation with a suitably gapped plasmid carrying the wild-type gene as shown (*13*). (c) Gene disruption can be accomplished by integration of a linear fragment containing an insertion or deletion containing a selectable marker (*16*). (d) Integrative gene disruption occurs when an internal fragment of a gene integrates by homologous recombination into the intact locus, splitting the gene into two partially duplicated but incomplete parts (*17*), one missing the amino-terminal coding region and the other missing the carboxyl terminal.

generic receptors in other organisms, as do cytoskeletal elements and enzymes of similar function. This generality strongly supports the notion that there has been functional conservation to at least the same degree. This argument has already been buttressed in several cases by the demonstration of actual appropriate func-

tion of mammalian genes in yeast cells (for example, mammalian *ras* genes complement yeast *ras* mutants) (*24*) or vice versa [for example, assembly of hybrid tubulins in animal cells) (*25*)]. Recently a human gene was cloned directly by complementation of a *cdc2* mutation in *S. pombe* (*26*). We think that conservation most strongly validates the use of the yeasts as models for the primary deduction of functional and mechanistic aspects of proteins and protein systems shared by eukaryotes.

Finally, it is probably necessary to say that just as one cannot uncritically make conclusions from lethality in yeast, so must one be cautious about concluding too much about the function of proteins in animals based only on studies of their homologs in yeast. We believe that the facility of gene manipulation will continue to allow molecular genetic studies in yeast to lead the way to understanding gene function in more complex organisms. Nevertheless, every major conclusion will have eventually to be tested directly in higher organisms. The point of any model system is to make these final tests possible, not to make them unnecessary.

The Community of Yeast Biologists

An important ingredient in the success of yeast studies as a scientific field is the attractiveness of the yeast community itself. Newcomers find themselves in an atmosphere that encourages cooperation. In keeping with a set of traditions that began with the phage group founded by Delbruck, Luria, and Hershey, not only are published strains and materials generally made available, but many (if not quite all) laboratories in the field routinely exchange strains, protocols, and ideas long before publication.

In conclusion, we believe that the yeasts are a nearly ideal model system for eukaryotic biology at the cellular and molecular level. The main reason is experimental tractability, especially in associating genes with proteins and functions in the cell, but the open and cooperative traditions of the community also play an important role.

References and Notes

1. J. N. Strathern, Ed., *The Molecular Biology of the Yeast Saccharomyces: Life Cycle and Inheritance* (Cold Spring Harbor Laboratory, Cold Spring Harbor, NY, 1981); *The Molecular Biology of the Yeast Saccharomyces: Metabolism and Gene Expression* (Cold Spring Harbor Laboratory, Cold Spring Harbor, NY, 1982).
2. P. Russell and P. Nurse, *Cell* **45**, 781 (1986).
3. A. Hinnen, J. B. Hicks, G. R. Fink, *Proc. Natl. Acad. Sci. U.S.A.* **75**, 1929 (1978); D. T. Stinchcomb, K. Struhl, R. W. Davis, *Nature* **282**, 39 (1979).
4. D. Botstein and R. W. Davis, in *Molecular Biology of the Yeast Saccharomyces: Metabolism and Gene Expression*, J. E. Strathern, Ed. (Cold Spring Harbor Laboratory, Cold Spring Harbor, NY, 1982), p. 607.
5. L. Clarke and J. Carbon, *Nature* **287**, 504 (1980); *Annu. Rev. Genet.* **19**, 29 (1985).
6. J. Broach and J. B. Hicks, *Cell* **21**, 501 (1980).
7. D. Meeks-Wagner and L. H. Hartwell, *Cell* **44**, 43 (1987). J. H. Thomas, thesis, Massachusetts Institute of Technology (1984).
8. M. D. Rose and G. R. Fink, *Cell* **48**, 1047 (1987).
9. P. J. Schatz, F. Solomon, D. Botstein, *Mol. Cell. Biol.* **6**, 3722 (1986).
10. T. Orr-Weaver, J. Szostak, R. Rothstein, *Proc. Natl. Acad. Sci. U.S.A.* **78**, 6354 (1981).
11. S. Scherer and R. W. Davis, *ibid.* **76**, 4951 (1979).
12. G. S. Roeder and G. R. Fink, *Cell* **21**, 239 (1980).
13. F. Winston, F. F. Chumley, G. R. Fink, *Methods Enzymol.* **101**, 211 (1982).
14. J. B. Hicks, J. Strathern, A. Klar, S. Dellaporta, *Genet. Eng.* **4**, 219 (1982); T. Orr-Weaver, R. Rothstein, J. Szostak, *Methods Enzymol.* **101**, 228 (1983); T. Orr-Weaver and J. Szostak, *Proc. Natl. Acad. Sci. U.S.A.* **80**, 4417 (1983).
15. D. Shortle, P. Novick, D. Botstein, *Proc. Natl. Acad. Sci. U.S.A.* **81**, 4889 (1984).
16. R. J. Rothstein, *Methods Enzymol.* **101**, 202 (1983).
17. D. Shortle, J. E. Haber, D. Botstein, *Science* **217**, 317 (1982).
18. J. D. Boeke, J. Trueheart, G. Natsoulis, G. R. Fink, *Methods Enzymol.* **154**, 164 (1987).
19. G. S. Payne and R. Schekman, *Science* **230**, 1009 (1985).
20. G. S. Payne, T. B. Hasson, M. S. Hasson, R. Schekman, *Mol. Cell. Biol.* **7**, 3888 (1987).
21. S. K. Lemmon and E. W. Jones, *Science* **238**, 504 (1987).
22. J. Huffaker, A. Hoyt, D. Botstein, *Annu. Rev. Genet.* **21**, 259 (1987).
23. J. Hicks, Ed., *Yeast Cell Biology* (Liss, New York, 1986).
24. T. Katoaka *et al.*, *Cell* **37**, 437 (1984); K. Tatchell, D. T. Chaleff, D. DèFeo-Jones, E. M. Sconick, *Nature* **309**, 523 (1984); F. Tamanoi, M. Walsh, T. Kataoka, M. Wigler, *Proc. Natl. Acad. Sci. U.S.A.* **81**, 6924 (1984); D. Brevia-

rio, A. Hinnebusch, J. Cannon, K. Tatchell, R. Dhar, *ibid.* **83**, 4152 (1986).

25. J. F. Bond, J. L. Fridovich-Keil, L. Pillus, R. C. Mulligan, F. Solomon, *Cell* **44**, 461 (1987).
26. M. G. Lee and P. Nurse, *Nature* **327**, 31 (1987).
27. UBI1, UBI2, UBI3, UBI4 proteins of yeast (*Saccharomyces cerevisiae*): E. Özkaynak, D. Finley, M. J. Solomon, A. Varshavsky, *EMBO J.* **6**, 1429 (1987). Ubiquitin, human and bovine: D. H. Schlesinger and G. Goldstein, *Nature* **255**, 423 (1975); _____ and H. D. Niall, *Biochemistry* **14**, 2214 (1975).
28. Actin, yeast (*Saccharomyces cerevisiae*): D. Gallwitz and I. Sures, *Proc. Natl. Acad. Sci. U.S.A.* **77**, 2546 (1980); R. Ng and J. Abelson, *ibid.*, p. 3912; W. Nellen, C. Donath, M. Moos, D. Gallwitz, *J. Mol. Appl. Genet.* **1**, 239 (1981). Actin, human cytoplasmic beta actin: I. Nakajima, H. Hamada, P. Reddy, T. Kakunaga, *Proc. Natl. Acad. Sci. U.S.A.* **82**, 6133 (1985).
29. Tubulin beta chain, yeast (*Saccharomyces cerevisiae*): N. F. Neff, J. H. Thomas, P. Grisafi, B. Botstein, *Cell* **33**, 211 (1983). Beta tubulin, human: S. A. Lewis, M. E. Gilmartin, J. L. Hall, N. J. Cowan, *J. Mol. Biol.* **182**, 11 (1985); M. G. S. Lee, S. A. Lewis, C. D. Wilde, N. J. Cowan, *Cell* **33**, 477 (1983).
30. HMGCoA reductase, human: K. L. Luskey and B. Stevens, *J. Biol. Chem.* **260**, 10271 (1985). HMGCoA, yeast: M. E. Basson, M. Thorsness,

J. Rine, *Proc. Natl. Acad. Sci. U.S.A.* **83**, 5563 (1986).
31. Cytochrome c, yeast: M. Smith *et al.*, *Cell* **16**, 753 (1979); K. Narita and K. Titani, *J. Biochem.* **65**, 259 (1969). Cytochrome c, human: H. Matsubara and E. L. Smith, *J. Biol. Chem.* **238**, 2732 (1963); *ibid.* **237**, PC3575 (1962).
32. Citrate synthase, yeast: M. Suissa, K. Suda, G. Schatz, *EMBO J.* **3**, 1773 (1984); D. P. Bloxham, D. C. Parmelee, S. Kumar, K. A. Walsh, K. Titani, *Biochemistry* **21**, 2028 (1982); S. Remington, G. Wiegand, R. Huber, *J. Mol. Biol.* **158**, 111 (1982). The comparison is with the porcine enzyme.
33. RAS protein, yeast: R. Dhar, A. Nieto, R. Koller, D. Dereo-Jones, E. M. Scolnick, *Nucleic Acids Res.* **12**, 3611 (1984).
34. Transforming protein 1, human: J. P. McGrath *et al.*, *Nature* **304**, 501 (1983); K. Shimizu *et al.*, *ibid.*, p. 497; D. J. Capon *et al.*, *ibid.*, p. 507; E. Taparowsky, K. Shimizu, M. Goldfarb, M. Wigler, *Cell* **34**, 581 (1983).
35. Glucose transporter, human: M. Mueckler *et al.*, *Science* **229**, 941 (1985). Glucose transporter, yeast: J. L. Celenza, L. Marshall-Carlson, M. Carlson, *Proc. Natl. Acad. Sci. U.S.A.* **85**, 2130 (1988).
36. We thank J. Mulholland and C. Watanabe for the homology searches and C. Morita for preparation of figures.

5

Dictyostelium discoideum: A Model System for Cell-Cell Interactions in Development

Peter Devreotes

The life cycle of *Dictyostelium* consists of distinct growth and developmental phases. Development is characterized by a stage in which 10^5 single amoebae aggregate to form a multicellular organism (Fig. 1). This spontaneous process is organized by extracellular adenosine 3',5'-monophosphate (cAMP), which binds to surface receptors that mediate chemotaxis and oscillatory cell-cell signaling. The signaling derives from the capacity of each cell to synthesize and secrete cAMP in response to stimulation by cAMP. Within aggregation centers the concentration of cAMP spontaneously oscillates; each peak initiates a wave of cAMP that propagates through the cell monolayer. The leading edge of each passing wave provides a gradient that briefly orients the chemotactically sensitive cells toward the aggregation center. Cells move up the gradient for several minutes until the peak of the wave reaches the position of the cell. The cells then move in random directions until the next wave elicits another coordinated movement step. About 50 movement steps result in the formation of a multicellular structure (*1–3*). Within the multicellular structure, the signaling system continues to act as a developmental timer and to play a role in morphogenesis and pattern

formation. Signaling controls the classes of genes expressed at each stage of development, ensuring that cells differentiate into two types, vacuolated nonviable stalks and durable spores, in a pattern along the axis of the structure (*4–12*).

Biochemistry and Genetics

Free-living *Dictyostelium* amoebae are easily grown and maintained (*13*). The wild-type strains grow to high densities on bacterial lawns with doubling times of 3 hours. About 10^9 cells are typically harvested from a single agar dish (10 cm in diameter). The wild-type strains can also be grown in bacterial suspensions or in axenic media containing particulate nutrients. Axenic strains are available that grow with doubling times of 8 hours in a simple, inexpensive medium of glucose, peptone, and yeast extract. Shaken cultures grow to a density of 10^{10} cells per liter, so that a 100-liter culture yields up to 10^{12} identical cells (*14*). The axenic strains will also grow with doubling times of 24 hours in defined liquid media.

The developmental program is initiated when cells are deprived of nutrients. Environ-

P. Devreotes is in the Department of Biological Chemistry, Johns Hopkins University School of Medicine, Baltimore, MD 21205. This chapter is reprinted from *Science* 245, 1054 (1989).

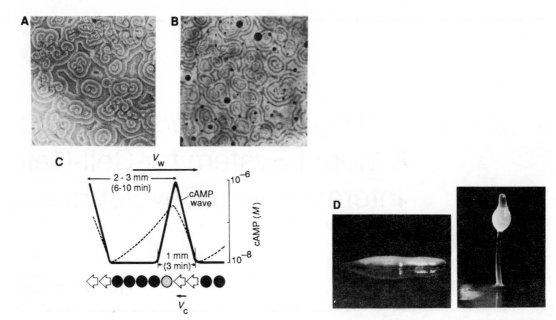

Fig. 1. Stages of *Dictyostelium* development. (A through C) Chemotaxis and cAMP signaling mediate aggregation. (A) Dark-field photograph of aggregating cells at the 5-hour stage of development. Territories containing approximately 10^6 cells are 1 to 2 cm in diameter. (B) Fluorographic image of cAMP waves within a monolayer of aggregating cells. (C) Dynamics of signal relay and chemotaxis. The solid line represents the cAMP concentration; the dotted line represents the adaptation process. Symbols represent a single radial line of cells: open arrows, cells moving toward the center; shaded and filled circles, randomly oriented cells; and arrow vectors, the speed and direction of motion of the cAMP wave ($V_w = 300$ mm/min) and the moving cells ($V_c = 20$ mm/min). [(A), (B), and (C) are adapted from (66) with permission of AAAS](D) Morphogenesis and differentiation. (Left) Migrating sluglike stage (approximately 2 to 3 mm in length). (Right) Fruiting body (approximately 2 to 3 mm tall) with a mass of spores at the end of a vertical stalk.

mental parameters that affect the rate of development include cell density and temperature and, when cells are spread on a surface, the wetness of the surface. When these parameters are held constant, the timing of stages of the 24-hour developmental program is reproducible to within 1 hour (15).

Growth and development are completely separate, and it is a simple matter to switch between the two modes. When cells are depleted of nutrients, growth ceases and development is initiated. However, differentiation is completely reversible until late stages of the program; on reintroduction of nutrients, cells "erase" their developmental markers and resume growth (15). Thus, once developmental mutations are selected, they can be propagated and maintained, and the phenotypes can be expressed again when development is reinitiated. Mutant cell lines can also be stored indefinitely as spores or as frozen amoebae.

The genome of *Dictyostelium* consists of about 40,000 kb, about 1% of the size of the human genome, and is arranged on seven chromosomes (16). Although the cells have a true diploid phase, formed from opposite mating types as in yeast, all of the major biological phenomena being studied are expressed when the cells are haploid. Thus, nonlethal recessive mutations affecting processes of interest can be directly scored. It is relatively easy to mutagenize cells, and selection of mutants is limited only by the screening or selection procedure (17).

Parasexual diploids form by spontaneous cell fusion at a frequency of 10^{-5} in a population of identical cells. Recessive sensitivity markers can be used to select against the haploid cells and allow survival of diploids (17). Thus, if a strain containing a *tsg* marker (cells die at 27°C) is mixed with a strain containing a different *tsg* marker or with a *bsg* (cells cannot feed on *Bacillus subtilis*)-marked strain, then approximately one in 10^5 cells will grow at 27°C or on *B. subtilis* at 27°C, respectively. These diploids can be destabilized, resulting in the production of haploids with a random segregation of the chromosomes. There is a low frequency of mitotic crossing over, which can be used for finer structure mapping (*17*). Mutations can be assigned to complementation and linkage groups (over 100 loci have been mapped) by use of tester strains marked with recessive resistance markers. A series of wild isolates and divergent strains are available for restriction fragment length polymorphism mapping studies, and numerous cloned genes have been assigned to linkage groups (*18*). This information is useful when complementation groups and suspected mutated cloned genes are assigned to the same linkage group.

Transformation of *Dictyostelium* with exogeneous DNA is a routine procedure (*19–21*). Efficiencies of 10^{-5} to 10^{-3} are readily obtained with either calcium phosphate precipitation or electroporation. Both transient (up to 3 days) and stable cells lines can be constructed. Establishment of a stable cell line requires about 3 hours of culture manipulation followed by approximately 1 week of incubation for the appearance of foci. The available transforming plasmids contain a neomycin transferase gene flanked by an actin promoter and a variety of terminators. Most vectors integrate into the genome in tandem repeats from 10 to 300 copies, although nonintegrating vectors have also been developed by including regions

of endogenous plasmids (*22*). Multiple endogenous plasmids occur in *D. discoideum* and other species of slime molds; these plasmids are stably carried at a few to several hundred copies and may facilitate the construction of new vectors (*23*). Vectors have been used for expression of heterologous genes, for analysis of promoter function, and for antisense and homologous recombination experiments.

A number of gene products from other cells have been expressed in *Dictyostelium* with retention of functional properties. Because the cells appear to carry out mammalian-like glycosylation, the system may be suitable for expression and large-scale production of mammalian gene products (*24*); it may be especially useful for receptors and other membrane proteins that are difficult to express in bacteria and may not function in yeast. In addition, developmentally regulated *Dictyostelium* gene products can be expressed during the growth stage of the life cycle when they are normally absent (Table 1).

The antisense approach can often be applied in *Dictyostelium* to mimic a mutation. Antisense cell lines that express less than 0.5% of the amount of gene product found in the wild type have been constructed for discoidin (an analog of fibronectin), D2 (a serine esterase), myosin heavy chain, and surface cAMP receptors (*25–28*) (Fig. 2). These cell lines display strong phenotypes correlated with the specific loss of the targeted gene product and have helped delineate the roles of these proteins in

Table 1. Expression of functional proteins in *Dictyostelium*.

Protein	Source	Developmental stage of expression (hours)	Function assayed	References
Luciferase	Firefly	0, 12, 15	Luminescence	(*60*)
Glucoronidase	Mammalian	12, 15	Fusion protein	(*61*)
$G_\alpha 1$	*Dictyostelium*	0	Multinucleated cells	(*40*)
$G_\alpha 2$	*Dictyostelium*	0	Rescue of *Frigid A*	(*40*)
cAMP receptor	*Dictyostelium*	0	cAMP binding	(*27*)
β-Adrenergic receptor	Hamster	0	Isoproterenol binding	(*62*)
Ras	*Dictyostelium*	0	Multitip phenotype	(*63*)
Ras	Mammalian	0	Multitip phenotype	(*64*)
Phosphodiesterase	*Dictyostelium*	0	Aggregation	(*65*)

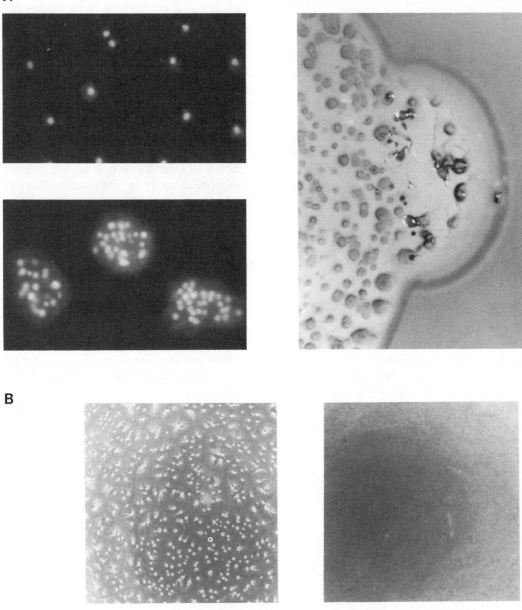

Fig. 2. Mutations with phenotype. (**A**) Disruption of myosin heavy chain by homologous recombination. (Left) Multinucleate morphology of cells expressing the NH_2-terminal portion of myosin. Control (upper) or mutated (lower) cells were allowed to attach to slides, then fixed and stained with 4',6-diamidino-2-phenylindole. Individual nuclei are $3\,\mu m$ in diameter. (Right) Reversion of the phenotype. Transformed cells were grown on a bacterial lawn in the absence of drug selection. The left portion shows a region where the cells have depleted the bacteria and formed mounds that did not develop further. A reversion event, recorded in the center of the photograph, is visible as an outgrowth (1 cm across) from the leading edge of the feeding fruit. The reversion gave rise to cells that grew faster, formed normal fruiting bodies, and contained intact myosin heavy chain. [(A) is adapted from (*29*) with permission of AAAS] (**B**) Antisense transformation mimics receptor mutation. Vectors were designed to express full-length antisense RNA of the cAMP receptor in growing cells. The control transformants (no receptor sequence in vector) (left) differentiated as wild-type cells, aggregating with streams of cells entering the centers. Antisense transformants (right) remained as a uniform cell monolayer, shown here at the 7.5-hour stage of development. A 5 by 5 cm section of each monolayer is shown. [(B) is adapted from (*27*) with permission of AAAS]

development. The success of the antisense approach does not appear to depend on the abundance of the targeted protein. Whereas surface cAMP receptors constitute only a small fraction of the cell protein (0.02%), discoidin and myosin heavy chain constitute about 2 and 0.5%, respectively, of cell protein.

In several instances, such as for myosin heavy chain and α-actinin, transformation by homologous recombination with deleted copies of the gene or cDNA has led to integration into and disruption of the endogenous gene (29–31). In the case of myosin, cells were transformed with a construct containing the 5' end of the gene with a nonfunctional promoter. A single crossover occurred and the cells began to express the NH_2-terminal half of myosin, heavy meromyosin. This construct was excised spontaneously with reversal of the phenotype when cells were removed from selection (29) (Fig. 2). In the case of α-actinin, cells were transformed with an internal fragment of the coding sequence. Again a single crossover occurred, resulting in two nonfunctional copies of the gene, and the gene product was either not synthesized or was lost by degradation (30). The studies of myosin heavy chain have been extended by transforming cells with regions flanking the coding sequence separated by the neomycin transferase gene. In several instances, this type of construct resulted in a double crossover and irreversible disruption of the gene (31).

These and other recent observations contain the seeds for growth of a powerful genetic system. It is possible to complement "null" mutants with cloned genes so that the wild-type phenotype is restored (32). Moreover, a thymidine synthetase mutant has been complemented with a clone from a genomic library (33). Mutant complementation with genomic libraries will probably become a more widely used technique as improvements in transformation efficiency are achieved. It should be possible to introduce mutations into endogenous genes and to clone new genes by complementation of mutants. *Dictyostelium* is especially well suited for studies of signal transduction, motility and chemotaxis, cell-cell communication, gene expression, and pattern formation, and because of the similarity of many functions and genes between *Dictyostelium* and mammalian cells, these studies may provide general insights.

Signal Transduction

Transmembrane signaling systems in *Dictyostelium* seem to be essentially the same as those in mammalian cells (Fig. 3). Surface cAMP receptors have many of the features of receptors that are linked to G proteins. Like bovine rhodopsin and the β-adrenergic receptor, the

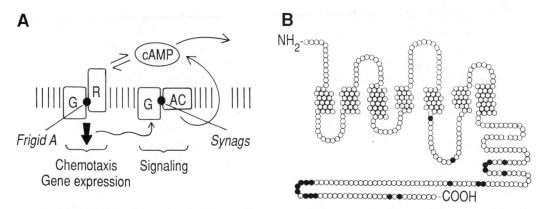

Fig. 3. Transmembrane signaling system. (**A**) Diagram of cAMP signaling system, including interactions among receptor (R), G proteins (G), and adenylate cyclase (AC). (**B**) Model of the cAMP receptor. The seven hydrophobic domains are arranged as α-helices in the lipid bilayer. The proposed extracellular domains are shown above the α-helices, and the intracellular domains, including the serine-rich COOH-terminus (serines are filled), are shown below these transmembrane helices. [(B) is adapted from (27) with permission of AAAS]

cAMP receptor has seven putative membrane-spanning domains and a serine- and threonine-rich COOH-terminus (*27*). The affinity of receptors for cAMP is reduced by guanine nucleotides, and a primary effector, adenylate cyclase, is stimulated by guanosine triphosphate in vitro (*34–36*).

These guanine nucleotide-dependent activities can be independently affected by mutation. A series of mutants, designated *Frigid*, that fail to enter the developmental program even when provided with exogenous cAMP has been isolated (*37*). In several mutants, belonging to the complementation group *Frigid A*, guanine nucleotides do not influence cAMP binding. A second series of mutants, designated *Synag*, has also been characterized. In these mutants, adenylate cyclase and the cAMP signaling response are no longer sensitive to guanine nucleotides, although repeated stimulation with cAMP still is able to induce chemotaxis and gene expression (*38*). Cells contain two homologs of the α-subunits of mammalian G proteins that are 45% identical and a β-subunit that is 63% identical to their mammalian counterparts at the amino acid level (*39*). *Frigid A* cells carry deletions or mutations in one of the two G protein α-subunits, Gα2 (*40*). A defect in a *Synag* mutant has not been associated with a mutation in a G protein subunit. Homologs of G protein α and β subunits have also been found in yeast, where they participate in the α/a-mating factor response (*41*). Thus G protein-linked transmembrane signaling systems appeared early in the evolution of eukaryotic cells.

The cAMP receptors undergo a process of desensitization similar to that observed for related mammalian receptors. Desensitization involves both down-regulation and adaptation (*42*), and the adaptation properties of several receptor-mediated cellular responses, such as guanosine 3′,5′−monophosphate (cGMP) accumulation, myosin phosphorylation, and adenylate cyclase activation, have been studied. Cells respond to increases in the fractional occupancy of surface receptors, and when occupancy is held constant, responses subside within a few minutes. The magnitude of a response is proportional to the fractional increase in receptor occupancy, and this is true for the initial challenge with cAMP or for any subsequent increment in the amount of cAMP. When the stimulus is withdrawn, adaptation decays and cells regain sensitivity in about 15 min. Adaptation is probably due to a rapid, extensive ligand-induced phosphorylation of the receptors. The kinetics and ligand-concentration dependence of phosphorylation correlate closely with the adaptation process (*43*). This reversible phosphorylation cycle appears to be an essential feature of the biological oscillator during aggregation.

Signal transduction through surface cAMP receptors participates in a multitude of developmental functions. It appears that there is a family of related surface cAMP receptor subtypes. The existence of such families of receptor subtypes seems to be a common theme within the class of G protein-linked receptors. Each subtype may be responsible for one or a few of the cAMP-mediated functions (*44*). Expression of antisense RNA of the cAMP receptor subtype first characterized may block expression of all the receptor subtypes because the antisense cell lines are defective in most aspects of the developmental program (Fig. 2).

Motility and Chemotaxis

The movement and chemotactic responses of *Dictyostelium* cells closely resemble those of amoeboid-like cells in higher eukaryotes. The cascade of physiological and biochemical responses that occurs in response to cAMP in *Dictyostelium* is similar to that triggered by chemoattractants in leukocytes and macrophages (*45*). In the absence of stimuli, *Dictyostelium* cells move at between 7 to 10 μm/min and extend pseudopods every 30 s at a rate of about 30 μm/min. The cells contain many of the contractile and cytoskeletal proteins found in higher eukaryotes, including actin, myosin, gelsolin, tubulin, and α-actinin. Actin seems to play a central role in motility and chemotaxis. Filamentous actin (F-actin) is present in the advancing pseudopods and forms a meshwork at the sides and posterior of the cell; its extent of polymerization is transiently increased upon addition of chemotactic stimuli (*46*). Myosin heavy chain is found in punctate regions of the cytoplasm and is concentrated in the posterior ectoplasm (*47*); it redistributes to the cell periphery in activated cells.

A large number of mutants defective in chemotaxis have been isolated, and although only a few have been characterized, they have helped in defining which components are essential for motility and chemotaxis. For example, *Frigid A* mutants are unable to undergo chemotaxis, suggesting that a G protein is essential for this process (*48*). Although in *Synag 7* mutants the activation of adenylate cyclase is defective, their chemotactic response is normal, suggesting that intracellular cAMP does not participate in this process. *Streamer F* mutants lack cGMP-specific phosphodiesterase, so that there is a persistent elevation of intracellular cGMP (*49*). As a result, the cells remain elongated for several minutes after removal of the external gradient of chemoattractant (*50*). The heavy chain of conventional myosin, α-actinin, and gelsolin (which have all been disrupted) do not appear to be essential for motility and chemotaxis. Cells also contain shorter forms of myosin heavy chains, which may be involved in motility (*51*). Conventional myosin appears to play a role in cytokinesis, as cells lacking the heavy chain become multinucleate (*29*) (Fig. 2).

Cell-Cell Communication

The cell-cell signaling system in *Dictyostelium* appears to be unlike that of any other member of the biological kingdom. It is a simple method of communication within a population of identical cells, all of which secrete and respond to the same molecule. Similar communication systems may be used in other cell types, however, possibly during embryogenesis or regeneration when groups of identical cells must spontaneously organize.

The cAMP signaling system does not supplant other mechanisms of cell-cell interaction. Since all species of slime mold appear to use cAMP signaling in later development but do not form interspecies structures, there must be

additional modes of cell-cell recognition. Specific cell adhesion molecules (CAMs) found in *Dictyostelium* may serve this function (*52*). The best characterized CAM is the protein referred to as GP80 or "contact sites A," which shares regions that are homologous to CAMs from higher eukaryotic cells, such as L-CAM and N-CAM in birds and mammals (*53*). *Dictyostelium* also contains an analog of fibronectin, discoidin, which mediates cell adherence to the substrate. This protein contains the RGDX (Arg-Gly-Asp-X) sequence commonly found in adhesion proteins (*54*).

Gene Expression

Most of the known components of the sensory system, such as surface cAMP receptors, adenylate cyclase, and the G protein subunits, are subject to developmental regulation. The periodic cAMP signaling, in turn, regulates the expression of these components as well as that of other developmentally controlled gene products. Receptor-mediated events also appear to regulate late cell type-specific differentiation, pattern formation, and morphogenesis. Characterization of mutants such as *Frigid A* and the *Synags* indicate that signal transduction pathways that include G proteins regulate these events (Fig. 3). The effectors of these G proteins are unknown, although they may include phospholipase C.

Cells must be exposed to cAMP in a specific regimen to induce expression of each of the gene classes. Low, constant concentrations of exogenous cAMP inhibit the early developmental program, whereas repeated application of cAMP at 6-min intervals accelerates the process. In contrast, late cell type-specific gene expression depends on continuous application of cAMP.

These observations suggest that the receptor-mediated responses that activate early gene expression adapt, whereas those that activate the late cell–type specific genes do not. Thus, the cAMP oscillator, regulated by receptor-mediated phosphorylation (and dephosphorylation), may be an essential part of the developmental timer.

Morphogenesis and Pattern Formation

As cells that are initially identical aggregate to form the multicellular "slug" structure, a simple pattern appears along its axis (Fig. 1). A sharp demarcation separates cells within the posterior three-fourths of the slug, which express "prespore" markers, from those at the anterior, which do not express these markers. The anterior cells later differentiate into stalk cells and are designated "prestalk" cells. A variety of proportioning mutants that affect the ratio of cell types have been isolated.

This morphological pattern seems to be achieved through the establishment of a pattern by chemotactic cell sorting and through subsequent position-dependent differentiation. Cells that have undergone periods of starvation, that enter the developmental program from early stages of the cell cycle, or that have been grown on glucose-deficient media preferentially sort to the anterior region. Subsequent prespore marker accumulation is confined to the posterior region of the slug, and later, stalk cell differentiation takes place at the tip and ventral side of the structure, indicating that there are position-dependent developmental cues (*3*).

The tip of the slug displays properties of an embryonic "organizer." If a slug is severed just behind the tip, the tissue connected to the tip continues to migrate. However, migration of the remainder ceases, the cell mass re-forms a mound, and a new tip appears before morphogenesis continues. Excised tips, grafted to host slugs, define new axes and thereby cause several smaller slugs to form and separate. These organizational properties of the tip appear to derive from its capacity to generate oscillatory cAMP signaling (*55, 56*).

In addition to cAMP, other compounds such as adenosine, differentiation-inducing factor (DIF), and ammonia have been implicated as morphogens in the developing system (*57–59*). Adenosine appears to antagonize the actions of extracellular cAMP, and its production in anterior regions has been postulated to inhibit the formation of additional tips and to prevent anterior cells from expressing prespore

genes. DIF has been identified as 1-(3,5-dichloro-2, 6-dihydroxy-4-methoxyphenyl)-1-hexanone, a lipid-like compound that is essential for stalk cell differentiation. Certain marker mRNAs have been shown to be dependent on DIF for expression and to be expressed only in anterior cells. In this sense DIF displays the properties of a morphogen, a compound described in theories of embryonic pattern formation as being able to induce cell differentiation in a region of a responsive field of cells. Ammonia appears to promote spore cell formation by counteracting the effects of DIF.

Conclusions

The mechanisms of signal transduction in eukaryotic microorganisms appear to be similar to those in mammalian cells, and thus genetics can be applied to study relevant signal transduction phenomena. The role of signal transduction in processes such as motility and chemotaxis are most easily addressed in *Dictyostelium*. The relation of the signaling properties of individual cells to the behavior of aggregates of cells is more clearly understood in *Dictyostelium* than in other organisms. However, there are major apparent differences between *Dictyostelium* development and that of embryos. In *Dictyostelium*, a cell mass forms from the aggregation of a population of independent cells; in an embryo, a highly ordered cell mass forms from the division of a single cell. Whether the rules for the steps of morphogenesis and pattern formation that follow, and appear similar in each case, are more fundamental than these differences will determine the future utility of *Dictyostelium* as a model for development.

References and Notes

1. P. N. Devreotes, in *The Development of* Dictyostelium discoideum, W. Loomis, Ed. (Academic Press, San Diego, 1982), pp. 117–168.
2. G. Gerisch, *Annu. Rev. Biochem.* **56**, 853 (1987).
3. P. Schaap, *Differentiation* **33**, 1 (1986).

4. _____ and R. van Driel, *Exp. Cell Res.* **159**, 388 (1985).
5. A. Kimmel, *Dev. Biol.* **122**, 163 (1987).
6. M. Oyama and D. D. Blumberg, *Proc. Natl. Acad. Sci. U.S.A.* **83**, 4819 (1986).
7. B. Haribabu and R. P. Dottin, *Mol. Cell. Biol.* **6**, 2402 (1986).
8. S. Mann and R. A. Firtel, *ibid.* **7**, 458 (1987).
9. J. A. Cardelli *et al.*, *Dev. Biol.* **110**, 147 (1985).
10. W. J. Kopachik, B. Dhokia, R. R. Kay, *Differentiation* **28**, 209 (1985).
11. J. H. Morrissey, K. M. Devine, W. F. Loomis, *Dev. Biol.* **103**, 414 (1984).
12. P. Schaap, J. E. Pinas, M. Wang, *ibid.* **111**, 51 (1985).
13. M. Sussman, *Methods Cell Biol.* **28**, 9 (1987).
14. P. Devreotes, D. Fontana, P. Klein, J. Sherring, A. Theibert, *ibid.*, p. 299.
15. D. R. Soll, *ibid.*, p. 413.
16. A. Kimmel and R. Firtel, in *The Development of* Dictyostelium discoideum, W. Loomis, Ed. (Academic Press, San Diego, 1982), p. 234.
17. W. F. Loomis, *Methods Cell Biol.* **28**, 31 (1987).
18. D. Welker *et al.*, *Genetics* **112**, 27 (1986).
19. W. Nellen, C. Silan, R. Firtel, *Mol. Cell Biol.* **4**, 2890 (1984).
20. W. Nellen *et al.*, *Methods Cell Biol.* **28**, 67 (1987).
21. D. Knetch, S. Cohen, W. Loomis, H. Codish, *Mol. Cell. Biol.* **6**, 3973 (1986).
22. R. A. Firtel *et al.*, *ibid.* **5**, 3241 (1985).
23. N. Farrar and K. Williams, *Trends Genet.* **4**, 343 (1988).
24. E. Henderson, in *The Biology of Glycoproteins*, R. J. Ivatt, Ed. (Plenum, New York, 1984), pp. 371–443.
25. T. E. Crowley, W. Nellen, R. H. Gomer, R. A. Firtel, *Cell* **43**, 633 (1985).
26. D. A. Knecht and W. F. Loomis, *Science* **236**, 1081 (1987).
27. P. S. Klein *et al.*, *ibid.* **241**, 1467 (1988).
28. S. Rubino, S. K. O. Mann, R. T. Hori, R. A. Firtel, *Dev. Biol.* **131**, 27 (1989).
29. A. De Lozanne and J. A. Spudich, *Science* **236**, 1086 (1987).
30. W. Witke, W. Nellen, A. Noegle, *EMBO J.* **6**, 4143 (1987).
31. D. Manstein *et al.*, *ibid.* **8**, 923 (1989).
32. M. Foure, G. Podogorsky, J. Franke, R. Kessin, *Dev. Biol.* **131**, 366 (1989).
33. J. Dynes and R. A. Firtel, *Proc. Natl. Acad. Sci. U.S.A.*, in press.
34. P. J. M. Van Haastert, *Biochem. Biophys. Res. Commun.* **124**, 597 (1984).
35. A. Theibert and P. N. Devreotes, *J. Biol. Chem.* **261**, 15121 (1986).
36. P. J. M. Van Haastert, *ibid.* **262**, 7700 (1987).
37. M. Coukell, S. Lappano, A. M. Cameron, *Dev. Genet.* **3**, 283 (1983).
38. P. Lilly *et al.*, *Bot. Acta* **101**, 123 (1988).
39. M. Pupillo, G. Pitt, A. Kumagai, R. Firtel, P. Devreotes, *Proc. Natl. Acad. Sci. U.S.A.*, in press.
40. A. Kumagai, M. Pupillo, R. Gundersen, P. Devreotes, R. Firtel, *Cell* **57**, 265 (1989).
41. M. Nakafuku, H. Itoh, S. Nakamura, Y. Kaziro, *Proc. Natl. Acad. Sci. U.S.A.* **84**, 2140 (1987).
42. P. J. M. Van Haastert, *J. Biol. Chem.* **262**, 7705 (1987).
43. R. Vaughan and P. Devreotes, *ibid.* **263**, 14538 (1988).
44. K. Saxe, R. Johnson, A. Kimmel, P. N. Devreotes, unpublished data.
45. _____ and S. H. Zigmond, *Annu. Rev. Cell Biol.* **4**, 649 (1988).
46. S. McRobbie, *CRC Crit. Rev. Microbiol.* **13**, 335 (1986).
47. Y. Fukui and S. Yumura, *Cell Motil. Cytoskeleton* **6**, 662 (1986).
48. F. Kesbeke, E. Snaar-Jagalska, P. Van Haastert, *J. Cell Biol.* **107**, 521 (1988).
49. P. J. M. Van Haastert, M. M. Van Lookeren Campagne, F. M. Ross, *FEBS Lett.* **147**, 149 (1983).
50. F. M. Ross and P. C. Newell, *J. Gen. Microbiol.* **127**, 339 (1981).
51. J. Spudich, personal communication.
52. S. Bozzaro, R. Merkl, G. Gerisch, *Methods Cell Biol.* **28**, 359 (1987).
53. A. Noegel, G. Gerisch, J. Stadler, M. Westphal, *EMBO J.* **5**, 1473 (1986).
54. W. Springer, D. Cooper, S. Barondes, *Cell* **39**, 557 (1984).
55. R. L. Clark and T. L. Steck, *Science* **204**, 1163 (1979).
56. A. J. Durston and F. Vork, *J. Cell Sci.* **36**, 261 (1979).
57. P. Schaap and M. Wang, *Cell* **45**, 137 (1986).
58. H. R. Morris, G. W. Taylor, M. S. Masento, K. A. Jermyn, R. R. Kay, *Nature* **328**, 811 (1987).
59. J. Schindler and M. Sussman, *J. Mol. Biol.* **116**, 161 (1977).
60. P. K. Howard, K. G. Ahern, R. A. Firtel, *Nucleic Acids Res.* **16**, 2613 (1988).
61. S. Datta, R. Gomer, R. A. Firtel, *Mol. Cell. Biol.* **6**, 811 (1986).
62. R. Dixon, personal communication.
63. C. Reymond, W. Nellen, R. A. Firtel, *Proc. Natl. Acad. Sci. U.S.A.* **82**, 7005 (1985).
64. A. Kumagai and R. Firtel, personal communication.
65. M. Faure, G. Podgorski, J. Franke, R. Kessin, *Proc. Natl. Acad. Sci. U.S.A.* **85**, 8076 (1988).
66. K. J. Tomchik and P. N. Devreotes, *Science* **212**, 443 (1981).
67. Supported in part by NIH grants GM28007 and GM34933.

6

Xenopus laevis in Developmental and Molecular Biology

Igor B. Dawid and Thomas D. Sargent

Perhaps the best known experiment in embryology is the Spemann and Mangold experiment on embryonic induction, defining what these researchers called the "organizer" (*1, 2*). Induction is a widespread and fundamentally important phenomenon in biology; in its broadest terms it describes any interaction between cells or groups of cells that affects differentiation. As such, induction also occurs in adult organisms, but the term is usually used in the context of embryogenesis, when the processes that generate new tissues and cell types are most active.

How different tissues with their great morphological and functional diversity are formed from the comparatively simple egg is the basic question of developmental biology. Inductive interactions constitute one of the two general developmental mechanisms—cytoplasmic localization of information in the egg being the other—that are thought to be instrumental in setting up regional differences in the embryo, which result in a complex organized structure. Although induction events occur in the development of all animals, this phenomenon has been studied most extensively in amphibians, the phylogenetic class in which it was discovered.

The original work involved newts, but more recently *Xenopus laevis* has become the animal of choice for studies of induction as well as many other aspects of embryogenesis, in particular at the interface of molecular and developmental biology.

The advantages of *Xenopus* as an experimental animal include its easy husbandry, the fact that it is a vertebrate, the accessibility of embryonic material from the earliest stage onward, and the comparatively large size of the egg and embryo that facilitates physical manipulations. These advantages, in spite of the limitation of the almost total inapplicability of classical genetics, have stimulated a great deal of research on *Xenopus* over the past three decades.

In this article we discuss three areas in which this system has made important contributions. These are (i) the role of localized cytoplasmic information and of inductive interactions in the establishment of the polarity and initial tissue differentiation in the embryo, and in the nature and molecular basis of embryonic induction; (ii) the study of genes for RNA components of the ribosome and the control of their expression; and (iii) the productive use of the *Xenopus* oocyte as a "super test tube" in a broad range of studies on translation and transcription.

I.B. Dawid and T.D. Sargent are members of the Laboratory of Molecular Genetics, National Institute of Child Health and Human Development, National Institutes of Health, Bethesda, MD 20892. This chapter is reprinted from *Science* 240, 1443 (1988).

Fig. 1. Induction of a secondary axis in the amphibian embryo. Gastrulation starts with an invagination at one side, forming the dorsal lip, shown at the right of the schematized gastrula at the top. Cells originating from the area of the dorsal lip organize the dorsal axis, as shown in a cross section of a neurula below. Transplantation of a small segment of tissue including the dorsal lip into the ventral side of another embryo, as indicated on the top right, leads to the formation of a double-axis embryo. The second axis, including mesodermal and neural derivatives, is formed mostly from host tissue, that is, it is induced by the transplant. [Adapted from (*1*)]

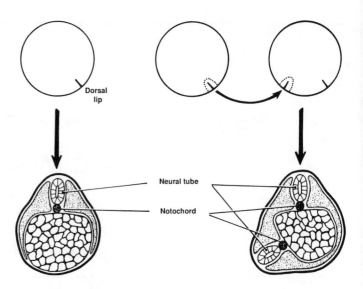

The Spemann-Mangold Organizer Concept

Amphibian gastrulation begins with cell migrations at the dorsal side of the embryo. Cells moving up along the blastocoel roof form the presumptive dorsal mesoderm (chordamesoderm), and the ectoderm overlying this tissue develops into the central nervous system (CNS). As the earliest externally visible sign of gastrulation the dorsal lip forms at a subequatorial position in conjunction with these migrations, marking the future dorsal side of the embryo. The Spemann-Mangold experiment involved the transplantation of the dorsal lip from one embryo into the ventral side of another, leading to the development of two dorsal axes in the host (*1–3*). Figure 1 illustrates such an experiment: a cross section of a host embryo is shown at the late neurula stage with two neural tubes, two notochords, and duplicated somites. In external morphology a second head or tail is formed with almost complete duplication of the embryo as a "Siamese twin" in certain cases. Most of the tissue in the duplicated axis, including mesodermal as well as neural derivatives, originates from the host, that is, it is induced by the implanted dorsal lip, which was therefore named the organizer (*1, 2*).

Fascination with the organizer concept gradually turned to frustration as attempts to isolate a factor with organizing activity met with varied difficulties (*4*). Although we are far from understanding the relevant phenomena and therefore not in a position to explain these difficulties from a historical perspective, it appears that a major reason for a lag in progress for many years, in addition to insufficient molecular techniques, was the way in which the problem had been defined (*5*). Our present view is that organizing a second axis is a complex phenomenon that may be separated into at least two components, mesoderm induction and neural induction, with the mesoderm induction occurring at an earlier stage than the organizer effect. Substantial progress in this area has come, therefore, not only from advances in technology, but also from a dissection of the problem into components that could be handled experimentally.

Determination of Polarity

In *Xenopus*, one embryonic axis is determined by the structure of the egg, the second is determined after fertilization. The egg is radially symmetrical around the animal-vegetal axis, which is defined by the center of the pigmented and unpigmented halves of the egg. This axis, which is set up during oogenesis, defines the future anterior-posterior polarity of the tadpole (Fig. 2); it is thus an example of a developmen-

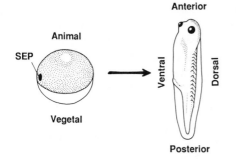

Fig. 2. Relation between the polarity of the fertilized egg and that of the tadpole; SEP, sperm entry point.

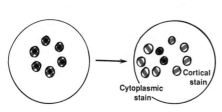

Fig. 3. Cytoplasmic movement before first cleavage as visualized by differential staining. A spot pattern was imprinted on the vegetal surface of a fertilized egg by staining through a grid. Two dyes were used: One stains the cortex (fluoresceinated potato lectin) and a second stains subcortical cytoplasm (Nile blue). The egg was embedded in gelatin, which allows normal development but immobilizes the cortex. About 1 hour after fertilization, the cytoplasm had rotated relative to the cortex by about 30°, as visualized by the staining pattern. Movement did not substantially distort the pattern. [Drawn after Vincent *et al.* (7)].

tal outcome predetermined by cytoplasmic localization of information in the egg. The second major axis of the tadpole is fixed only after fertilization. Sperm entry can occur anywhere in the animal hemisphere, and the sperm entry point (SEP) normally defines the future ventral side. The way in which this polarity determination comes about has been explained primarily through the work of Gerhart and colleagues (6). A movement of cytoplasm relative to the cortex takes place between fertilization and first cleavage, and it is the direction of this movement that is the most accurate predictor of the future dorsal-ventral polarity of the embryo (Fig. 3) (7, 8). This movement, which can be abolished by ultraviolet (UV) irradiation or cold shock, may be driven by a mechanism that is dependent on microtubules (9). Eggs treated with UV irradiation that do not undergo cytoplasmic movement develop into defective embryos lacking a dorsal axis and all dorsal structures including CNS, notochord, and skeletal muscle (9, 10); yet irradiated eggs can be completely rescued if they are rotated before first cleavage, thereby generating gravity-driven cytoplasmic movement (9). This is an important point since no substance is added to the egg to achieve

rescue; whatever substance may be affected can be regenerated by the egg as a consequence of cytoplasmic movements.

The spatial cue generated by cytoplasmic movement is translated into inductive capacity. Eggs treated with UV irradiation, destined to become axis deficient, can also be rescued after cleavage has been initiated by the transplantation of vegetal dorsal blastomeres from a normal embryo into an irradiated host (Fig. 4). Not only is rescue achieved, but dorsal structures such as CNS and notochord are formed from the host, not the graft (11, 12). Thus, the implanted normal cells do not simply replace structures that had been destroyed, but they induce the differentiation of host cells that are therefore capable of responding to the appropriate signal. This result suggests that in normal embryos vegetal dorsal blastomeres are the source of an inductive signal that generates dorsal structures by influencing the fate of other cells. The dorsal blastomeres acquire this inductive capability as a consequence of cytoplasmic movements during the first cleavage cycle. At what time, in what way, and by which molecular mechanism are these inductive signals generated, transmitted, and received?

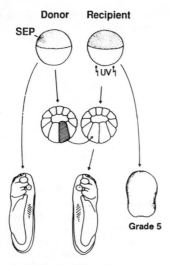

Donor   Recipient

SEP

↕UV↕

Grade 5

Fig. 4. Ultraviolet irradiation of fertilized eggs results in defective embryos lacking dorsal structures (Grade 5). Such embryos could be rescued by transplantation of two normal vegetal-dorsal blastomeres at the 64-cell stage. A normal dorsal axis is formed on the implanted side entirely from the cells of host embryo. [Reprinted from (*11*), with permission of Academic Press]

Mesoderm Induction

The Nieuwkoop experiment

The initial evidence for an inductive interaction responsible for mesoderm differentiation in *Xenopus* was obtained by experiments illustrated in Fig. 5. Animal region explants (animal caps), when cultured in standard buffered salt solution, differentiate only along epidermal lines; yet, when placed in contact with vegetal (future endodermal) tissue, such caps will form a variety of mesodermal tissues including notochord, muscle, kidney, and blood (*13*). These results were interpreted as suggesting that normal mesodermal development, which arises from centrally located cells, the so-called marginal zone, also requires an inductive signal; but this conclusion does not follow directly. The fact that a blastula animal cap can be induced toward mesoderm does not prove that mesoderm in a normal embryo, arising from different cells, requires induction. There are good reasons to believe, however, that it does. One is the rescue of UV-irradiated embryos by blastomere transplant (Fig. 4), in which dorsal mesoderm forms from the same

cells as in a normal embryo but would fail to differentiate from these cells without the inductive signal. A second reason is an experiment in which the cells of *Xenopus* embryos were dissociated and widely dispersed, thus preventing interactions. Cells divide for many hours under these conditions. The cells were kept dispersed through the period of cleavage and blastula development, when induction is presumed to occur, later reaggregated for additional culture, and then assayed for gene activation. It was found that muscle-specific genes, such as the α-actin gene, were not expressed, whereas keratins, specific for epidermal differention, were expressed at normal levels (*14*). This result suggests that cell interactions during blastula stages are required for the differentiation of muscle, which is a major derivative of dorsal mesoderm.

Studies with XTC mesoderm-inducing activity

Inducing factors have been obtained from various sources over many years (*15*). A breakthrough in this area came from the recent discovery by Smith of the powerful inducing effect of conditioned medium of XTC cells (XTC-CM) (*16*). This cell line was generated some time ago from a metamorphosing tadpole (*17*); the tissue of origin of the XTC cells is not clear, nor is it understood why cells from such a late stage should secrete inducing factor. These unknowns, and the fact that the active principle has not yet been purified to homogeneity, have not diminished the value of XTC-CM as a reproducible source of a soluble, highly effective inducing activity.

Animal caps induced by endodermal tissue or by XTC-CM differentiate into the whole range of mesodermal derivatives (*13, 16*). How these different tissues are generated is a critical question, especially if one considers that the normal embryo produces this range of tissues in an orderly way along its dorsal-ventral axis. Thus, it is insufficient to discuss mesoderm induction as such without considering the establishment of dorsoventral polarity. One result of studies already completed with XTC-CM is that

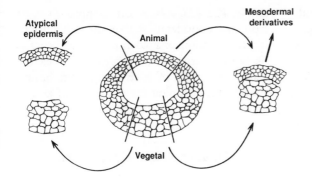

Atypical epidermis

Animal

Mesodermal derivatives

Vegetal

Fig. 5. An animal explant (animal cap) derived from a *Xenopus* blastula (middle), and cultured in salt solution, differentiates along the epidermal pathway, which is a major fate of the cells of this region in normal embryogenesis; vegetal tissue cultured alone yields little differentiation. Culture of an animal cap in contact with vegetal tissue leads to varied mesodermal structures (right) that develop from the animal explant (*13*).

a concentration-dependent effect occurs in which high levels induce mostly dorsal tissues (notochord and muscle), whereas lower levels lead to the differentiation of ventral mesodermal derivatives such as kidney and blood (*18*). Any conclusion from this result is preliminary since the possible involvement of multiple factors has not been excluded. But because the active principle in XTC-CM appears to be a single component, the result suggests that there is no need to postulate separate dorsal and ventral mesoderm-inducing factors; a graded distribution of a single factor might suffice. Yet one cannot postulate that the nature of the induced tissue depends entirely on a specific concentration of factor along the gradient. Such a mechanism would not be precise enough to assure a properly organized embryo, and interactions between neighboring induced cell groups may have to be invoked.

The nature of mesoderm inducer: relation to TGF-β2

The relation of mesoderm-inducing activity to growth factors has been implied by several recent experimental results, connecting the field of embryonic induction to a large and highly active area of exploration. Slack *et al.* (*19*) showed that fibroblast growth factor (FGF) can induce ventral mesoderm as well as small amounts of muscle; Kimelman and Kirschner (*20*) observed the muscle-inducing effect of FGF to be potentiated by transforming growth factor β (TGF-β); and Rosa *et al.* (*21*) found that TGF-β2 alone effectively induces muscle (Fig. 6). Although none of these

heterologous factors was as effective as the homologous XTC-CM, the activity of XTC-CM in muscle induction was inhibited by antibodies to TGF-β2 but not by antibodies to TGF-β1 (*21*). These results suggest that the active principle in XTC-CM is structurally related but not identical to mammalian TGF-β2.

TGF-β is a large family of factors with a variety of biological functions (*22*). The family includes a gene named *Vg1* (*23*), whose messenger RNA (mRNA) is accumulated in a localized manner in the vegetal region of *Xenopus* oocytes (*24*). The product of this gene is a good candidate for a component of the mesoderm induction system.

A summary of polarity determination and induction

Although the molecules and mechanisms involved in these processes are not fully understood, the following hypothetical scenario seems reasonable at this time. Factors localized in the egg (as exemplified by, but not limited to, *Vg1*) are displaced through cytoplasmic movements to bestow distinct properties on

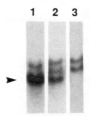

1 2 3

Fig. 6. Muscle induction in animal caps by XTC-CM and TGF-β2. The accumulation of α-actin mRNA has been used as a measure of muscle differentiation; muscle is a major derivative of dorsal mesoderm (*21*). The arrowhead points to α-actin mRNA; the other bands are cross-hybridizing cytoskeletal actin mRNAs. Lane 1, animal cap induced by XTC-CM; lane 2, animal cap induced by TGF-β2 (200 ng/ml); lane 3, uninduced control (*52*).

different vegetal blastomeres, which subsequently induce the mesoderm in the marginal zone. Dorsal vegetal blastomeres induce dorsal mesoderm, including the Spemann-Mangold organizer, which in turn induces the neural plate during gastrulation. Interactions between the different induced tissues are likely to ensure the emergence of an ordered global structure. This scenario is a simplification and is only meant as an aid to visualization of events, rather than as an established series of mechanistically understood processes.

Future Work on Cytoplasmic Localization and Embryonic Induction

The cell and molecular biology of mesoderm induction has entered a phase of rapid exploration. We may expect clarification of questions on the identity of inducers in the near future. The issue of establishment of dorsal-ventral polarity will be an important focus. Furthermore, attention will be focused on receptors for inducer molecules, an issue of interest in terms of transduction of the signal, but especially as an approach to the problem of competence, that is, the ability—highly regulated in development—of certain cells to respond to an inducing signal.

Although we have emphasized induction as an important phenomenon in early development, localization of information in the egg contributes greatly to specification of the embryo [for a general discussion, see (25)]. One approach to this question is the search for localized macromolecules in the egg, some of which may be regulatory in nature, for ex-

ample, *Vg1* (23, 24). Another approach starts from the consideration that the cell-autonomous activation of epidermal keratin genes is likely to be controlled by prelocalized factors (26). The study of such factors can be approached by defining the sequences of keratin genes that control their activation.

Neural induction and the organizer

While mesoderm is established by interactions during cleavage and blastula, neural induction takes place during gastrulation by the effect of migrating dorsal mesoderm on the overlying ectoderm (Fig. 7) (27). The CNS develops from the induced ectoderm. As such, neural induction may be thought of as a separate event that follows mesoderm induction, yet the situation is somewhat confusing because of overlapping ranges of inducing ability and competence. In fact, the classical organizer induction generates both dorsal mesoderm and CNS in the host (1–3). The cellular and molecular basis of neural induction will be amenable to detailed investigation when this phenomenon can be isolated and analyzed as a distinct entity.

Xenopus in Molecular Biology: Key Advances in Gene Isolation and Expression

The *Xenopus* system was used in the first isolation of a gene from any eukaryote (28), the initial studies on gene amplification (29), the earliest example of accurate transcription of a cloned eukaryotic gene (30), and the first isola-

Fig. 7. Neural induction during gastrulation. The drawing represents a cross section through a midgastrula stage. The small arrows pointing from the chordamesoderm to the dorsal ectoderm symbolize signal transfer in neural induction. Abbreviations: Blast., blastocoel; and Arch., archenteron.

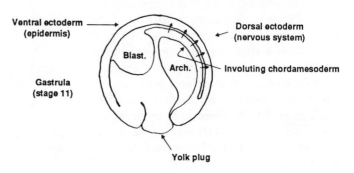

tion, cloning, and detailed characterization of a eukaryotic transcription factor (*31*).

Ribosomal RNA genes

Paradoxically, in view of the general lack of information about the genetics of *Xenopus*, it was probably the existence of the anucleolate mutation that caused investigators to focus on the ribosomal DNA (rDNA) of this animal. This mutation allowed the conclusion that the nucleolus is the site of rRNA synthesis (*32*), containing the rRNA genes (*33*). Soon it became clear that the ribosomal genes of *Xenopus* are tandemly repeated and, because of their high copy number and their distinctive base composition, could be physically separated from the bulk chromosomal DNA (*28*). *Xenopus* genes encoding components of the ribosome became the prototype for repeated gene families.

In the analysis of the structure and regulation of expression of rRNA genes, *Xenopus* has been a prime system. A complex arrangement of control regions has been identified (*34*). Each gene (repeat unit) has a basic promoter, a series of enhancer elements that have additive effects, and several duplicated promoter elements that can initiate transcription in vitro, and under some conditions in vivo (Fig. 8). Transcription termination on rDNA has proved even more surprising than initiation. It was believed at first that the 3' end of mature 28*S* rRNA corresponds to the termination site. This proved inaccurate: transcription extends

far into what was originally called the non-transcribed spacer, generating transcripts that are rapidly processed to the mature molecules (Fig. 8).

5S RNA genes

These genes, which have been isolated and extensively characterized by Brown, have yielded many insights into questions of transcriptional mechanisms. Achievement of accurate expression of 5*S* RNA genes (*30*) allowed a detailed analysis of the cis-regulatory regions of the 5*S* DNA. The unexpected result pointed to an intragenic control region (*35*), a feature now generally recognized in genes transcribed by polymerase III (*36*).

TFIIIA, one of three factors required for 5*S* RNA gene transcription, was isolated and cloned primarily through the efforts of Roeder and colleagues (*31*) and has become a prototype for eukaryotic transcription factors. Recognition of a repeated structural feature in TFIIIA, the so-called zinc finger (Fig. 9), led to significant further insights (*37*). The zinc finger is a structural motif in which four residues coordinate a Zn^{2+} ion that is presumed to help fold the protein domain into a configuration suitable for DNA binding. The zinc finger motif has since been recognized in numerous known or suspected DNA-binding regulatory proteins (*38*); it is a second major motif of DNA-binding proteins in addition to the helix-turn-helix motif recognized earlier (*39*). Detailed analysis of the binding of TFIIIA to

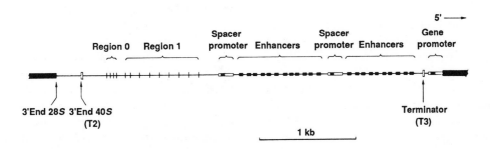

Fig. 8. Summary of regulatory regions in ribosomal RNA genes of *Xenopus*, showing promoters, enhancers, terminators, and major 3' termini (*31*). [From (*53*), with permission]

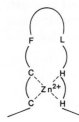

Fig. 9. The zinc finger motif. Protein domain recognized originally in TFIIIA (*37*) and found in many proteins (*38*). Conserved amino acids are shown in one-letter code. Abbreviations for the conserved amino acids follow: C, Cys; F, Phe; H, His; and L, Leu. The number of residues between conserved amino acids are: Cys- Cys, 2 or 4; Cys-Phe, 3, rarely 2; Phe-Leu, 5; Leu-His, 2, rarely 3; His-His, 3, rarely 4. Coordination of a zinc ion is hypothetical in most cases.

the internal control region of 5S DNA and exploration by mutagenesis of the protein regions involved (*40*) have led to a detailed understanding of the DNA-protein interaction.

The most interesting aspect of the 5S RNA system, however, is the existence of oocyte-specific and somatic cell-specific genes and their differential regulation. In a comparatively simple, well-understood system, this property provides a model for the general question of differential gene regulation, one of the basic questions in biology. Although the molecular mechanisms of differential 5S RNA gene activity are not understood at present, the relevance of stable transcription complexes, and of changing factor concentrations has been stressed (*41*). Brown (*42*) has suggested that stable transcription complexes may represent a general mechanism in establishment or preservation of the differentiated state, a hypothesis that should provide stimulation for useful experimentation.

Regulation of class II genes in embryogenesis

In the past 2 years progress has been made in this system toward understanding the control of genes transcribed by polymerase II. Two highly regulated genes, GS17, which is expressed only during gastrulation (*43*), and α-actin, a muscle-specific gene (*44*), have been cloned and introduced as purified DNA into fertilized eggs. Both of these genes are correctly regulated by the embryo; cloned GS17 is transcribed at the correct time because of the presence of a gastrulation-specific enhancer element located about 700 bases upstream from the initiation site (*45*). The α-actin genes are also controlled by elements residing in the upstream flanking region (*44*). Similar results have been obtained with a cloned epidermal keratin gene (*46*). The ability of the embryo to incorporate exogenous DNA into its regulatory circuitry is an important advantage in mapping cis-acting elements that regulate the transcriptional responses to temporal, positional, or inductive developmental cues.

The Uses of *Xenopus* Oocytes as an Expression System

The frog oocyte translates injected mRNAs with great efficiency, processes the resulting proteins, and distributes them into the correct compartment. The oocyte also transcribes accurately many, although not all, genes injected into its large nucleus. These powerful techniques were developed through the efforts of Gurdon (*30, 47*) and are now used widely. In particular, translation in the oocyte has broad applicability for characterization of mRNA and protein products and as an aid in cloning of complementary DNA (*48*). Transcription of genes injected into the oocyte nucleus has not been used as widely as mRNA translation, but the *Xenopus* oocyte provided an important early expression system for 5S and ribosomal RNA genes (*30, 34*). Recently, the oocyte provided a functional transcription factor assay. Injection of nuclear extracts from sea urchins stimulated the transcription of sea urchin histone genes in the frog oocyte (*49*), and injection of a purified protein factor allowed the expression of an introduced *Drosophila* heat-shock gene without actual heat shock (*50*).

A corollary of transcriptional studies is the work on RNA processing and transport that takes advantage of the ability to introduce genes, RNA precursor molecules, and protein or ribonucleoprotein factors into the nucleus and to determine both the nature and the localization of the products (*51*). Thus, the *Xenopus* oocyte continues to provide investigators with new opportunities in a variety of applications.

References and Notes

1. H. Spemann and H. Mangold, *Wilhelm Roux Arch. Entwicklungsmech. Org.* **100**, 599 (1924).
2. O. Mangold, *Hans Spemann* (Wissenschaftliche Verlagsgesellschaft, Stuttgart, ed. 2, 1982).
3. R. L. Gimlich and J. Cooke, *Nature* **306**, 471 (1983); M. Jacobson and U. Rutishauser, *Dev. Biol.* **116**, 524 (1986); J. Smith and J. M. W. Slack, *J. Embryol. Exp. Morphol.* **78**, 299 (1983).
4. L. Saxén and L. Toivonen, in *A History of Embryology* T. J. Horder, J. A. Witkowski, C. C. Wylie, Eds. (Cambridge Univ. Press, Cambridge, 1986), pp. 261–274.
5. I. B. Dawid, *Trends Biochem. Sci.* **12**, 34 (1987).
6. J. Gerhart, M. Danilchik, J. Roberts, B. Rowning, J.-P. Vincent, in Gametogenesis and the Early Embryo, J. G. Gall, Ed. (Liss, New York, 1986), pp. 305–319.
7. J.-P. Vincent, G. F. Oster, J. C. Gerhart, *Dev. Biol.* **113**, 484 (1986).
8. J.-P. Vincent and J. C. Gerhart, *ibid.* **123**, 526 (1987).
9. S. R. Scharf and J. C. Gerhart, *ibid.* **79**, 181 (1980); *ibid.* **99**, 75 (1983).
10. P. Grant and J. F. Wacaster, *ibid.* **28**, 454 (1972); G. M. Malacinski, A. J. Brothers, H.-M. Chung, *ibid.* **56**, 24 (1977).
11. R. L. Gimlich and J. C. Gerhart, *ibid.* **104**, 117 (1984).
12. R. L. Gimlich, *ibid.* **115**, 340 (1986).
13. S. Sudarwati and P. D. Nieuwkoop, *Wilhelm Roux Arch. Entwicklungsmech. Org.* **166**, 189 (1971); O. Nakamura and K. Kishiyama, *Proc. Jpn. Acad.* **47**, 407 (1971); P. D. Nieuwkoop, *Adv. Morphog.* **10**, 1 (1973); H. Grunz and L. Tacke, *Wilhelm Roux's Arch. Dev. Biol.* **195**, 467 (1986); J. B. Gurdon, S. Fairman, T. J. Mohun, S. Brennan, *Cell* **41**, 913 (1985).
14. T. D. Sargent, M. Jamrich, I. B. Dawid, *Dev. Biol.* **114**, 238 (1986).
15. H. Tiedemann, in *Biochemistry of Differentiation and Morphogenesis,* Colloquium Mosbach, L. Jaenicke, Ed. (Springer-Verlag, Berlin, 1982), vol. 33, pp. 275–287; J. Born et al., *Biol. Chem. Hoppe-Seyler* **366**, 729 (1985).
16. J. C. Smith, *Development* **99**, 3 (1987).
17. M. Pudney, M. G. R. Varma, C. J. Leake, *Experientia* **29**, 466 (1973).
18. J. Cooke, J. C. Smith, E. J. Smith, M. Yaqoob, *Development* **101**, 893 (1987). A similar idea was proposed by E. C. Boterenbrood and P. D. Nieuwkoop [*Wilhelm Roux Archiv. Entwichlungsmech. Org.* **173**, 319 (1973)].
19. J. M. W. Slack, B. G. Darlington, J. K. Heath, S. F. Godsave, *Nature* **326**, 197 (1987).
20. D. Kimelman and M. Kirschner, *Cell* **51**, 869 (1987).
21. F. Rosa et al., *Science* **239**, 783 (1988).
22. M. B. Sporn, A. B. Roberts, L. M. Wakefield, B. De Crombrugghe, *J. Cell. Biol.* **105**, 1039 (1987).
23. D. L. Weeks and D. A. Melton, *Cell* **51**, 861 (1987).
24. M. R. Rebagliati, D. L. Weeks, R. P. Harvey, D. A. Melton, *ibid.* **42**, 769 (1985); D. A. Melton, *Nature* **328**, 80 (1987).
25. E. H. Davidson, *Gene Activity in Early Development* (Academic Press, New York, ed. 3, 1986).
26. M. Jamrich, T. D. Sargent, I. B. Dawid, *Genes Dev.* **1**, 124 (1987).
27. B. I. Balinsky, *Introduction to Embryology* (Saunders, Philadelphia, ed. 4, 1975).
28. D. D. Brown and C. S. Weber, *J. Mol. Biol.* **34**, 661 (1968); *ibid.*, p. 681; M. L. Birnstiel, J. Speirs, I. Purdom, K. Jones, U. E. Loening, *Nature* **219**, 454 (1968); I. B. Dawid, D. D. Brown, R. H. Reeder, *J. Mol. Biol.* **51**, 341 (1970).
29. D. D. Brown and I. B. Dawid, *Science* **160**, 272 (1968); J. G. Gall, *Proc. Natl. Acad. Sci. U.S.A.* **60**, 553 (1968).
30. D. D. Brown and J. B. Gurdon, *Proc. Natl. Acad. Sci. U.S.A.* **74**, 2064 (1977).
31. D. R. Engelke, S.-Y. Ng, B. S. Shastry, R. G. Roeder, *Cell* **19**, 717 (1980); A. M. Ginsberg, B. O. King, R. G. Roeder, *ibid.* **39**, 479 (1984).
32. D. D. Brown and J. B. Gurdon, *Proc. Natl. Acad. Sci. U.S.A.* **51**, 139 (1964).
33. H. Wallace and M. Birnstiel, *Biochim. Biophys. Acta* **114**, 296 (1966).
34. R. H. Reeder, *Cell* **38**, 349 (1984); T. Moss, *ibid.* **30**, 85 (1982); P. Labhart and R. H. Reeder, *ibid.* **45**, 431 (1986); J. Windle and B. Sollner-Webb, *Mol. Cell. Biol.* **6**, 1228 (1986); *ibid.*, p. 4585; R. H. Reeder et al., *Nucleic Acids Res.* **15**, 7429 (1987); K. Mitchelson and T. Moss, p. 9577 (1987); R. F. De Winter and T. Moss, *J. Mol. Biol.* **196**, 813 (1987); for a broad review on rRNA transcription see B. Sollner-Webb and J. Tower [*Annu. Rev. Biochem.* **55**, 801 (1986)].
35. D. F. Bogenhagen et al., *Cell* **19**, 27 (1980); S. Sakonju et al., *ibid.*, p. 13.
36. E. P. Geiduschek and G. P. Tocchini-Valentini, *Annu. Rev. Biochem.*, **57**, 873 (1988).
37. J. Miller, A. D. McLachlan, A. Klug, *EMBO J.* **4**, 1609 (1985).
38. J. M. Berg, *Science* **232**, 485 (1986); U. B. Rosenberg et al., *Nature* **319**, 336 (1986); A. Ruiz i Altaba, H. Perry-O'Keefe, D. A. Melton, *EMBO J.* **6**, 3065 (1987); H. Blumberg, A. Eisen, A. Sledziewski, D. Bader, E. T. Young, *Nature* **328**, 443 (1987); J. T. Kadonaga et al., *Cell* **51**, 1079 (1987).

39. C. O. Pabo and R. T. Sauer, *Annu. Rev. Biochem.* **53**, 293 (1984).

40. K. E. Vrana, M. E. A. Churchill, T. D. Tullius, D. D. Brown, *Mol. Cell. Biol.*, **8**, 1684 (1988).

41. D. F. Bogenhagen, W. M. Wormington, D. D. Brown, *Cell* **28**, 413 (1982); A. P. Wolffe and D. D. Brown, *ibid.* **51**, 733 (1987).

42. D. D. Brown, *ibid.* **37**, 359 (1984).

43. P. A. Krieg and D. A. Melton, *EMBO J.* **4**, 3463 (1985).

44. T. J. Mohun *et al.*, *ibid.* **5**, 3185 (1986); C. Wilson *et al.*, *Cell* **47**, 589 (1986).

45. P. A. Krieg and D. A. Melton, *Proc. Natl. Acad. Sci. U.S.A.* **84**, 2331 (1987).

46. E. Jonas, A. M. Snape, T. D. Sargent, *Development*, **106**, 399 (1989).

47. J. B. Gurdon *et al.*, *Nature* **233**, 177 (1971).

48. D. A. Melton, *Method. Enzymol.* **152**, 288 (1987).

49. R. Maxson *et al.*, *Mol. Cell. Biol.* **8**, 1236 (1988); J. Mous, H. Stunnenberg, O. Georgiev, M. Birnstiel, *ibid.* **5**, 2765 (1985).

50. C. Wu *et al.*, *Science* **238**, 1247 (1987).

51. H. G. Stunnenberg and M. L. Birnstiel, *Proc. Natl. Acad. Sci. U.S.A.* **79**, 6201 (1982); G. M. Gilmartin, F. Schaufele, G. Schaffner, M. L. Birnstiel, *Mol. Cell. Biol.* **8**, 1076 (1988); M. Zasloff, *Proc. Natl. Acad. Sci. U.S.A.* **80**, 6436 (1983); P. de la Peña and M. Zasloff, *Cell* **50**, 613 (1987).

52. F. Rosa, unpublished data.

53. R. H. Reeder, unpublished data.

54. We thank D. Brown, J. Cooke, J. Smith, and F. Rosa for unpublished material and R. Reeder for Fig. 8.

7

The Nematode
Caenorhabditis elegans

Cynthia Kenyon

The nematode *Caenorhabditis elegans* was originally selected for study as the result of a deliberate search for a multicellular organism that could be analyzed with the ease and resolution characteristic of studies of microorganisms (*1*). Features of the *C. elegans* life cycle facilitate genetic analysis, and features of its development and anatomy make it possible to analyze multicellular processes in the living animal at the level of individual cells. Since then a concerted effort has been made by members of the field to develop the system for studies of neurobiology, development, and cell biology. A detailed genetic map and methods for genetic analysis have been compiled (*1–3*), a complete cell-by-cell description of the development and anatomy of *C. elegans* has been achieved (*4–7*), and now a library of ordered cosmid clones representing the genome is nearing completion (*8*).

Caenorhabditis elegans is well suited to genetic studies for several reasons. First, it takes about 3 days to do a genetic cross in *C. elegans*; this is about the time required for genetic crosses in yeast. In addition, unlike many closely related nematode strains, *C. elegans* reproduces by self-fertilization. Self-fertilization allows new mutations to become homozygous automatically, without requiring brother-sister matings. The ability to freeze strains allows mutant stocks to be maintained indefinitely. Finally, genes defined by mutation can be cloned easily by transposon tagging or by injection of cloned DNA sequences from the genetic region of interest.

Each cell in the developing nematode is visible under the light microscope (Fig. 1). Furthermore, the entire cell lineage, from egg to adult, is essentially the same in each animal and is known precisely (*4, 5*) (Fig. 2). This means that any process involving the behaviors of individual cells in a multicellular context can be analyzed at the level of the single cell.

Finally, development of *C. elegans* appears to involve mechanisms that are conserved throughout the animal kingdom. When nematode lineages were found to be invariant, it seemed possible that much of nematode development would be controlled by mechanisms internal to individual lineages, and not by cell-extrinsic signaling, which plays such an important role in the development of higher organisms. Because the majority of *C. elegans* lineages are invariant, showing that cell-cell communication occurs is problematic. In order to show that cellular interactions influence cell fate, one must be able to alter a cell's behavior by changing its environment, for example by ablating other cells with a laser microbeam. By using this and other approaches, many ex-

C. Kenyon is an assistant professor in the Department of Biochemistry and Biophysics, University of California, San Francisco 94143. This chapter is reprinted from *Science* **240**, 1448 (1988).

amples of cellular interactions have now been discovered; it is clear that cell-cell communication plays a central role in *C. elegans* development (*9*). Furthermore, molecules that regulate *C. elegans* development have been found to contain homology to familiar vertebrate growth factor precursors and cell surface receptors (*10*), and to evolutionarily conserved homeodomains (*11*). Thus, with *C. elegans* it is possible to use genetics and single-cell analysis to dissect regulatory mechanisms of general significance and also to gain insights into how these mechanisms are modified to produce different kinds of animals.

We will first describe nematode development, then discuss experimental approaches available, and finally mention some current topics in *C. elegans* development, neurobiology, and cell biology. More comprehensive reviews are also available (*12–17*).

Overview of *C. elegans* Development

Caenorhabditis elegans adults are about 1 mm in length and live in the soil, feeding on microorganisms. They are sexually dimorphic,

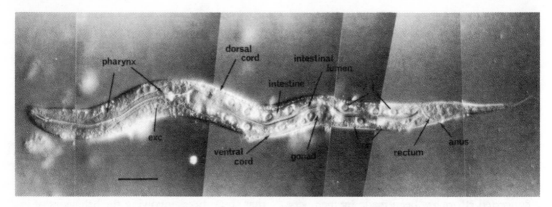

Fig. 1. A live, newly hatched hermaphrodite, as seen with Nomarski optics. Abbreviations: exc, excretory cell; i, intestinal nuclei; and vcn, ventral cord neurons. Scale bar, 20 μm [from Sulston and Horvitz (*4*)].

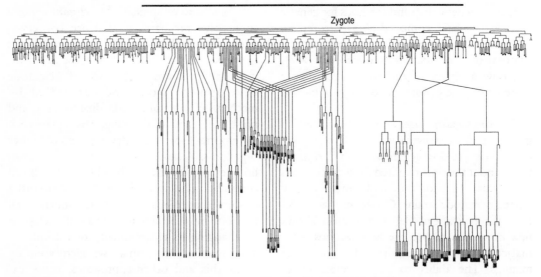

Fig. 2. The cell lineage of the *Caenorhabditis elegans* hermaphrodite. The rapid early divisions are embryonic, and generate the 558-cell juvenile. Fifty-five cells continue to divide after hatching, adding cells to existing tissues and also producing the reproductive system.

with hermaphrodites having two X chromosomes and males only one. Hermaphrodites have a total of 959 somatic nuclei (5), and a relatively simple 80,000-kb genome (18). After fertilization, the egg undergoes a series of asymmetric cleavages that produce six cells called founder cells. One founder cell gives rise to the germ line, another the intestine, and one makes only body muscle cells. The other founders generate predominantly ectodermal or mesodermal cell types. The cells divide rapidly for several hours, then the animal elongates and, 14 hours after fertilization, hatches. The juvenile proceeds to the adult stage via four molts. Of the cells generated during embryogenesis, roughly 10% divide after hatching. These postembryonic divisions generate additional epidermal cells, muscles, neurons, and nearly all the reproductive structures.

Techniques Used for Studying *C. elegans*

Genetic analysis

The isolation of mutants in *C. elegans* is a straightforward procedure because of its small size and short generation time, and because self-fertilization leads automatically to the segregation of recessive mutations in the homozygous form. The ease of isolating homozygotes also makes it possible to identify suppressors, secondary mutations that restore a mutant phenotype to wild type. Suppression analysis can be used to identify genes that interact in a pathway (19–21).

To isolate suppressors, the investigator simply mutagenizes a mutant strain and looks for wild-type animals among its progeny (for dominant suppressors) or its grandchildren (for recessive suppressors). Because it is often easy to see wild-type animals in the midst of mutants, a very large number of animals (as many as 10^8 in some cases) can be examined at one time. The ability to screen large numbers allows the isolation of rare suppressors, such as mutations that correct defective protein-protein interactions. Genes involved in sex determination, dosage compensation, cell lineage, and behavior have been identified by virtue of their being recessive suppressors of other mutations.

To map and complement new mutations, and to construct genetic regulatory pathways, mutations must be moved into other genetic backgrounds. For this, males are required. In *C. elegans*, males are produced at low frequency by spontaneous meiotic nondisjunction of the X chromosome, or at high frequency in certain mutants (22). XO animals are males, and will mate with the XX hermaphrodites. One half of the outcross progeny are males that now carry genes from the hermaphrodite and can be used to move these genes into other hermaphrodites. About 700 genes have now been identified and mapped to one of the five autosomes or the X chromosome (23). In addition, genetic analysis has been enhanced by the isolation of chromosome deficiencies, duplications, and translocations (2).

Given the prevalence of cell signaling systems in development, an important question is whether or not gene activity is cell-autonomous; that is, whether a gene's influence is confined to or extends beyond those cells that express it. This question can be answered in vivo by means of genetic mosaic analysis. Genetic mosaics are animals containing clones of mutant cells in a wild-type background. In *C. elegans*, they can be produced by the rare, spontaneous loss of chromosome fragments carrying wild-type genes in an otherwise mutant animal (3). Mutant cells are then examined to see whether they can develop normally despite the absence of the gene function of interest. As an example, the technique has been used to show that acetylcholinesterase, which inactivates the neurotransmitter acetylcholine at neuromuscular junctions, is made in muscle (24).

Single-cell analysis

The ability to analyze multicellular processes at the level of single cells is a great advantage of working with *C. elegans*. The complete cell lineage has been determined by direct observation of living animals with Nomarski optics (*4, 5*). Single-cell resolution is feasible because the animal is transparent, and because all but a few cells behave in the same way in every animal throughout development. Cell divisions, migrations, and morphogenesis can be observed and monitored precisely in the developing organism. Individual cells can be ablated with a laser microbeam in living animals in order to test for cellular interactions (*25*). To determine how genes influence cells in vivo, mutants can be compared to wild-type animals throughout development, cell by cell. The point at which mutant cells first begin to behave differently from their wild-type counterparts can be identified, and the alternative fates expressed by the mutant cells—changes in division patterns, types of descendants, cell death, cell migration behaviors, altered sequences of morphogenesis—can be determined exactly by following cell lineages. This resolving power has allowed the identification of cell lineage and migration mutations whose phenotypes would have been uninterpretable otherwise. Furthermore, because the whole animal is accessible, it is possible to learn how gene functions are distributed among all cells in the developing organism.

Molecular genetics

A variety of new methods have streamlined molecular analysis in the nematode. In particular, J. Sulston and A. Coulson of the Medical Research Council (Cambridge, England) are assembling an ordered library of cosmid clones representing the genome. Overlapping cosmids are identified by the presence of common restriction fragments (*8*). At this time, cosmids representing about 95% of the genome have been analyzed. The map positions of sets of overlapping clones are known for about 50% of the genome (*26*). This means that for 50% of mapped genes, a set of overlapping cosmids in the region is available on request.

For cloning genes in locations not yet associated with sets of overlapping cosmids, the relevant set can generally be identified by find-ing a genetically linked DNA probe, as described below. Furthermore, because procedures for DNA transformation have been developed (*27*), genes are now being cloned simply by injecting cosmid clones known to map in the genetic region of interest into mutants and looking for phenotypic rescue (*28*). A. Fire has shown that the coupling of the β-galactosidase gene with the control sequences of a muscle-specific gene produces a transgenic animal with appropriate tissue-specific expression (*29*). Thus control sequences may be identified in this way.

The discovery of transposable elements, and their variable numbers and positions in different strains of *C. elegans* has further enhanced molecular genetic methods (*18, 30*). Strain-specific transposons occur approximately once every 120 kb (four cosmid lengths). These create restriction site polymorphisms, and thus become physical markers for the genetic map. By finding polymorphic transposons linked to the gene of interest, it is possible to identify previously unmapped sets of overlapping cosmids likely to contain the gene. Even more powerful is the ability to create mutations by insertion of the transposon into the gene itself, thereby tagging the gene. The frequency of transposition is very high in certain strains, allowing mutants to be selected easily (*18, 31*). A number of genes involved in development, neurobiology, and muscle structure have now been cloned by transposon tagging (*10, 32*).

It is also possible to start with conserved molecules from other organisms, and obtain mutants in *C. elegans* to study their in vivo functions at the cellular level. Valuable initial information can be obtained about possible gene functions with antibodies (*33*) or in situ hybridization to RNA (*34–36*) to determine which cells express the gene of interest. The mapped cosmid library can be used to determine the genetic location quite precisely. To identify mutants, it is possible to examine known mutations and deletions in the region, or else isolate new mutations in the region, and identify those that alter the gene product. An essential function of a cloned myosin heavy chain gene has been identified by this approach (*37*).

Control of Cell Division and Differentiation

Embryonic cell patterning

Cellular asymmetries that arise early in development establish the future body plan of the animal. The first embryonic cleavage produces the cell AB (which generates most of the ectoderm) and the cell P1 (which generates the germ line, intestine, and much mesoderm). When these two blastomeres are separated from one another, they produce very different types of progeny, suggesting that regulatory molecules segregate asymmetrically at the first division (5, 38). In support of this interpretation, germ line–specific granules can be seen to move asymmetrically during the first few cleavages, until they enter the appropriate germ line precursor cell (39).

Mechanisms for establishing asymmetries during the first embryonic cleavages are under investigation with genetic approaches and probes for the cytoskeleton. Studies with cytoskeletal inhibitors suggest that microfilaments, but not microtubules, are involved in the establishment of these early asymmetries (39, 40). Mutations that cause the AB and P1 blastomeres to behave similarly have been isolated as a means of identifying genes involved in generating embryonic asymmetry. Products of the *par* (*par*tition-defective) genes are particularly interesting, because in *par* mutants some or all manifestations of asymmetry are absent (41).

Caenorhabditis elegans embryos have classically been regarded as highly mosaic in character; that is, intracellular segregation of regulatory information has been thought to lead to embryonic cell patterning. However, recent experiments by J. Priess have shown that an extensive network of cellular interactions influences the fates of at least two-thirds of the cells in the embryo (42). In normal embryos, the daughters of the AB blastomere, ABa and ABp, generate different lineages and cell types. If the positions of ABa and ABp are switched, their fates are switched, and each cell generates the descendants normally produced by the other. Differences between the ABa and ABp

lineages appear to arise by cellular interactions that occur at a later time during embryogenesis. By ablating individual cells with a laser microbeam, Priess has discovered an example of classic embryonic induction: cells derived from P1 induce three AB-derived cells to generate the anterior portion of the pharnyx (esophagus). Mutations in the gene *glp-1* (*g*erm *l*ine *p*roliferation) produce an animal specifically lacking the anterior pharynx, suggesting that the gene participates in this inductive interaction (43). The same gene also functions postembryonically. Here *glp-1* is required for developing germ cells to respond to a mitogenic signal from a nearby somatic cell.

Postembryonic cell patterning

How complex patterns of cell types form within initially uniform arrays of cells is a fundamental problem in development. Several examples in *C. elegans* are under investigation (20). Among the more intensively studied are the development of the vulva and the patterning of the antero-posterior body axis. The vulva is an epidermal structure generated by three ventrally located epidermal cells. If a cell in the overlying gonad (called the anchor cell) is ablated, the vulva is not formed; thus a signal from the anchor cell induces vulva formation (44). Laser ablation experiments also show that any of six epidermal cells is capable of participating in vulva formation, and that each can adopt any of three alternative vulval cell fates. Which fate each cell adopts depends on its distance from the anchor cell, and also on signals from other epidermal precursor cells (45). An extensive search for mutations affecting vulva formation has led to the identification of 22 genes required for vulva development (46). These fall into several classes, including genes involved in the generation of the anchor cell, the generation of vulval precursor cells, the decision process that determines which of three fates the precursor cells adopt, and the generation of the corresponding vulval cell lineages. One of the genes involved in the choice between alternative precursor fates, *lin-12* (*lin*eage abnormal), is similar to the precursor of vertebrate epidermal growth factor and the low density lipoprotein (LDL) receptor. Like *glp-1*, *lin-12* mutations influence the fates of additional cells known to participate in cellular interactions. These two proteins share extensive sequence similarity and both appear to function as transmembrane receptors for developmental signals in a variety of cell types in the animal *(10)*.

Many antero-posterior differences in body pattern arise after hatching, and genes influencing patterning in the head *(16)*, anterior body region *(47)*, and posterior body region *(48)* have been identified. Raising or lowering the dosage of some of these genes can shift the boundaries between position-specific patterns toward the head or the tail, suggesting that these genes function to position cell types within the animal. The best characterized of these genes is *mab-5*, (*male ab*normal), which promotes posterior specialization in many cell types *(48)*. The cells affected by *mab-5* are related only by position, not by lineage or cell type, suggesting that localized cell-extrinsic signals may be involved in patterning the body axis. The *mab-5* gene contains a homeobox similar to that of the *Drosophila* homeotic gene *Antennapedia (36)*, again suggesting conservation of developmental mechanisms.

Temporal control of development

A set of genes called heterochronic genes controls the timing of developmental events *(49)*. In heterochronic mutants, lineage patterns expressed at one developmental period appear at another period. In precocious mutants, cells in early larval stages express lineages characteristic of later stages. In retarded mutants the opposite is true, and cells adopt fates characteristic of earlier stages. The gene *lin-14* appears to play a regulatory role; high levels of *lin-14* activity specify cell fates characteristic of early larval stages, whereas low levels specify late cell fates *(50)*. Most heterochronic mutations affect many cell types, but one, *lin-29*, affects only one tissue, the lateral epidermis *(49)*. This gene may regulate lineage patterns in specific cells in response to more global regulatory signals.

Sex determination

Developing cells not only respond to spatial and temporal cues, but also to information that controls sexual development. Analysis of sex determination mutants has defined a complex regulatory network *(21)*. In nematodes, as in Drosophila, the ratio of X chromosomes to autosomes (X : A) determines sex; a ratio of 0.5 leads to male development and a ratio of 1.0 to hermaphrodite development. However, the developmental machinery can discriminate between very fine differences in the X : A ratio; animals with a ratio of 0.67 are males, while those with a ratio only slightly higher, 0.75, are hermaphrodites *(51)*. It will be interesting to see how such a fine discrimination can be made, and whether the mechanism has general significance. A number of sex determination genes have been placed into a formal regulatory pathway *(52–54)*; upstream genes affect both sex determination and dosage compensation (a process that equalizes levels of X-linked gene expression in the two sexes), while downstream genes have more restricted effects. Much of the complexity of the pathway is thought to reflect a requirement for fine tuning. For example, the genes *fog-2* (*f*eminization *o*f *g*erm line *(55)* and *tra-2* (*tra*nsformer) *(56)* permit XX animals to undergo a brief period of male gametogenesis, so that some sperm are produced and the animals become hermaphrodites instead of females. Together the sex determination genes appear to establish the state of the gene *tra-1* *(57)*; if *tra-1* is on, somatic cells adopt hermaphrodite-specific fates, otherwise they adopt male fates.

The Nervous System

The nervous system of *C. elegans* contains a set of only 302 neurons, but this set is complex, and consists of 118 distinguishable cell types *(58)*. A variety of neurotransmitters are represented among *C. elegans* neurons, including acetylcholine, GABA (γ-aminobutyric acid) serotonin, dopamine, octopamine, FMRF-amide, and others. In order to facilitate studies of neurogenesis and behavior, the entire nervous system has been reconstructed from 20,000 electron micrographs taken of serial sections (6, 58). Some portions of the nervous system can be reconstructed quickly, which permits the isolation of mutations affecting synapse formation and synaptic specificity (58). Reconstructions of developing animals are currently under way; knowing the order in which processes extend and make interconnections will be quite informative. For example, neurons that first extend processes have been identified, and laser ablation studies have shown that they function to guide subsequent axons to their targets (59).

Knowledge of the complete structure should allow us to analyze behavior in a unique way, by simulating how it might be generated from the known neuronal network. For example, M. Chalfie and co-workers have identified neurons that mediate a response to mechanical stimulation by analyzing the wiring diagram and then ablating cells likely to mediate aspects of the behavior. Because of its small size, it is not yet possible to do neurophysiological experiments in C. elegans, so it is not possible to follow the flow of information directly. However, the structure and function of the nervous system of the larger nematode Ascaris appears to be essentially the same as the C. elegans nervous system, and models about the C. elegans mechanosensory response are consistent with electrophysiological studies in Ascaris (60).

Many mutations affecting neurogenesis have been isolated by screening for animals with alterations in behaviors or in such cellular properties as transmitter production, drug sensitivity, or lineage alterations (58). Because the C. elegans lineages are known, and because many neural cell types can be distinguished, it is possible to ask how a neuronal cell type is specified genetically. For example, genes that specify the mechanosensory response have been identified genetically. The gene lin-32 is required for many epidermal precursor cells to generate neuroblasts instead of additional epidermal cells (61), including neuroblasts that produce mechanosensory receptors. The gene unc-86 (uncoordinated) is required at a later stage for the generation of complex neural cell

lineages that generate mechanosensory receptors; in unc-86 mutants, many neuroblasts undergo simple stem cell divisions instead of generating complex lineages (62). Further analysis of this gene may reveal whether (and if so how) cell-intrinsic events generate complex lineages. The gene mec-3 (mechanosensory) which has a homeo box and may therefore regulate gene expression, initiates differentiation of the mechanosensory neurons (11). Candidates for genes regulated by mec-3 include the gene mec-7, a β-tubulin gene required for a mechanosensory cell-specific form of microtubules (63).

A total of 131 cells normally undergo programmed cell death during C. elegans development. The majority of cell deaths occur within neural cell lineages, and specifically require activity of the genes ced-3 (cell death) and ced-4 (64). In the absence of either gene function, cells that would normally die differentiate into functional neurons (64, 65). A different class of genes that includes mab-5 (48), specify some, but not all, cell deaths. This class of genes may be used to activate the death program, which can be considered a state of differentiation, in particular cells.

Caenorhabditis elegans is particularly well suited for learning how neural processes are guided to their destinations. Guidance mutants can be identified by staining fixed animals with neuron-specific antibodies, or better, they can be isolated by looking directly at axon placement in live animals. To see axons in living animals, Nomarski optics can be used to examine mutants in which processes are visible (11, 66), or else the animals can be soaked in fluorescent dyes that are taken up by many sensory neurons (67). Many guidance genes have been identified (16), and some of these are being analyzed at the molecular level.

In C. elegans, a small number of mesodermal and neural cells migrate for long distances. These migrations are easy to see in the light microscope in normal, untreated animals, and so the isolation and characterization of mutations affecting cell migration is possible. The problem of cell migration is not unlike that of axon guidance, so it is not surprising that many mutations affect both. So far, mutations in approximately 30 genes disrupt the movements of cells or axons along the body wall (16). A particularly interesting set includes three genes affecting both cell migration and axon outgrowth along the dorso-ventral axis. One of these, unc-6, is required for all such migrations, while unc-5 is required only for dorsalward movement, and unc-40 is required only for ventralward migration. At least two of these genes are homologous to genes encoding known extracellular matrix proteins (68).

Muscle Assembly and Function

An important question in cell biology is how proteins such as muscle components assemble into machines capable of generating force. In C. elegans, muscle assembly and function is being dissected by means of the genetic approaches that have been so successful in understanding the related problem of bacteriophage morphogenesis. Mutations affecting muscle function can be isolated by looking for animals that move abnormally; mutants with remarkably subtle changes in the smooth sinusoidal pattern of movement can be detected. The development and structure of muscle fibers can be examined with polarizing optics in living animals and also by electron microscopy of fixed individuals. By analyzing uncoordinated (unc) mutants, about 30 genes encoding myosin isoforms, paramyosin, actins, and additional, possibly new, muscle components have been identified [reviewed in (69)]. The sequence of myosin heavy chain, first elucidated in C. elegans, has refined our conception of the two major myosin domains, the globular head region and the coiled-coil rod region. In particular, the repetitive coiled-coil sequence in the rod provided a structural basis for the periodicity of the thick filament. Fine structure genetic analysis is revealing structural and functional properties of the myosin domains. For example, all unc mutations known to affect the rod inhibit assembly, indicating that the rod functions in this process. All of these rod mutations have been deletions; one possibility for the failure to isolate missense mutations in this region is that the coiled-coils contain sufficient

structural redundancy to withstand small perturbations. Surprisingly, some dominant mutations in the myosin head region also block muscle assembly, suggesting that this domain can also influence the assembly process (69). Body-wall thick filaments have been found to contain two myosin isotypes that have different locations and functions. The sequences responsible for these positional and functional differences can now be identified by constructing hybrid myosin genes and introducing them into the animal by DNA transformation.

The unc collection contains additional genes thought to encode thick filament proteins. One of these is a 500-kD protein with protein kinase homology (70). In addition, three genes with mysterious functions affect the dense body (the nematode equivalent of the Z line). Two of these are thought to bind actin filaments, and the third may anchor the dense body to the adjacent hypodermis. These gene products could help us understand how muscle filaments are positioned within a sarcomere, and how the force generated within the sarcomere is transduced to the body wall.

Other interesting topics, beyond the scope of this review, include aging (71); sperm morphogenesis (72); the dauer larva, an alternative developmental form that appears under conditions of stress (73); construction of the elaborate collagen-based cuticle; aspects of neuronal ultrastructure (74); and the role of the cytoskeleton in the elongation of the embryo (75). Also under study are aspects of DNA and RNA metabolism, including transposable elements (30) and the provocative phenomenon of trans-splicing (76), which is the post-transcriptional joining of separate RNA molecules.

Conclusions

Caenorhabditis elegans provides the experimental biologist with a powerful experimental system for the study of development and other complex systems in a multicellular animal. Its virtues are that the cellular plan and developmental lineages are essentially invariant and both have been determined completely; genetic analysis is extensive and now can exploit all the techniques of modern molecular genetics. In principle, in this organism the analysis of the genes and how their products work to implement the developmental process could be carried to completion.

References and Notes

1. S. Brenner, *Genetics* 77, 71 (1974); *British Med. Bull.* 29, 269 (1973).
2. R. Herman, in *The Nematode Caenorhabditis elegans*, W. B. Wood, Ed. (Cold Spring Harbor Laboratory, Cold Spring Harbor, NY, 1988), pp. 17–45.
3. _____, *J. Neurogenet.* 5, 1 (1989).
4. J. E. Sulston and H. R. Horvitz, *Dev. Biol.* 56, 110 (1977); J. Kimble and D. Hirsh, *ibid.* 70, 396 (1979).
5. J. E. Sulston, E. Schierenberg, J. G. White, J. N. Thomson, *ibid.* 100, 64 (1983).
6. J. G. White, E. Southgate, J. N. Thomson, S. Brenner, *Philos. Trans. R. Soc. London* 275B, 327 (1976).
7. D. G. Albertson and J. N. Thomson, *ibid.*, p. 299; J. E. Sulston, D. G. Albertson, J. N. Thomson, *Dev. Biol.* 78, 542 (1980).
8. A. Coulson, J. Sulston, S. Brenner, J. Karn, *Proc. Natl. Acad. Sci. U.S.A.* 83, 7821 (1986).
9. S. W. Emmons, *Cell* 51, 881 (1987).
10. I. Greenwald, *ibid.* 43, 583 (1985); G. Seydoux and I. Greenwald, *ibid.*, 57, 1237 (1989); J. Yochem and I. Greenwald, *ibid.*, 58, 553 (1989); J. Austin and J. Kimble, *ibid.*, 58, 565 (1989).
11. J. Way and M. Chalfie, *ibid.*, 54, 5 (1988).
12. P. W. Sternberg and H. R. Horvitz, *Annu. Rev. Genet.* 18, 489 (1984).
13. I. Greenwald, *Philos. Trans. R. Soc. London* 312B, 129 (1985).
14. C. Kenyon, *ibid.*, p. 21.
15. J. Hodgkin, *Annu. Rev. Genet.* 21, 133 (1987).
16. E. M. Hedgecock, J. G. Culotti, D. H. Hall, B. D. Stern, *Development* 100, 365 (1987).
17. W. B. Wood, Ed., *The Nematode Caenorhabditis elegans* (Cold Spring Harbor Laboratory, Cold Spring Harbor, NY, 1988).
18. S. Emmon, in *ibid.*
19. J. Hodgkin, K. Kondo, R. H. Waterston, *Trends Genet.* 3, 325 (1987).
20. R. Horvitz, in (*17*).
21. J. Hodgkin, in *ibid.*
22. _____, H. R. Horvitz, S. Brenner, *Genetics* 91, 67 (1979).
23. M. Edgley, Caenorhabditis Genetics Center, personal communication.
24. R. K. Herman and C. K. Kari, *Cell* 40, 509 (1985); C. D. Johnson, J. B. Rand, R. K. Herman, B. D. Stern, R. L. Russell, *Neuron*, 1, 165 (1988).
25. J. E. Sulston and J. G. White, *Dev. Biol.* 78, 577 (1980).
26. J. Sulston, personal communication.
27. A. Fire, *EMBO J.* 5, 2673 (1986).
28. I. Hope, R. Hoskins, A. Spence, J. Sulston, D. Thierry-Mieg, personal communication.
29. A. Fire, personal communication.
30. R. K. Herman, and J. E. Shaw, *Trends Genet.* 3, 222 (1987).
31. P. Anderson, personal communication.
32. D. G. Moerman, G. M. Benian, R. H. Waterson, *Proc. Natl. Acad. Sci. U.S.A.* 83, 2579 (1986).
33. H. Okamoto and J. N. Thomson, *J. Neuroscience* 5, 643 (1985).
34. M. R. Klass, *Int. Rev. Cytol.* 102, 1 (1986).
35. W. B. Wood, P. Meneely, P. Schedin, L. Donahue, *Cold Spring Harbor Symp. Quant. Biol.* 50, 575 (1985).
36. M. Costa, M. Weir, A. Coulson, J. Solston, C. Kenyon, *Cell*, 55, 747 (1988).
37. R. Waterston and A. Fire, personal communication.
38. J. S. Laufer, P. Bazzicalupo, W. B. Wood, *Cell* 19, 569 (1980).
39. S. Strome and W. B. Wood, *ibid.* 35, 15 (1983).
40. A. A. Hyman and J. G. White, *J. Cell Biol.* 105, 2123 (1987).
41. K. J. Kemphues, J. R. Priess, D. G. Morton, N. Cheng, *Cell* 52, 311 (1988).
42. J. R. Priess and J. N. Thompson, *ibid.* 48, 241 (1987).
43. J. R. Priess, H. Schnabel, R. Schnabel, *Cell* 51, 601 (1987); J. Austin and J. Kimble, *ibid.*, p. 589.
44. J. Kimble, *Dev. Biol.* 87, 286 (1981).
45. P. Sternberg, *Nature* 335, 551 (1988); _____ and H. R. Horvitz, *Cell* 44, 761 (1986).
46. E. L. Ferguson and H. R. Horvitz, *Genetics* 110, 17 (1985).
47. H. R. Horvitz, P. W. Sternberg, I. S. Greenwald, W. Fixsen, H. M. Ellis, *Cold Spring Harbor Symp. Quant. Biol.* 48, 453 (1983).
48. C. Kenyon, *Cell* 46, 477 (1986).
49. V. Ambros and H. R. Horvitz, *Science* 226, 409 (1984).
50. _____, *Genes Dev.* 1, 398 (1987).
51. J. E. Madl and R. K. Herman, *Genetics* 93, 393 (1979).
52. A. M. Villeneuve and B. J. Meyer, *Cell* 48, 25 (1987).

53. J. Hodgkin, *Annu. Rev. Genet.* **21**, 133 (1987).
54. M. K. Barton, T. B. Schedl, J. Kimble, *Genetics* **115**, 107 (1987).
55. T. B. Schedl and J. Kimble, *ibid.* **119**, 43 (1988).
56. T. Doniach, *ibid.* **114**, 53 (1986).
57. J. Hodgkin, *Genes Dev.* **1**, 731 (1987).
58. M. Chalfie and J. White, in (*17*).
59. R. Durbin, thesis, University of Cambridge, England (1987); W. M. Walthal and C. Chalfie, *Science* **239**, 643 (1988).
60. M. Chalfie, J. E. Sulston, J. G. White, E. Southgate, J. N. Thomson, S. Brenner, *J. Neurosci* **5**, 956 (1985); A. O. Stretton, R. E. Davis, J. D. Angstadt, J. E. Donmoyer, C. D. Johnson, *Trends Neurosci.* **8**, 294 (1985).
61. C. Kenyon, E. Hedgecock, M. Chalfie, unpublished data.
62. M. Chalfie, H. R. Horvitz, J. E. Sulston, *Cell* **24**, 59 (1981).
63. L. Gremke, Ph.D. thesis, Northwestern University (1986); C. Savage and M. Chalfie, personal communication.
64. H. M. Ellis and H. R. Horvitz, *Cell* **44**, 817 (1986).
65. L. Avery and H. R. Horvitz, *ibid.* **51**, 1071 (1987).
66. E. Hedgecock, personal communication.
67. _____, J. G. Culotti, J. N. Thomson, L. A. Perkins, *Dev. Biol.* **111**, 158 (1985).
68. C. Leung-Hagesteijn, N. Ishii, B. Stern, J. Culotti, E. Hedgecock, personal communication.
69. R. H. Waterston, in (*17*).
70. G. Benian, D. Moerman, R. Waterston, personal communication.
71. D. B. Friedman and T. E. Johnson, *Genetics* **118**, 75 (1988).
72. J. Kimble and S. Ward, in (*17*).
73. D. Riddle, in (*17*).
74. L. A. Perkins, E. M. Hedgecock, J. N. Thomson, J. G. Culotti, *Dev. Biol.* **117**, 456 (1986).
75. J. R. Priess and D. I. Hirsh, *ibid.*, p. 156.
76. M. Krause and D. Hirsh, *Cell* **49**, 753 (1987).
77. I would like to express the deep gratitude that we in the nematode field feel toward Sydney Brenner for initiating the study of *C. elegans*. I thank M. Costa, R. Horvitz, A. Kamb, C. Loer, M. Rykowski, K. Sawin, D. Waring, and L. Wrischnik for discussions and comments on the manuscript and S. Miner for help in preparing the manuscript. I thank R. Waterson for discussions about muscle.

8

Parasitic Protozoans and Helminths: Biological and Immunological Challenges

Adel A. F. Mahmoud

Parasitic protozoans and helminths represent two major groups of infectious agents that are responsible for considerable morbidity and mortality in human and animal populations. These agents have a wide geographic distribution and, as multiple infections are common in endemic areas, the total number of infections in humans far exceeds the world population. The distribution of some of these infections is expanding because of the increased prevalence of immunosuppressive conditions such as malignancies, chemotherapeutic medications, and retroviral infections. Approximately 1 billion people are infected by ascaris; 600 million by malaria-causing plasmodia, and 300 million each by schistosomes and filariae (*1*). While information on morbidity and mortality is less complete, it is estimated that malaria alone causes 1 million deaths among children yearly and that other infections such as schistosomiasis and filariasis cause chronic debilitating conditions that are associated with loss of productivity and reduced life-span. Parasitic protozoans such as *Pneumocystis carinii, Toxoplasma gondii,* and *Cryptosporidium* are emerging as major opportunistic infections in those infected with HIV or in those with other immunosupressive conditions (*2*).

In spite of this enormous public health impact, a staggering contrast exists between our knowledge of protozoan and helminthic infections and our understanding of infections caused by bacteria and viruses. Details of the biology, biochemistry, and molecular biology of viruses and bacteria have been examined extensively while similar studies of protozoans and helminths have lagged. Although there is no specific scientific reason for this disparity, the complex structure of protozoans and helminths and the fact that they are endemic mainly in less developed countries may provide partial explanation. Whatever the reasons for this gap, the past two decades have witnessed the scientific rediscovery of these parasites.

The ability of protozoan and helminthic organisms to establish themselves in multiple hosts and to overcome host defense mechanisms has become a central scientific challenge (*3*). These organisms are characterized by complex life cycles in which changes between free living and parasitic stages occur and each stage represents a different biochemical and antigenic structure. The discovery of antigenic variation in trypanosomes has

Adel A. F. Mahmoud is professor and chairman, Department of Medicine, Case Western Reserve University School of Medicine and University Hospitals, Cleveland, OH 44106. This chapter is reprinted from *Science* **246**, 1015 (1989).

resulted in a considerable shift in interest in parasitic protozoans (*4*). In fact, investigations of the underlying mechanisms of antigenic variation and the molecular basis for antigen switching in trypanosomes are bridging the gap between the status of studies on protozoans and the expanding horizon of molecular biology and are adding to our understanding of gene structure and organization in eukaryotes in general.

The scientific community has also focused on the expanding medical challenge posed by these organisms. For example, multiple drug resistance is emerging as the single most serious problem in malaria chemotherapy (*5*). Continued analyses of parasite proteins are needed to develop species-specific diagnostic tests, which are lacking for many parasite species (such as *Leishmania*) that differ in their pathogenicity. Furthermore, we still have no effective and safe chemotherapeutic agents against certain parasitic infections such as *Trypanosoma cruzi* and *Cryptosporidium*.

This chapter will highlight some of the most important and exciting issues in research on parasitic protozoans and helminths. The goal is to use the major biological differences between these two groups of organisms to contrast mechanisms of adaptation to parasitism. We will use information gained from representative examples of parasitic protozoans (*Plasmodium* and *Trypanosoma*) and of a parasitic helminth (*Schistosoma*).

Because of the relative ease of culture of protozoans in vitro, there has been considerable progress in our knowledge of their biology and molecular organization. In contrast, maintaining the life cycle of worms in the laboratory necessitates passaging the organisms in experimental animals. To obtain significant numbers of schistosomes, a colony of the snail intermediate host has to be maintained and the parasite must also go through a phase in mammalian hosts such as mice or hamsters. No in vitro culture systems have as yet been developed to maintain and expand the population of a parasitic worm in vitro. Despite these difficulties, there also have been recent advances in the characterization of gene expression in parasitic helminths.

Biology of Parasitic Protozoans and Helminths

Parasitism involves a specialized and dependent mode of life; as such, all infectious organisms are parasites (*3*). The life cycles of protozoans and helminths often involve differentiation through several morphologically and antigenetically distinct stages (*6*). Protozoans and helminths that parasitize humans gain access to their bodies via multiple routes, including ingestion, direct penetration of intact skin, or the bites of insect vectors. They inhabit specific locations within their human host. Because of their small size (4 to 400 μm), protozoans may reside intracellularly (*Leishmania, Toxoplasma,* and *Plasmodium*); intravascularly (*Trypanosoma*) or in tissues such as brain (*Toxoplasma*); or they may be located in the lumen of the gut (*Entaemoeba*). The larger size of the multicellular parasitic helminths (several millimeters to several meters) partially dictates their final habitat in the human host. For example, the small schistosomes (1–2 cm in length) are found in the portal venous or vesical circulation whereas the large tape worms (several meters in length) parasitize the lumen of the gut.

Humans usually serve as the definitive host for many parasitic protozoans and helminths; however, the concept of biological vectors that also are intermediate hosts adds a special feature to this group of infectious agents (*7*). Passage through intermediate hosts may simply be a mode for transmission but often it is significant for the life cycle of the parasite. For example, sexual multiplication of plasmodia during malarial infections occurs only in the mosquito vector whereas asexual reproduction, essential for expansion of the parasite population, occurs in infected humans (*8*). The situation is reversed in schistosomiasis: sexual reproduction of the worms occurs in infected humans whereas asexual multiplication takes place in the intermediate host, the snail (*9*).

Although there are many fundamental biologic differences between unicellular and multicellular organisms, a broad distinction may be made between microparasites and macroparasites (*10*). Microparasites (including bacteria, viruses, and protozoans) exhibit high rates of direct reproduction in the host, however, macroparasites (helminths) generally have no direct reproductive capabilities within their definitive host (*11*). It follows therefore that protozoans increase their population inside their human host without the necessity of passage through the environment or another intermediate host. For example, many malaria sporozoites may be introduced in a human host during the bite of one infected mosquito. After differentiation and multiplication in liver cells (7 to 14 days), several thousand merozoites emerge to begin the erythrocytic phase, in which the parasites invade and mature within host red cells. Cycles of multiplication and reinvasion of host erythrocyte continue at regular intervals (1 to 3 days) and may lead to accumulation of enough parasites to result in significant morbidity (*12*).

In contrast, parasitic helminths (in general) do not multiply within the human host. Schistosome infection in humans is initiated via penetration of intact skin by the infective stage, cercariae, which are found in contaminated bodies of fresh water. The likelihood that a cercaria will mature into a worm of either sex varies with the host species but in susceptible hosts is never greater than 40 to 60%. The exact mechanisms underlying the inability of all cercariae to mature into adult worms are not known, but the phenomenon may relate to innate immunity exhibited by each host species (*13*). For a human host to acquire a pair of male and female worms, exposure to more than two cercariae is, therefore, necessary. Even if a pair of the opposite sex were present, mature adult worms residing in humans are incapable of increasing their population by any multiplication mechanism within the human host; they can only produce eggs, which do not mature into adult worms in humans. To increase the number of worms in a human host, exposure to additional cercariae is necessary (*14*). The schistosomes increase their population in the environment by multiplication in the snail intermediate host. This occurs when mature adult worms in humans mate and produce eggs that have to pass with excreta to the outside environment. If these eggs reach fresh water they hatch, releasing motile organisms that invade the tissues of snails where they multiply and differentiate into the infective cercariae.

This fundamental difference between protozoan and helminthic organisms also has certain implication in describing and understanding factors that are involved in regulating host-parasite relationships. For example, mathematical models for the patterns of transmission of these organisms must take into account these differences in life cycle (*15*). Prevalence models largely fit microparasites; they describe the fractions of the human population that are either susceptible, infectious, or immune. Helminthic infections, in contrast, are better fit by density models because of the aggregated nature of the infection in human populations (certain individuals carry a high load of parasites) and the unreliability of prevalence as an index of intensity of infection (the number of parasites per infected person). Thus, the distribution of helminthic infections in any given community does not follow a statistically normal pattern. Most infected individuals harbor a low worm burden and only a small proportion carry the brunt of heavy worm counts; this pattern corresponds to a negative binomial relationship.

The difference in life cycle is also reflected in the different impact that macroparasites have on the host immune responses (*12, 16*). Both protozoans and helminths pose a significant challenge to the host because of the multiplicity of antigens they present to the immune system, but the rate of growth and multiplication of parasitic protozoans is considerably faster than in helminthic infections. While the consequences of this difference are not fully understood, it is reflected in differences in the ability of the host to mount an effective and protective immune response. For example, primary infection with *Leishmania tropica* is followed by immunity to reinfection (*17*), and injection of irradiated malaria sporozoites into experimental animals results in acquisition of resistance against subsequent challenge (*12*). In contrast, acquired resistance to helminthic infection, if it ever exists in nature, is a slow process that takes several years to be established. Evidence for acquired immunity against schistosomes and most other helminthic infections is based on observations of changes in age-specific intensity of infection in humans (Fig. 1). In most endemic communities that have been studied, intensity of schistosome infection increases with age, reaching a peak in individuals 15 to 20 years old. In older people, a dramatic reduction in the intensity of infection is seen. This decrease in adults has been considered strong evidence for acquired immunity in humans, although it takes a decade or so to be clinically significant (and other explanations such as changes in the pattern of exposure to infected water, are possible) (*18*). Additional observations in experimental animals and in infected individuals after chemotherapy strongly argue for a role of acquired immunity in regulating intensity of infection in helminthiases (*19*).

Invasion and Adaptation to Parasitism

Invasion of cells or internal organs of the host is one of the most significant biological characteristics of parasites in general. For protozoan and helminthic infections, this process takes on special importance because of their size and the complexity of their life cycle. In protozoan infections, invasion of only a very restricted set of cell types may occur, as in the case of *Plas-*

modium and *Leishmania* parasites, other organisms, (such as *Toxoplasma*) are less discriminating, invading various host cells including muscle, brain, and mononuclear phagocytes.

Malaria provides an example of a stage-dependent specificity of invasion (*12*). The sporozoites are the first stage of the parasite in humans; they only invade liver cells. The next stage, merozoites, invade only host erythrocytes; the specificity of this process is dependent on receptor molecules on the erythrocyte surface (*20*). A suggestion that merozoites of *P. vivax* require the presence of the Duffy blood group substance on the erythrocyte surface was obtained from studies performed many years ago. Some black individuals who were inoculated with *P. vivax* for therapeutic purposes failed to acquire infection while Caucasians receiving similar inocula did. These results were correlated to the absence or presence, respectively, of Duffy blood group substance on their erythrocytes. Evidence was later obtained from in vitro studies that Duffy-negative erythrocytes are resistant to invasion by *P. vivax*. (*21*). This resistance was found to be species-dependent since Duffy-negative erythrocytes are susceptible to invasion by *P. falciparum* (*20*). The ability of *P. falciparum* to invade is related to the presence of glycophorin

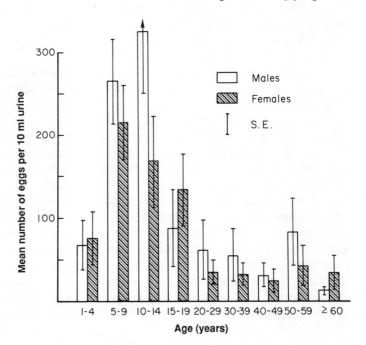

Fig. 1. An example of evidence used to indicate the existence of acquired immunity against helminthic infection. The histogram represents mean counts of *Schistosoma haematobium* eggs in 10-ml urine samples of the population in an endemic area (Kilole, Coast Province, Kenya) (*71*). The sharp decrease in egg counts in individuals over 30 years may reflect acquisition of immunity.

A on the erythrocyte membrane (22). Infection of red cells lacking glycophorin A or glycophorin A and B is considerably reduced. Nevertheless, the fact that several strains of *P. falciparum* can still invade such cells suggests that the parasite may use other membrane receptors (22, 24). Furthermore, invasion has been shown to be dependent on sialic acid residues on the erythrocyte surface receptors; erythrocytes were made less sensitive to invasion by prior treatment with neuroaminidase.

Invasion by the merozoites proceeds through the formation of a junction between the anterior end of its surface and the cell membrane (25). The junction moves around the parasite, creating a vacuole through which the merozoite invades the host erythrocyte (26). Subsequent to invasion, the membranes of infected cells are altered significantly, including the appearance of electron-dense knobs on the erythrocyte surface. The significance of this process is not clear, although knob formation has been implicated in the pathogenesis of cerebral malaria by inducing red cell sequestration (27). This effective, though biologically simple, intra-erythrocytic localization of the parasite provides excellent protection against multiple host defense mechanisms and allows the organisms to grow, multiply, and proceed to infect another cohort of red cells.

A different model for evasion of host protective responses, antigenic variation, has been described in trypanosomiasis (28). The phenomenon of antigenic variation in trypanosomes is the key adaptive mechanism for parasitism; it is, however, not restricted to protozoan organisms as it has been observed in *Borrelia, Neisseria,* and several viruses (29). During chronic trypanosomiasis in humans, waves of parasites appear in peripheral blood. This is particularly true in *Trypanosoma brucei gambiense* infection, which causes the chronic Gambian form of sleeping sickness. These waves of peripheral blood parasites are serologically distinct; each represents a different antigenic variant and results in the formation of variation-specific antibodies. In trypanosomes, switching of surface glycoproteins is, however, not related to the presence of antibodies (30) suggesting that the process is due to programmed gene rearrangements that are independent of host or environmental factors. However, if the detected rate of antigenic switch in vitro (10^{-5} to 10^{-7} per cell division) is extrapolated to the situation in vivo, rapid exhaustion of the parasite repertoire or of variable surface glycoproteins would occur. Since this does not happen in vivo, other mechanisms besides antigenic variation may be involved in controlling the orderly appearance of cohorts of parasites in the blood stream (4).

In spite of their size (approximately 200 to 600 μm), the infective stages of some worms are capable of penetrating intact human skin while transforming from free-living to parasitic organisms (31). For schistosomes, the infective stage is the free-living cercariae that are shed from the snail intermediate host into bodies of fresh water at ambient temperature. Each multicellular cercaria is bound by a syncytium that is limited by a traditional trilaminate plasma membrane. Upon encountering a mammalian host, cercariae attach to the skin and evacuate the contents of their pre- and post-acetabular penetration glands, which facilitates entry into the dermal and subdermal tissues (32). The process of invasion by the cercariae sets in motion a series of changes in the organisms that are aimed at adaptation to the environment of the mammalian hosts. The cercariae are transformed into schistosomula, the first parasitic stage in humans or other mammals, which can live in physiologic salt concentrations at 37°C. The most dramatic morphological change occurs in the surface membrane of the schistosomulum; the cercarial trilaminate membrane is converted into a heptalaminate complex structure composed of two closely apposed lipid bilayers (33). The surface of the schistosomulum (and all subsequent stages of maturing larvae and adults in mammalian hosts) is completely enclosed by the heptalaminate membrane, which shields the internal cellular structures of the parasite. The syncytium is connected by microtubule-lined intracytoplasmic processes to subtegumental cells. These cells contain multilaminate bodies that are thought to form the precursors of the heptalaminate surface membrane (34). The change from free-living cercariae to the parasitic schistosomula is usually completed within 3 hours, and can be observed in vitro. The biological relevance of the membrane transformation has recently been demonstrated in studies with inhibitors of microtubule organization (35). Colchicine or vinblastine treatment of transforming organisms resulted in accumulation of significant numbers of multilaminate bodies in the subtegumental cells (Fig. 2). The surface of such treated schistosomula was limited by a trilaminate membrane (Fig. 3). In vivo survival of colchicine-treated organisms was significantly reduced, demonstrating that heptalaminate membrane formation is a prerequisite adaptation to the parasitic mode of life for schistosomula.

Strategies for Vaccine Development

Induction of immunity is a key strategy for control of infectious diseases in general. Nevertheless, the task of producing vaccines against any of the clinically significant parasitic protozoan and helminthic infections has proven to be a tremendous challenge. One reason, which relates to their biologic and epidemiologic characteristics (36), is that a vaccine against a microparasite such as *Plasmodium* has to be almost 100% effective in the target population, in this case children less than 1 year of age. Furthermore, a significant proportion of this target population has to be vaccinated and protected if such a strategy is to succeed in either eradication of infection or reduction of

its incidence to very low levels. In contrast, characteristics of parasitic helminths may allow the implementation of vaccination strategies in which vaccines are less than 100% protective. The observation that most helminthic parasites do not multiply within their definite host, that a small proportion of infected individuals carry a high number of parasites, and that morbidity is related to the number of parasites suggest that a vaccine preparation less than 100% protective may be of major biological and clinical significance for controlling this group of infectious agents. Once the group of individuals with high intensity of infection can be identified, induction of partial protection in these individuals could dramatically decrease their contaminating effect on the environment and consequently reduce transmission to uninfected individuals. Furthermore, partial protection of heavily infected subjects would reduce their chances of becoming ill.

Until recently, induction of immunity against parasitic protozoan and helminthic infections has not been successfully achieved in humans or experimental animals except in limited studies with whole irradiated organisms or their crude extracts. Practically, there is only one example of consistently successful induction of protection in humans against a parasitic infection. For decades, it has been known that individuals who developed a scar from a primary infection with the protozoan *L. tropica* were immune to subsequent exposure on the same parasite. A more organized approach to induction of resistance to *L. tropica* by means of cultured organisms and measured doses is cur-

Fig. 2. Effect of colchicine treatment on the transformation of cercariae of *S. mansoni* into schistosomula. Transmission electron micrographs of **(A)** untreated schistosomula (final magnification, 5100X) and **(B)** Schistosomula incubated during transformation and for 3 hours thereafter in colchicine (Final magnification 5700X). While very few multilaminate bodies were observed in the subtegumental cells of control organisms, large accumulations of these bodies were demonstrated in the perinuclear cytoplasm of cells in colchicine-treated schistosomula (**C**). These observations suggest that during transformation multilaminate bodies are guided or transported to the surface via microtubuleline intracytoplasmic processes.

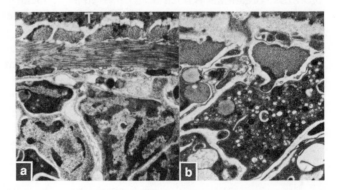

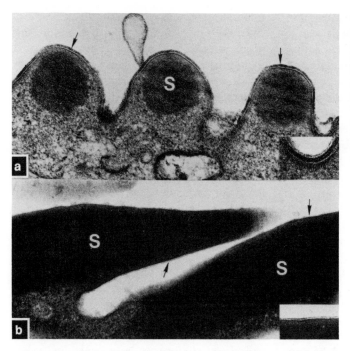

Fig. 3. Effect of colchicine treatment on the structure of the surface membrane during cercariae transformation into schistosomula. Final magnifications: upper and lower panels, 69,000; insets, 99,500X. Panel a shows the surface of control schistosomula. "S" is a spine covered by heptalaminate membrane; the details of its structure are shown in inset. Panel b indicates spines "S" covered by a trilaminate membrane in a colchicine-treated sample; details are shown in the inset.

rently being used to protect humans (*37*). Other approaches to vaccination such as the use of irradiated infective stages of plasmodia, schistosomes, and hookworms have been successfully employed in inducing resistance in experimental animals and (in the case of malaria) in humans. It is usually necessary to attenuate, but not kill, the infective stages. For example, dead sporozoites are not internalized by cells and do not induce protective immunity (*38*).

The limitation of irradation and similar techniques is related to the unavailability of sufficient materials. For example, it is a monumental task to develop an insectory for breeding enough mosquitoes to prepare irradiated malaria sporozoite vaccine for a small village in an endemic area. Furthermore, the multiplicity of antigens contained in these irradiated preparations adds another complicating factor since the corresponding host immune responses may not all be protective and may lead to untoward immunopathological or immunosuppressive consequences. Currently, a major approach to defining biologically relevant antigens involves production of monoclonal antibodies against stages of the parasite life cycle that may be involved in eliciting protective responses. In some instances

(see below) these monoclonal antibodies have been shown to adoptively transfer partial protection to recipient animals and may, therefore, be helpful in characterizing the corresponding antigens.

Since malaria is the most clinically important parasitic protozoan disease in humans, the possibility of vaccine production has attracted considerable attention. For example, monoclonal antibodies against the circumsporozoite antigen of the malaria sporozoites (infective stage to humans) and against the erythrocytic parasites and gametocytes (infective stages to mosquitoes) have been produced (*39*). These antibodies were used to protect mice against subsequent challenge, prevent clinical disease, or block fertilization of the male and female gametocytes in the stomach of the insect vector. Extensive efforts have been directed at isolation, characterization, and testing of malaria antigens identified by these monoclonal antibodies. Furthermore, immunologically important epitopes of candidate malaria vaccines have been synthesized chemically or produced by recombinant techniques and have recently been evaluated in humans. However, to date no single preparation of a defined antigen or combinations thereof has resulted in uniform response in humans. Developing a vaccine for malaria is further complicated by the lack of a systematic knowledge of the nature of protective host responses, the nature of the effector mechanisms of resistance, and the possible heterogeneity of the causative organisms. Furthermore, most studies on vaccination protocols have used adjuvants, which adds another complicating dimension because of their known side effects and limitations of their use in humans.

The circumsporozoite (CS) protein of *P. falciparum* has received considerable attention since it was first recognized as being involved in immunity after animals are exposed to irradiated parasites (*40*). Sera from such immunized mice precipitated a 44-kD antigen (the CS protein) that reacts with protective monoclonal antibodies and constitutes the major protein on the parasite surface. The structure of CS protein and its encoding DNA sequence in several mammalian and human malaria parasite species have been identified (*41*). In *P. falciparum,* the CS protein contains 412 amino acids. The central region of the molecule is made of approximately 40 repeats of the sequence Asn-Ala-Asn-Pro (NANP). Repeats have been found in other malaria species but their detailed structure is species specific. The CS protein also has two highly conserved regions flanking the central repeats.

Knowledge of the structure of the CS protein has facilitated a detailed examination of the host immune response to the different regions of the molecule and therefore the ability to construct subunit candidate vaccines. In humans and experimental animals, the immunodominant B-cell epitope is the repeat region of the molecule (*42*). Antibodies to this region block sporozoite invasion of hepatocytes in vitro and adoptively transfer resistance in mice.

The humoral immune response to the repeat region may not be a sufficient protective mechanism for construction of a vaccine. Other cellular elements (such as cytotoxic T cells) and mediators (such as interferon γ) may be involved (*43*). Only partial resistance can be achieved in animals adoptively transferred with antibodies to the repeat region and active im-

munization of humans with subunit repeat vaccine was only successful in two of nine individuals (*43*). Evidence also has accumulated that indicates a role for T cell responsiveness in vaccine-induced immunity to malaria. Immunity induced by malaria sporozoites is adoptively transferred by T cells from immunized to normal recipients. In addition, sporozoite immunization does not protect athymic nude mice (*44*). Recently systematic examination of T cell responses in immunized animals and in infected humans have been conducted. In one study (*45*), mice that had been immunized repeatedly with irradiated sporozoites were injected with antibodies to either CD8 or CD4 determinants (to deplete the corresponding T cells) and were then exposed to live sporozoites. The absence of CD8$^+$ but not CD4$^+$ cells resulted in abrogation of acquired immunity (*45*).

Examination of the T cell response and its genetic restriction in immunized mice has resulted in identification of the Th2R region (amino acids 326 to 345) as a helper T cell site. Similarly, human T cell epitopes of the CS proteins have been examined in studies with overlapping synthetic peptides corresponding to different areas of the molecule. In one such study, the ability of 29 synthetic peptides (covering the entire CS protein) to stimulate peripheral blood mononuclear cells of individuals from a *P. falciparum* endemic area was examined (*46*). In a significant number of individuals, recognition of residues corresponding to the Th2R region of the molecule occurred. In another study, approximately 70% of samples tested reacted with synthetic peptides corresponding to Th2R and Th3R regions of three strains of *P. falciparum* (*47*). In addition, human T cell reactivity to Th3R correlates with resistance to *P. falciparum* (*48*). An outcome of these mapping studies is the recognition that T cell epitopes of the CS protein are located in the polymorphic segments of the molecule. Examination of intra-species variation of the CS gene in several *P. falciparum* isolates demonstrated substitutions in biologically important regions such as Th2R, suggesting that strain variability in CS protein is significant (*49*). These findings indicate potential serious complications for vaccine production against malaria.

The task of developing defined vaccines for helminthic infections may be far more challenging. However, convincing evidence for the existence of acquired resistance against schistosomiasis, filariasis, and other helminthic infections in humans has been obtained (50). The development of specific anti-eosinophil sera and in vitro systems to examine antibody-dependent cell-mediated cytotoxicity against *Schistosoma mansoni* schistosomula and larval stages of other helminths demonstrated the pivotal role of eosinophils as well as antibodies and complement in host resistance (51). Other effector mechanisms, particularly those resulting in activation of the mononuclear phagocytes, are also involved (52). The increased understanding of effector mechanisms has led to several approaches to induction of resistance against multicellular parasites in experimental models. Irradiated organisms, protective monoclonal antibodies, single purified or recombinant antigens, and anti-idiotypic and anti-antiidiotypic antibodies have all been shown to induce significant in vivo resistance against schistosomiasis (53). It has to be noted, however, that protection achieved has only been partial. Even with the use of several adjuvants, no single antigen preparation has resulted as yet in complete protection against any helminth infection. The basis for this limitation is unknown.

Isolation and characterization of defined helminthic antigens with potential protective effects has been dependent either on examinations of crude parasite extracts or the development of monoclonal antibodies capable of adoptively transferring resistance to normal recipients (54). In studies with crude extracts of *S. mansoni* schistosomula, significant protection against subsequent challenge was observed (55). The antibody response in animals vaccinated with extracts has been used to identify individual active antigens in the crude preparations. Protective *S. mansoni* antigens of 97, 68, and 43 kD have been isolated (55).

The purified 68-kD schistosome glycoprotein induced significant protection by reducing worm load by 22 to 66% upon challenge of immunized mice (55); this level of resistance was induced without the use of adjuvants. Administration of the 68-kD antigen to mice did not result in sensitization to subsequent injection of schistosome eggs, which suggests that the use of such antigen in infected animals or humans will not lead to exacerbation of egg-related immunopathology. Immunolocalization studies have revealed that the 68-kD antigen is associated with cytoplasmic granules in the head and pre-acetabular glands of cercariae and schistosomula (55). The DNA sequences encoding this antigen and, thus, its amino acid sequence are now being determined.

Other protective schistosome antigens have been isolated; some have been cloned and their immunologic epitopes have been mapped (56). Synthetic peptides corresponding to highly hydrophillic or mobile areas of the 28-kD antigen have been used to examine the B and T cell epitopes (57). Two peptides were identified as major IgG targets including one (corresponding to amino acids 24–43) that is the target for the IgG2a antibody that mediates eosinophil-dependent cytotoxicity against schistosomula in vitro. As in the case of malaria subunit vaccines, these studies pave the way to examine in detail the most optimal structure of candidate anti-helminthic vaccines. Furthermore, approaches such as purification of schistosome paramyosin either immunologically or biochemically or the parasite gluthathione transferase have demonstrated the potential protective effect of these molecules (58). In other studies, the possibility of utilization of anti-idiotypic as well as anti-anti-idiotypic antibodies to induce significant protection has been reported (59).

Similarly, in lymphatic filariasis, it has been demonstrated that a crude extract of microfilariae of *Brugia malayi* induces resistance to subsequent injection of the parasite larvae in mice or to exposure of rodents of the genus *Meriones* (called jirds) to the infective stage (60). The limited nature of antibody response in immunized animals led to partial purification of several putative protective antigens. One of these is an ~60 kD antigen that has recently been examined in detail; a recombinant cDNA clone has been isolated (61). Insertion of this clone in vaccinia has been achieved, and the recombinant virus induces significant resistance to jirds.

Induction of resistance to a multicellular organism such as the schistome by a combination of several pathways is essential because of the complex nature of the parasite and because of the partial protection induced by each pathway. We now have to explore the possibilities of multiple antigens and custom-made adjuvants as complementary strategies for the development of effective vaccines.

Gene Regulation

While this review has focused on some features that characterize adaptation to parasitism in protozoan and helminthic infections and new insights gained in vaccine production, it should be noted that considerable progress has been made in understanding the molecular aspects of gene regulation in these organisms.

Parasitic protozoans and helminths provide excellent models for studying cellular differentiation. Because of the obligate parasitic nature of these organisms, they are programmed to switch different stages of their life cycle between free-living forms in the environment and parasitic forms in one or more hosts. In many examples switching means a change in morphological, biochemical, and antigenic structure of the organisms. Switching has to be regulated at the molecular level, often via elaborate mechanisms. Studies of gene regulation in protozoan and helminthic organisms have demonstrated some intriguing similarities and many differences and have uncovered an unexpected wealth of information relevant to eukaryotic gene regulation.

Switching of antigenic variants in trypanosomes involves control of the expression of the antigenetically distinct variable surface glycoproteins (VSGs), such that only one variant is transcribed at a given time. There are approximately 1000 VSG genes per trypanosome genome (*62*). Switching is accomplished primarily through gene conversion,

in which an unexpressed VSG gene is transferred to a telomeric "expression site" located on another chromosome (*63*).

Another facet of gene expression in trypanosomes that is the focus of considerable attention is trans-splicing (*64*). Messenger RNAs in trypanosomes consist of two exons that are transcribed from separate genes and are spliced post-transcriptionally. At the 5' end of each mature mRNA is a capped, 39-nucleotide (nt) sequence called the mini-exon or spliced leader (*65*). The mini-exon becomes joined to a main exon that was cut out of a polycistronic unit. The conclusion that mRNA maturation in trypanosomes occurs by trans-splicing rather than by the conventional cis-splicing, mechanisms was based on the observation that free 100-nt "intron" (or "transon") is released from poly A-fractionated trypanosome RNA after treatment with a debranching extract of HeLa cells (*66*). This 100-nt piece indicates that trypanosome transcription proceeds through a Y-branched intermediate, rather than through the lariat structure observed during cis-splicing.

Only recently have comparable studies been possible in helminthic organisms. These organisms also use trans-splicing in the maturation of some mRNAs. A 22-nt spliced leader was reported on some, but not all, actin mRNA in the free-living nematode *C. elegans* (*67,68*). The leader RNA is not polyadenylated and is only found on a subset of messages; this is in contrast to the trypanosomes, where the spliced leader has been found on all messages.

An identical 22-nt spliced leader has been described in two medically important nematodes, *B. malayi* and *Ascaris* (*69*). The orientation of the 22-nt piece within the 5S genes of *B. malayi* and *Ascaris* is opposite to that seen in *C. elegans*; the effect of this difference on function is unknown. Although there is little primary sequence identity between the spliced RNAs in different nematodes or in other organisms such as trypanosomes, they can assume identical secondary structures; each is folded into three stem-loops (Fig. 4). Comparison of these loops with the previously

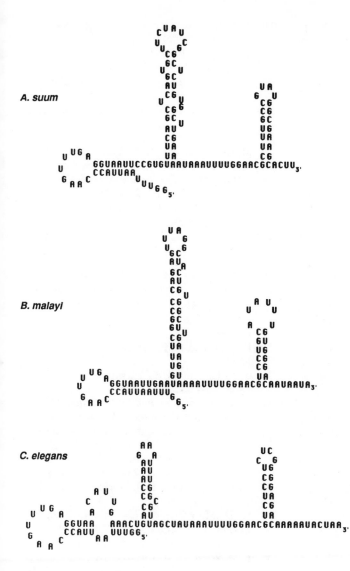

Fig. 4. Potential secondary structure of spliced leader RNAs from the free living nematode *Caenorhabditis elegans* as compared to those of two parasitic nematodes *Brugia malayi* and *Ascaria suum* (*69*). The computer program of Zuker and Stiegler (*72*) was used to generate the possible secondary structures. Comparison with structures of spliced leader RNAs of other parasitic protozoans has been described (*70*). Each contains three stem-loops and a 5′ splice site is next to the turn of the 5′ most loop. An Sm-binding sequence is located between stem-loops 2 and 3.

reported secondary structure of spliced leader RNAs of other organisms (*70*) suggest similarities with mRNA assembly and biogenesis in vertebrate cells.

Conclusion

This review concerned selected aspects of the biology, immunology, and molecular organization of parasitic protozoans and helminths. These organisms pose a significant challenge to the scientific community. Future research efforts need to address two major fundamental areas: examination of the molecular basis for adaptation to parasitism and development of newer control measures, including drugs and vaccines.

References and Notes

1. J. A. Walsh, in *Tropical and Geographical Medicine,* K. S. Warren and A. A. F. Mahmoud, Eds. (McGraw Hill, New York, ed. 2, 1990), p. 185.
2. P. Piot *et al., Science* **239,** 573 (1988); R. M. Selik, E. T. Sharcher, J. S. Curran, *AIDS* **1,** 175 (1987).
3. K. S. Warren, in *Tropical and Geographical Medicine,* K. S. Warren and A. A. F. Mahmoud,

Eds. (McGraw Hill, New York, ed. 2, 1990), p. 110; A. A. F. Mahmoud, in *The Oxford Companion to Medicine,* J. Walton, P. B. Beeson, R. B. Scott, Eds., (Oxford University Press, Oxford, 1986), p. 1005.

4. P. Barst and G. A. M. Gross, *Cell* **29,** 191 (1982); L. H. T. Van der Ploeg, *ibid.* **51,** 159 (1987).

5. L. J. Bruce-Chwatt, *Essential Malariology* (Heinmann, London, ed. 2, 1984); D. J. Wyler, *N. Eng. J. Med.* **308,** 875 and 934 (1983).

6. W. A. Petri, Jr. and J. I. Ravdin, *Curr. Opin. Infect. Dis.* **1,** 229 (1988); K. W. Warren, *Am. Rev. Publ. Health* **2,** 101 (1981); R. M. Anderson and R. M. May, *Adv. Parasitology* **24,** 1 (1985).

7. G. D. Schmidt and L. S. Roberts, Eds., *Foundation of Parasitology* (St. Louis, ed. 2, 1981).

8. D. J. Wyler, in *Principles and Practice of Infectious Diseases,* G. L. Mandell, R. Gorden Douglas, Jr., J. E. Bennet, Eds. (Churchill Livingstone Inc., New York, ed. 3, 1990), p. 2056.

9. A. A. F. Mahmoud, in *Principles and Practice of Infectious Disease,* G. L. Mandell, R. Gorden Douglas, Jr., J. E. Bennet, Eds. (Churchill Livingstone Inc., New York, ed. 3, in press), p. 2145.

10. R. M. Anderson and R. M. May, *Nature* **280,** 361 (1979); _____, *Adv. Parasitol.* **241,** 1 (1985).

11. A. A. F. Mahmoud, in *Tropical and Geographic Medicine,* K. S. Warren and A. A. F. Mahmoud, Eds. (McGraw Hill, New York, ed. 2, 1990), p. 368.

12. L. Miller and D. Warrell, *ibid.,* in press.

13. C. A. Peck, M. D. Carpenter, A. A. F. Mahmoud, *J. Clin. Invest.* **71,** 55 (1983).

14. A. A. F. Mahmoud and M. F. Abdel Wahab, in *Tropical and Geographical Medicine,* K. S. Warren and A. A. F. Mahmoud, Eds. (McGraw Hill, New York, ed. 2, 1990), p. 458.

15. G. F. Medley and R. M. Anderson, *Trans. Roy. Soc. Trop. Med. Hyg.* **79,** 532 (1985); R. M. Anderson, in vol. 2 of *Bailliere's Clinical Tropical Medicine and Communicable Diseases,* A. A. F. Mahmoud, Ed. (Bailliere's Tindall, Philadelphia, 1987), pp. 279–300.

16. D. G. Colley, in *Bailliere's Clinical Tropical Medicine and Communicable Diseases* (Vol. 2, No. 2), A. A. F. Mahmoud, Ed. (Bailliere's Tindall, Philadelphia, 1987), pp. 315–332; A. E. Butterworth *et al., Trans. Roy. Soc. Trop. Med. Hyg.* **79,** 393 (1985).

17. F. Y. Liew, in *Vaccination Strategies for Tropical Diseases,* F. Y. Liew, Ed. (CRC Press Inc., Boca Raton, FL, 1989), p. 239.

18. A. E. Butterworth, *Acta Trop.* **44,** 31 (1987).

19. A. E. Butterworth *et al., Trans. Roy. Soc. Trop. Med. Hyg.* **79,** 393 (1985).

20. T. J. Hadley, F. W. Klotz, L. H. Miller, *Annu. Rev. Microbiol.* **40,** 451 (1986).

21. L. H. Miller *et al., N. Engl. J. Med.* **295,** 302 (1976),

22. G. Pasvol, J. S. Wainscoat, D. J. Weatherall, *Nature* **297,** 64 (1982).

23. L. H. Miller *et al., J. Exp. Med.* **146,** 277 (1977); M. E. Perkins, *ibid.* **160,** 788 (1984).

24. T. J Hadley *et al., J. Clin. Invest.* **80,** 1190 (1987); M. E. Perkins and A. N. Holt, *Mol. J. Biochem. Parasitol.* **27,** 23 (1988).

25. G. Pasvol and R. J. M. Wilson, *Br. Med. Bull.* **38,** 133 (1982).

26. M. Aikawa *et al., J. Cell Biol.* **77,** 72 (1978).

27. A. Kilekian, *Proc. Natl. Acad. Sci. U.S.A.* **76,** 4650 (1979).

28. P. T. Englund, S. L. Hajduk, J. C. Marini, *Annu. Rev. Biochem.* **51,** 695 (1982); J. E. Donelson and A. G. Rice-Ficht, *Microbiol. Rev.* **49,** 107 (1985); M. R. Turner, *Adv. Parasitol.* **21,** 69 (1982).

29. P. Borat and D. R. Greaves, *Science* **235,** 658 (1987).

30. G. S. Lamont, R. S. Tucker, G. A. M. Cross, *Parasitology* **92,** 355 (1986).

31. R. F Sturrock, in *Bailliere's Clinical Tropical Medicine and Communicable Diseases* (Vol 2, No. 2), A. A. F. Mahmoud, Ed. (Bailliere's Tindall, Philadelphia, 1987), pp. 249–266.

32. M. A. Stirewalt, *Adv. Parasitol.* **12,** 115 (1974); _____, *Exp. Parasitol.* **13,** 395 (1983).

33. D. J. McLaren and D. J. Hockley, *Nature* **269,** 147 (1977).

34. K. D. Murrell *et al., Exp. Parasitol.* **46,** 247 (1978); J. C. Samuelson, J. P. Caulfield, J. R. David, *J. Cell Biol.* **94,** 355, (1982).

35. P. M. Wiest, A. M. Tartakoff, M. Aikawa, A. A. F. Mahmoud, *Proc. Natl. Acad. Sci. U.S.A.* **85,** 3825 (1988).

36. R. M. Anderson and R. May, *Nature* **315,** 493 (1985).

37. C. L. Greenblatt, in *New Developments with Human and Veterinary Vaccines,* A. Mizrahi, Ed. (Alan R. Liss, New York, 1980), pp. 259–285.

38. D. F. Clyde *et al., Am. J. Med. Sci.* **266,** 166 (1973); G. L. Spitalny and R. S. Nussenzweig, *Proc. Helm. Soc. Wash.* **39,** 506 (1972); L. H. Miller *et al., Science* **324,** 1349 (1986).

39. R. S. Nussenzweig and V. Nussenzweig, *Trans. Roy. Soc. London ser. B.* **307,** 117 (1984); W. R. Ballou *et al., Lancet* **2,** 1277 (1987).

40. V. Nussenzweig and R. S. Nussenzweig, *Cell* **42,** 401 (1985); P. Potocnjak *et al., J. Exp. Med.* **151,** 1504 (1980); V. F. de la Cruz, A. A. Lal, T. F. McCutchan, *J. Biol. Chem.* **262,** 11935

(1987).

41. J. B. Dame *et al., Science* **225**, 593 (1984).
42. W. R. Ballou *et al., ibid.* **229**, 996 (1985).
43. D. A. Herrington *et al., Nature* **328**, 257 (1987); H. M. E. Hinger *et al., J. Immunol.* **140**, 626 (1988).
44. M. F. Good, J. A. Berzofsky, L. H. Miller, *Annu. Rev. Immunol.* **6**, 663 (1988).
45. W. R. Weiss *et al., Proc. Natl. Acad. Sci. U.S.A.* **85**, 573 (1988).
46. M. F. Good *et al., ibid.,* p. 1199.
47. A. S. DeGroot *et al., J. Immunol.* **142**, 4000 (1989).
48. S. L. Hoffman *et al., ibid.,* p.1299.
49. V. F. de la Cruz, A. A. Lal, T. A. McCutchan, *J. Biol. Chem.* **262**, 11935 (1987).
50. A. E. Butterworth *et al., Trans. Roy. Soc. Trop. Med. Hyg.* **78**, 108 (1984); J. W. Kazura, H. G. Cicirello, K. Forsyth, *J. Clin. Invest.* **77**, 1985 (1986); R. M. Anderson, *Trans. Roy. Soc. Trop. Med. Hyg.* **80**, 686 (1986).
51. A. A. F. Mahmoud, K. S. Warren, P. A. Peters, *J. Exp. Med.* **142**, 805 (1975); A. A. F. Mahmoud, *J. Infect. Dis.* **145**, 613 (1981); E. Butterworth, D. W. Taylor, M. C. Veith, *Immunol. Rev.* **61**, 5 (1982).
52. J. J. Ellner and A. A. F. Mahmoud, *Rev. Infect. Dis.* **4**, 698 (1982); S. L. James, D. Lanar, E. J. Pearce, *Acta Trop.* **44**, 50 (1987).
53. A. A. F. Mahmoud, *Annals N.Y. Acad. Sci.,* in press; A. Capron *et al., Science* **238**, 1065 (1987); A. E. Butterworth, in *Bailliere's Clinical Tropical Medicine and Communicable Diseases* (Vol. 2, No. 2), A. A. F. Mahmoud, Ed. (Bailliere's Tindall, Philadelphia, 1987), pp. 465–483.
54. D. M. Zodda and S. M. Phillips, *J. Immunol.* **129**, 2326 (1982); D. A. Harn, M. Mitsuyama, J. R. David, *J. Exp. Med.* **159**, 1371 (1984); M. A. Smith, J. A. Clegg, D. Snary, E. J. Trejdosiewicz, *Parasitology* **84**, 83 (1982); J. M. Grzych, M. Capron, H. Bazin, A. Capron, *J. Immunol.* **129**, 2739 (1982).
55. C. H. King *et al., J. Immunol.* **139**, 4218 (1987); R. Blanton, M. Aikawa, A. A. F. Mahmoud, in preparation.
56. C. Dissous, J. M. Balloul, R. Pierce, A. Capron, in *Bailliere's Clinical Tropical Medicine and Communicable Diseases* (Vol. 2, No. 2), A. A. F. Mahmoud, Ed. (Bailliere's Tindall, Philadelphia, 1987), pp. 267–278.

57. C. Anriault *et al., J. Immunol.* **141**, 1687 (1988).
58. T. P. Flanigan, C. H. King, R. R. Lett, J. Nanduri, A. A. F. Mahmoud, *J. Clin. Invest.* **83**, 1010 (1989); S. L. James, *J. Immunol.* **134**, 1956 (1985); E. J. Pearce *et al., Proc. Natl. Acad. Sci. U.S.A.* **85**, 5678 (1988).
59. T. F. Kresina and G. R. Olds, *J. Clin. Invest.* **83**, 912 (1989); J. M. Grzych *et al., Nature* **316**, 6023 (1986); F. Velge-Roussel *et al., J. Immunol.* **142**, 2527 (1989).
60. J. W. Kazura, H. G. Cicirello, J. W. McCall, *ibid.* **136**, 1422 (1986).
61. T. W. Nilsen *et al., Proc. Natl. Acad. Sci. U.S.A.* **85**, 3604 (1988); K. G. Perrine, J. A. Denker, T. W. Nilsen, *Mol. Biochem. Parasitol.* **30**, 97 (1988).
62. L. H. T. van der Ploeg *et al., Nucleic Acids Res.* **10**, 5905 (1982).
63. E. Pays *et al., Cell* **34**, 371 (1983).
64. P. W. Laird, *Trends Genet.* **5**, 204 (1989).
65. J. C. Boothroyd and G. A. M. Cross, *Gene* **20**, 281 (1982).
66. W. J. Murphy, K. P. Watkins, N. Agabian, *Cell* **47**, 517 (1986); R. E. Sutton and I. Boothroyd, *ibid.,* p. 527.
67. M. Milhausen *et al., Cell* **38**, 721 (1984); P. Borst, *Annu. Rev. Biochem.* **55**, 701 (1986).
68. M. Krause and D. Hirsh, *Cell* **69**, 753 (1987); S. L. Bektesh and O. I. Hirsh, *Nucleic Acids Res.* **16**, 5692 (1988).
69. A. M. Takacs *et al., Proc. Natl. Acad. Sci. U.S.A.,* **85**, 7932 (1988); R. M. Ransohoff, J. A. Denker, A. M. Takacs, *Nucleic Acid Res.* **17**, 3773 (1989); T. W. Nilsen *et al., Proc. Natl. Acad. Sci. U.S.A.,* in press; P. A. Maroney, G. J. Hannon, T. W. Nilsen, *Mol. Biochem. Parasitol.* **35**, 277 (1989).
70. J. P. Bruzik *et al., Nature* **335**, 559 (1988); T. W. Nilsen *et al., Mol. Cell Biol.* **9**, 3543 (1989).
71. C. H. King *et al., Am. J. Trop. Med. Hyg.* **39**, 361 (1988).
72. M. Zuker and P. Stigler, *Nucleic Acids Res.* **9**, 133 (1981).
73. I am grateful to T. W. Nilsen for his critical comments and valuable discussions and to B. Jasny for her invaluable help in constructing the focus of this article. The editorial assistance of L. Ryan and K. Mong is gratefully acknowledged.

Drosophila melanogaster as an Experimental Organism

Gerald M. Rubin

Many problems in eukaryotic cell biology can be most easily studied in unicellular organisms, such as yeast, or in cell cultures derived from multicellular organisms. Other problems, however, currently can be studied meaningfully only in intact animals. This may be because we do not know how to mimic crucial aspects of the organismal environment in vitro, because cell-cell interactions play an important role, or because the process under study involves a behavior that is not currently understood in terms of the properties of individual cells. Examples include pattern formation in the embryo and the development and function of organ systems, such as the nervous system. *Drosophila's* intermediate level of complexity, in combination with its sophisticated genetics, makes it particularly well suited for the study of basic problems in metazoan biology.

Drosophila melanogaster, the Most Genetically Manipulable Metazoan

The fruit fly has a small size and a short life cycle—features that make feasible the raising of large numbers of individuals for the many generations required for genetic analysis (Fig. 1). It also has a small genome, 1/20 the size of a typical mammalian genome, which facilitates molecular genetic analysis (*1*). Other organisms share these features, however, and they are not the primary reasons for *Drosophila's* place in modern genetics. For this, credit must go to the hundreds of skillful and creative geneticists who have developed the wide range of tools available for use with this organism. Genetic studies with *Drosophila* began in T. H. Morgan's laboratory at Columbia University in 1909 (*2*). A year later, Morgan described the isolation of the first *Drosophila* mutant, *white eye*, and the observation that its inheritance was sex-linked (*3*). In 1915, Morgan, Sturtevant, Muller, and Bridges published their classic book *Mechanisms of Mendelian Inheritance* (*4*), establishing the relation between genes and chromosomes. Over the past 80 years, much of what we know about recombination, mutation, chromosome rearrangements, and other genetic phenomena has been discovered through the use of *Drosophila* as an experimental organism. It is important to remember that most of the techniques now used in genetic work with *Drosophila* were developed in the course of these efforts.

G.M. Rubin is a member of the faculty, Howard Hughes Medical Institute and Department of Biochemistry, University of California, Berkeley, CA 94720. This chapter is reprinted from *Science* **240**, 1453 (1988).

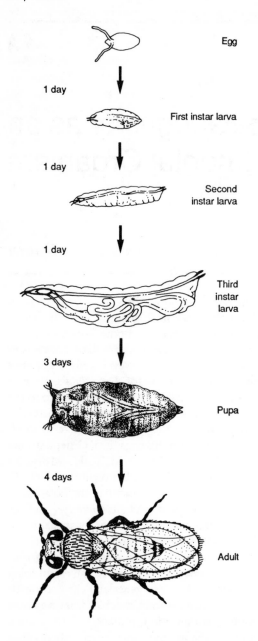

Egg

1 day

First instar larva

1 day

Second instar larva

1 day

Third instar larva

3 days

Pupa

4 days

Adult

Fig. 1. Diagram of the *Drosophila* life cycle. The approximate time span of each developmental stage at 25 °C is indicated.

characteristic for which an assay can be devised, and a wide variety of such screens have been, and are being, carried out. Most common are screens for mutations affecting either viability or aspects of adult or embryonic morphology (*5*). Morphological analysis has advanced to the level of individual cells by the use of antibodies that allow the scoring of specific cells or cell types (*6*). In addition, screens for mutations affecting oogenesis (*7*), sex determination (*8*), sensory perception (*9*), and learning (*10*) have been performed. Such genetic screens can provide an unbiased way to identify genes that function in forming a particular structure or in carrying out a particular process. No prior information about the biochemical nature of the structure or process is required. Thus, even processes that might be extremely complex, such as pattern formation in the developing embryo (*11*) or circadian rhythms (*12*), can be genetically dissected into identifiable components that can be individually approached. Mutation of many important genes is lethal to the organism. However, it is easy to maintain stocks of recessive mutations whose homozygosity would result in either death or sterility (*13–15*).

Mutations in over 3000 genes have been described and analyzed in *Drosophila* (*16*), and this number is growing rapidly. Sufficient numbers of individual flies can be examined to allow isolation of mutations in nearly all genes that contribute to the phenotype of interest. However, the number of flies that can be handled conveniently is usually not sufficient to look for phenotypes that cannot be generated by inactivation of a gene, but require a much rarer event such as a particular nucleotide change or amino acid change in the gene product, as is the case in certain types of second-site suppressor mutations (*17*).

Use of Mutations to Dissect Complex Processes into Discrete Steps

In *Drosophila* it is possible to screen systematically for all genes that can mutate to produce a given phenotype. This phenotype can be any

Use of Genetic Tools to Analyze Gene Function, Cell Lineage, and Cell Interaction

Once mutations that affect the process of interest have been identified, they can be studied by standard genetic methods. Genetic complementation tests can be used to estimate the number of different genes involved in the process (11). The phenotypes of individuals carrying mutations in more than one of the genes can be studied to infer hierarchies of gene function. Measuring the rates of meiotic recombination between a mutation and other genes can be used to map the relative chromosomal position of the mutated gene. In addition to these general methods, a set of extremely useful genetic tools, many unique to *Drosophila*, can be applied to the analysis of mutations. These same tools provide noninvasive approaches for studying many aspects of the development and function of normal individuals that are not easily addressed by classical manipulative techniques.

These approaches rely largely on the production of mosaics, individuals that bear clones of genetically altered somatic cells. These clones can be produced either by chromosome loss (18) or by induced mitotic recombination (19) and can be recognized by the presence of cell-autonomous genetic markers (20). A large number of cell-autonomous genetic markers have been identified, allowing analysis of nearly all parts of the animal. By using mosaic animals, it has been possible to construct genetic "fate maps," which plot the relative positions in the early embryo of precursor cells for structures that arise later in development (18). Use of this method has made it possible to address questions, such as "What parts of an animal have to be male for it to behave as a male in mating?" (21). In addition, analysis of mosaic individuals can be used to determine the cell-autonomy of gene action (19), founder cell numbers (22), growth patterns (23), and restrictions of cell fate (24).

A few general examples may illustrate some of these applications. Suppose a gene is known to be essential for the development of a particular cell type, "X." This gene might be required in cell type X itself, in neighboring cells that send an important inductive signal, or in distant cells that provide a required hormone or metabolite. These alternatives can be distinguished by asking what cells, in a mosaic animal, must have a wild-type copy of the gene

in order to permit normal development of cell type X. Mosaics are also used in the study of genes that, when mutated, result in the death of the animal. Such a gene could be required in every cell in the body or in just a particular cell type or organ. Clones of marked cells that are homozygous for the mutant gene can be induced, by mitotic recombination or chromosome loss, in a background of heterozygous cells. If the gene is a "housekeeping" gene required by each cell for its own viability, no marked cells will survive and no clones will be observed. Conversely, if marked clones are observed, then expression of the gene in the cells of the clone cannot be essential for the viability of either these cells or the organism. By examining a large number of clones, one can deduce which cells require the gene for the whole organisms to survive.

Mitotic recombination generally occurs at very low rates but can be increased greatly by x-irradiation. Thus the investigator can control the time at which marked clones are induced. By observing the marked cells later in development, one can learn much about patterns of cell growth and restrictions of cell fate. By generating such clones in flies carrying particular mutations, one can study the role of other genes in controlling these parameters. For example, cell clones induced after blastoderm formation in the posterior region of the wing never contribute to the anterior region, indicating that a lineage restriction has been established (25). However, when clones are similarly induced in animals carrying a mutation in the *engrailed* gene, no such lineage restriction is observed, indicating that *engrailed* gene function is required to establish or maintain this posterior-anterior lineage restriction (25).

High Resolution Cytogenetic Analysis Made Possible by Polytene Chromosomes

Cytogenetic analysis in *Drosophila* is greatly facilitated by the giant polytene chromosomes present in the salivary glands of third instar larvae (26). These interphase chromosomes contain more than 1000 strands of chromatin precisely aligned to produce a characteristic and highly reproducible banding pattern (Fig. 2). Polytene chromosomes are useful in correlating cytological and genetic maps. The positions of chromosomal inversions, deletions, and other rearrangements in particular mutant strains can be determined by analysis of polytene chromosomes by use of light microscopy. Moreover, genetic and molecular maps can be aligned by the technique of in situ hybridization (*27*) to place cloned DNA segments on the polytene chromosome map. Some of the bands that can be seen in the light microscope correspond to less than 10 kb of DNA, with the average of the approximately 5000 visible bands being about 25 kb. Thus, the resolution of cytogenetic analysis in *Drosophila* is orders of magnitude greater than in other animals and there is no large gap between what can be seen in the light microscope and what can be cloned in a single recombinant DNA molecule.

Many of the roles traditionally played by cytological maps of polytene chromosomes may be fulfilled instead by ordered molecular maps of cloned DNA segments now being constructed for many organisms (*28*), including *Drosophila* (*29*). However, such molecular maps are made from a particular strain and cannot be used readily when the analysis of different strains is required, as in the mapping of chromosome rearrangements or the positions of transposable elements in different populations.

From Genetic Function to DNA Clone

Many genetic loci in *Drosophila* have been identified by their phenotypic effects on development, physiology, or behavior. For a large fraction of these, the gene product is unknown and methods that depend only on knowledge of the phenotype must be used to molecularly clone the gene. Two approaches have been taken: one relies on cytogenetic and molecular mapping, the other on transposon mutagenesis. In the first approach, classical genetic mapping is used initially to determine the position of the gene relative to known genes and to the breakpoints of chromosomal deletions or other rearrangements. A large number of such rearrangements exist (*16*), and it is generally feasible to produce more for a chromosomal location of interest. Then the region of the genome containing the desired gene is isolated by chromosome walking (*30*) or chromosome microdissection (*31*). In chromosome walking overlapping segments of DNA are isolated, starting from the nearest known cloned DNA

Fig. 2. (A) The distal third of the X chromosome as seen in the polytene chromosomes of third instar salivary glands. (B) A metaphase spread made from cells of the brain of a third instar male. The region of the X chromosome corresponding to that shown in the top panel is indicated by the arrows. This region contains approximately 8000 kb of DNA.

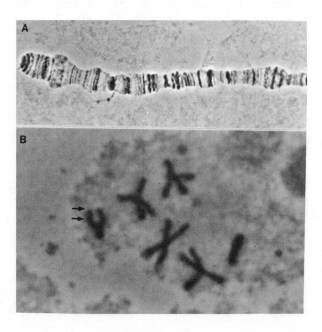

segment, by sequentially screening a genomic library, so that one extends the cloned region in an ordered way toward the gene of interest. In cloning by microdissection, the chromosomal region containing the desired gene is physically dissected from the polytene chromosomes of the salivary glands and used as a source of DNA for constructing a library in a bacteriophage λ vector.

Both of these methods yield DNA sequences from the region containing the gene of interest. The problem is then reduced to identifying the DNA sequences comprising the gene within the larger cloned region. In some cases, the region defined by the positions of chromosome rearrangements or restriction site polymorphisms may encompass only the gene of interest. In other cases, looking for DNA sequences that have a transcript accumulation pattern that matches the expected pattern of expression of the desired gene has permitted the localization of a gene in a chromosome walk. The most rigorous test, however, is to find whether a cloned DNA segment can complement the mutant defect when introduced into the fly genome by DNA transformation methods.

In the second approach, transposon tagging, transposable elements are used to "tag" a gene biochemically to aid in its cloning (*32*). Transposable elements are segments of DNA that move as discrete units from place to place in the genome (*33*). The insertion of a transposable element into a new genomic site often inactivates a gene located at that site. Indeed, a large fraction of spontaneous mutations in *Drosophila* appears to be due to transposable element insertions. P transposable elements have proven useful to molecular geneticists studying *Drosophila* as tools for transposon tagging because their mobility is under genetic control and can be manipulated experimentally (*34*).

A typical protocol for transposon tagging using P elements would be as follows (*35*). Males from a strain containing P elements (a P strain) are crossed to females from a strain lacking P elements (an M strain). In such a cross P element transpositions are induced in the germ line of the progeny. The progeny are bred and their offspring are screened or selected for new mutations in the gene of interest. In most cases, the new mutations will have resulted from P element insertions and the gene will be "tagged" with a P element. DNA corresponding to the mutant allele can then be retrieved from a genomic DNA library of the mutant strain by virtue of its sequence homology to the P element (*36*). Alternatively, strains containing one or only a few P elements can be constructed by P element-mediated transformation of M strains and used instead of naturally occurring P strains as the source of P elements in the initial cross (*37*). This approach has the advantage that P elements can be used that have been modified to contain genetic markers or other features to permit their easy identification and cloning.

The major advantage of transposon tagging as a cloning method is that only the approximate cytogenetic location of the desired gene need be known. Thus, it can be readily applied to most of the *Drosophila* mutations already identified. Moreover, screens can be initiated for new mutations that have a particular phenotype with P element insertion used as the mutagen, thereby greatly facilitating the subsequent cloning of those genes (*37*). A major disadvantage in all transposon tagging methods, however, is that P element insertion is not random but shows target site preferences. Thus, not all genomic sites are mutated with equal frequency and perhaps only one-half of all genes will be mutated at rates high enough to make their isolation by this method practical (*38*).

The collections of fly strains, each carrying an insertion of a genetically marked transposable element, that are generated by transposon mutagenesis and by gene transfer experiments are proving very useful for a variety of purposes. For example, with only 1000 such transposable element insertions any site in the genome would be expected to be within 100 kb of an element. Any gene of interest could then be mapped by recombination relative to the markers carried by the transposable element, localized to between two such elements, and isolated by a chromosomal walk between these elements.

From DNA Clone to Genetic Function

Often a segment of DNA will be isolated first, not on the basis of its corresponding to a particular genetic locus, but rather because it displays a desired biochemical property: the DNA might hybridize to RNA present in one tissue or in one developmental stage, it may be similar to another cloned gene from the same or a different organism, or it may encode an antigen that reacts with an antibody of interest. Each of these properties has been exploited with great success in *Drosophila* and other organisms to isolate genes that have a pattern of expression or DNA sequence that indicates they are performing a function of interest. For example, many *Drosophila* genes have been cloned because they contain a homeo box domain (*39*), or are homologous to a particular mammalian oncogene (*40*) or receptor (*41*). Other genes have been selected because they are expressed in the eye (*42*) or at the blastoderm stage of development (*43*), or because they encode an antigen that reacts with a monoclonal antibody that stains the surfaces of a subset of neurons (*44*).

The demonstration that a gene's product is present at the right time and place or has similarity to genes of known or suggestive biochemical function does not prove, however, that it plays a role in the process under study. To get beyond such "guilt by association" arguments one needs to know how the process is affected when the gene product is removed. Mutation of the gene offers a general method for removing its product. Once a gene has been cloned, and its position on the genetic map has been determined by in situ hybridization to polytene chromosomes, a variety of genetic methods (*15*) can be used to isolate mutations that map in the general area of the gene. Then, assuming that a mutation of the gene of interest produces a detectable phenotype, such a mutation can be identified from among all the mutations in the area by its ability to be complemented by the cloned gene (*45*), or to produce an alteration either in the DNA sequence of the gene (*46*) or in the abundance of the gene product (*47*). If mutation of the gene is lethal to the whole organism, it is nonetheless possible to ask what role, if any, it plays in individual cells by making clones of cells that are homozygous for the mutant gene in a background of heterozygous cells.

Introduction of Normal and Altered Genes Back into the Genome

The experimental utility of gene transfer methods is evident at two levels. The most straightforward and routine application is to determine whether a cloned segment of DNA encodes a product that is missing or altered in a particular mutant (that is, genetic complementation). A more sophisticated application of gene transfer involves the systematic manipulation of a gene and the determination of the biological consequences of the changes introduced. In *Drosophila*, stable gene transfer into the germ line can be achieved by using P transposable elements as vectors (*48*).

The strategy for using P elements as vectors for gene transfer is based on mimicking the events that take place during a cross between P and M strains. In such a cross, P elements on paternally contributed chromosomes enter the M strain egg and are induced to transpose at high rates. An analogous situation occurs if DNA containing an autonomous 3-kb P element is microinjected into an M cytotype embryo shortly after fertilization. This element can transpose from the injected DNA to the germ line chromosomes of the host embryo in a reaction that is catalyzed by a "transposase" protein encoded by the P element. Smaller P elements that lack the DNA sequences encoding this protein can also transpose if co-injected with the 3-kb element. Other DNA segments of interest can be transferred into the germline if they are inserted within such internally deleted P elements and then co-injected with the 3-kb element. Figure 3 shows a typical protocol for such a gene transfer experiment.

Because genes transferred by means of P element vectors are incorporated into the germ lines of their hosts, their function can be assayed in all cell types and developmental stages

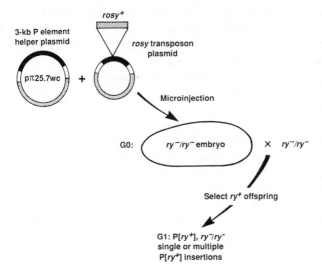

3-kb P element
helper plasmid

pπ25.7wc

+

rosy⁺

rosy transposon
plasmid

Microinjection

G0: ry⁻/ry⁻ embryo × ry⁻/ry⁻

Select ry⁺ offspring

G1: P[ry⁺], ry⁻/ry⁻
single or multiple
P[ry⁺] insertions

Fig. 3. Typical protocol for a gene transfer experiment. DNA of the plasmid pπ25.7wc, which carried a 3-kb P element, and DNA of the *rosy* transposon plasmid, which carried a wild-type *rosy* gene inserted into a small P element (to generate a *rosy* transposon), are co-injected into an embryo that is homozygous for a mutation in the *rosy* gene. The inclusion of the plasmid pπ25.7wc is necessary, since the *rosy* transposon can only transpose with the aid of protein factors encoded by the 3-kb "helper" P element. Approximately 10% of the injected embryos will survive to become fertile adults (G0 adults) and, in about one-third of these, transposition of the *rosy* transposon from the injected plasmid DNA to germline chromosomes will have occurred. Since transposition only occurs in the germline, the expression of the introduced *rosy* gene is not evident in the somatic tissues of the G0 adults, but if they are mated to ry^- individuals the expression of the *rosy* gene can be assayed in the next (G1) generation. In those offspring that show expression of the *rosy* gene (ry^+ offspring), single or multiple copies of the *rosy* transposon (P[ry^+]) are found inserted in the chromosomes and are stably inherited in future generations. The *rosy* gene (the structural gene for the enzyme xanthine dehydrogenase) affects eye color and thus is easily identified. Genes that have functions difficult to assay can be transferred by constructing a transposon that contains both the gene of interest and the *rosy* gene. Successful transfer of the entire transposon can be detected by identifying *rosy* gene function and then the ry^+ progeny can be assayed for the function of the second gene.

in subsequent generations. Although the transferred genes are not inserted at their normal chromosomal locations, they appear to be regulated properly and, in the majority of cases, exhibit correct tissue and temporal specificity of expression (*49*).

A cloned gene can be altered by mutagenesis in vitro and then put back into the genome where its function can be assessed. Specific changes in the amino acid sequence of the encoded protein can be engineered and the effects of these changes determined in the organisms's normal environment, where behavioral or developmental phenotypes caused by the altered gene can be monitored. In addition, mutations outside the protein-coding region of the gene can be made and assayed in vivo to determine which cis-acting DNA sequences control the tissue specificity and the developmental timing of gene expression.

More complex mutations can also be constructed in vitro, including mutations that cause a gene to be expressed in a cell type where it is not normally active (*50*). For example, suppose a gene encoding a cell surface protein specific to cell type A has been isolated. The protein coding portion of this gene could be

joined to the control region of a second gene whose expression is limited to another cell type, B, and the fusion gene introduced into the genome. Flies carrying the fusion gene should now express the protein on the surface of cell type B, directed by the fusion gene, as well as on the surface of cell type A (as a result of the unaltered copy of the gene present in the fly's genome). If this cell surface protein is involved in cell-cell recognition, specific developmental abnormalities might be expected.

One problem with this type of approach is that the incorrect expression of many important genes will be lethal to the organism. Although genetic tools exist for handling recessive lethal mutations in *Drosophila*, mutations like the one described earlier are expected to be dominant. The classical way to overcome this difficulty is to make expression of the mutant gene conditional. For example, fusions can be made to transcriptional promoters that are inducible either by environmental factors, such as heat shock (*51*), or by an exogenous trans-acting factor that can be supplied only by mating to a specially engineered strain (*52*). An alternative solution would be to make fusions to promoters that are specific for nonessential cell types. For example, flies without eyes are viable under laboratory conditions and therefore fusions made to a promoter for a photoreceptor-specific gene (for example, rhodopsin) would be expected to be viable even if the fusion product resulted in death of the photoreceptor cells or otherwise disrupted eye development.

Other potential applications of gene transfer remain to be developed. For example, many genes have been cloned by virtue of their differential expression in certain cell types or developmental stages. However, these genes frequently do not correspond to known mutations, and thus the phenotype of an individual lacking the gene function cannot be discerned. Although with effort mutations in such a gene can be induced by classical means, a convenient and reliable method that utilizes the cloned copy of the gene to inactivate the corresponding chromosomal copy would be particularly useful.

Methods based on homologous recombination, such as those used in yeast (53), are presently not available in *Drosophila*. It is possible that methods based on the production of an antisense RNA (54) can be adapted to *Drosophila* (55), although it is not clear that this method can completely inactivate a gene.

Embryological and Biochemical Methods Supplement Genetic Approaches

In addition to the genetic methods described above, it is possible to apply most of the techniques of experimental embryology to *Drosophila*. The embryo develops outside the mother and can be easily observed with a variety of techniques with varying degrees of resolution and invasiveness (56). Many gross morphological changes can be seen in living embryos by use of light microscopy, while at the other extreme fixed embryos can be readily stained, allowing one to see cellular details at high resolution. Methods for nuclear and cell transplantation (57), microsurgery (58), laser ablation (59), cell and organ culture (60), biochemical cell marking (61), immunocytology (62), and in

situ hybridization (63) have been developed and widely applied.

Large quantities of *Drosophila* can be readily obtained at each stage of its life cycle. This facilitates biochemical analysis and has made it possible to obtain a wide variety of purified proteins. However, the difficulty in obtaining sufficient quantities of specific tissues or cell types for biochemical analysis is often a limiting factor. Extracts of embryos or cell cultures have been used for in vitro transcription (64) and RNA splicing studies (65). The ability to combine such biochemical studies with genetic analysis should be particularly useful in understanding how gene expression is controlled during development.

Use of *Drosophila* to Study the Development and Function of the Nervous Stystem

A number of scientists working with *Drosophila* are now using the embryological and genetic methods outlined here to investigate how the nervous system develops. Electrophysiological methods, including intracellular recording (66) and patch clamping (67), have been used in a variety of preparations. Given the small size of *Drosophila* neurons, however, it has been possible to map neuronal circuits in only a few cases (68). A large number of mutations affecting the electrical properties of nerve cells have been isolated (69), including those in the structural genes for channels (70). Much has been learned from analysis of the electrophysiological effects produced by these mutations. Even lethal mutations have been amenable to study in nerve cells cultured from the homozygous embryos (71). Much information about the relationship of structure to function has been gained by studying genes encoding ion channels and receptors that have been expressed in heterologous systems, such as *Xenopus* oocytes (72). However, questions concerning the regulated expression, localization, and modification of such important molecules can only be fully answered if they are studied in their natural environment. *Drosophila* can be expected to play an important role in these studies. In addition,

Table 1. Proteins that are highly similar in *Drosophila* and vertebrates.

Class	Protein	Reference
Cytoskeleton	Actin	*(76)*
	Myosin	*(77)*
	Tropomyosin	*(78)*
	Tubulin	*(79)*
	Spectrin	*(80)*
Neuronal function	Acetylcholinesterase	*(81)*
	Choline acetyl transferase	*(82)*
	Acetylcholine receptor	*(83)*
	Sodium channel	*(84)*
Transcription factors	Homeo box proteins	*(85)*
	Zinc finger proteins	*(86)*
Second messenger systems	Calmodulin	*(87)*
	Protein kinase C	*(88)*
	Protein kinase A	*(89)*
Oncogenes	Src	*(90)*
	Abl	*(91)*
	Myb	*(92)*
	Rel (dorsal)	*(93)*
	Int-1 (wingless)	*(94)*
Growth factors and receptor	TGF-β	*(95)*
	Insulin receptor	*(96)*
	EGF receptor	*(97)*
Cell and substrate adhesion	Fibronectin receptor-like	*(98)*
	Laminin	*(99)*
	NCAM*-like (amalgam)	*(100)*

*Neuronal cell and adhesion molecule.

sensory transduction *(9)*, neuronal pathfinding *(73)*, various behaviors *(74)*, biological rhythms *(12)*, as well as learning and memory *(10)* are being analyzed in *Drosophila* by combined genetic and molecular approaches.

Concluding Remarks

When the techniques described here are used on an organism with the intermediate level of complexity of *Drosophila*, many challenging problems are accessible to experimental analysis. Will the answers obtained be relevant to mammalian organisms such as ourselves? The high degree of homology of a wide range of fundamental molecules suggests that many, perhaps nearly all, of the basic mechanisms used for the development and function of multicellular organisms were established before the evolutionary divergence of the progenitors of flies and humans some 500 million years ago. Representative examples of the high degree of

primary sequence similarity observed between many *Drosophila* and vertebrate proteins are given in Table 1. This homology extends to functional properties. For example, the *Drosophila* insulin receptor binds to and is activated by bovine insulin *(75)*. Mammalian organisms are more complex than *Drosophila*. However, there is no compelling reason not to believe, and much circumstantial evidence to support, the contention that most of this complexity is achieved by reiteration and adaptation of common, evolutionarily ancient processes.

Although other experimental organisms may equal or exceed *Drosophila* in the facility of a particular experimental area, only flies contain in one system the potential for the application of the tools of classical genetics, cytogenetics, molecular genetics, biochemistry, electrophysiology, cell cultures, and other cell biological techniques. *Drosophila*'s unique ability to support such a multidisciplinary approach, combined with its intermediate level of complexity, ensures that this animal will continue to play an important role in biological research for many years to come.

References and Notes

1. The genome of *Drosophila melanogaster* is 165,000 kb. This is approximately 50× the *Escherichia coli* genome or about 1/20 the human genome. For a review of *Drosophila* genome structure see A. C. Spradling and G. M. Rubin, *Annu. Rev. Genet.* 15, 219 (1981).

2. For historical reviews of early genetic work on *Drosophila* see C. P. Oliver, in *The Genetics and Biology of* Drosophila, M. Ashburner and E. Novitski, Eds. (Academic Press, London, 1976), vol. 1A, p. 1; A. H. Sturtevant, *A History of Genetics* (Harper & Row, New York, 1965).

3. T. H. Morgan, *Science* 32, 120 (1910).

4. _____, A. H. Sturtevant, H. J. Muller, C. B. Bridges, *The Mechanism of Mendelian Heredity* (Holt, New York, 1915).

5. The first genetic screens were carried out by the "fly group" at Columbia University [see T. H. Morgan, *Proc. Am. Philos. Soc.* 54, 143 (1915)]. See (*11*) for an example of a recent genetic screen.

6. See R. Bodmer *et al.*, *Cell* 51, 293 (1987).

7. A. Bakken, *Dev. Biol.* 33, 100 (1973); M. Gans, C. Audit, M. Masson, *Genetics* 81, 863 (1975); R. B. Rice and A. Garen, *Dev. Biol.* 43, 277 (1975); J. D. Mohler, *Genetics* 85, 259 (1977); K. V. Anderson and C. Nüsslein-Volhard, in *Pattern Formation: A Primer in Developmental Biology*, G. M. Malacinski and S. Bryant, Eds. (Macmillan, New York, 1984), p. 269.

8. Reviewed in B. S. Baker and J. M. Belote, *Annu. Rev. Genet.* 17, 345 (1983); B. S. Baker, R. N. Nagoshi, K. C. Burtis, *Bioessays* 6, 66 (1987).

9. W. L. Pak, in *Neurogenetics: Genetic Approaches to the Nervous System*, X. O. Breakefield, Ed. (Elsevier, New York, 1979), p. 67; O. Siddiqi, *Trends Genet.* 3, 137 (1987); M. Heisenberg and R. Wolf, *Vision in* Drosophila (Springer-Verlag, Berlin, 1984).

10. T. Tully, *Trends Neurosci.* 10, 330 (1987); Y. Dudai, *Annu. Rev. Neurosci.* 11, 537 (1988).

11. C. Nüsslein-Volhard and E. Wieschaus, *Nature* 287, 795 (1980); _____, H. Kluding, *Wilhelm Roux's Arch. Dev. Biol.* 193, 267 (1984); G. Jurgens, E. Wieschaus, C. Nüsslein-Volhard, H. Kluding, *ibid.*, p. 283.

12. For review see J. C. Hall and M. Rosbash, *Annu. Rev. Neurosci.* 11, 373 (1988).

13. The number of different stocks that can be conveniently maintained by any one individual is limited to several hundred, however, since it is not possible to store genetic stocks in a frozen state and all stocks must be maintained by serial passage. For review of some basic genetic methods see T. Grigliatti, in (*14*), p. 39; A. S. Wilkins, *Genetic Analysis of Animal Development* (Wiley, New York, 1986).

14. D. B. Roberts, Ed., Drosophila: *A Practical Approach* (IRL Press, Oxford, 1986).

15. M. Ashburner, Drosophila: *Laboratory Manual* (Cold Spring Harbor Laboratory, Cold Spring Harbor, NY, in press).

16. D. L. Lindsley and E. H. Grell, *Carnegie Inst. Washington Publ.* 627 (1968).

17. However, a few allele-specific suppressors have been isolated in *Drosophila*; see W. A. Harris and W. S. Stark, *J. Gen. Physiol.* 69, 261 (1977).

18. A. H. Sturtevant, *Z. Wiss. Zool.* 135, 323 (1929); A. Garcia-Bellido and J. R. Merriam, *J. Exp. Zool.* 170, 61 (1969); Y. Hotta and S. Benzer, *Nature* 240, 527 (1972).

19. C. Stern, *Genetic Mosaics and Other Essays* (Harvard Univ. Press, Cambridge, MA, 1968); H. J. Becker, in *The Genetics and Biology of* Drosophila, M. Ashburner and E. Novitski, Eds. (Academic Press, London, 1976), vol. 1C, p. 1020.

20. P. Lawrence, P. Johnston, G. Morata, in (*14*), p. 229.

21. Y. Hotta and S. Benzer, *Proc. Natl. Acad. Sci. U.S.A.* 73, 4154 (1976).

22. J. R. Merriam, in *Genetic Mosaics and Cell Differentiation*, W. J. Gehring, Ed. (Springer-Verlag, Berlin, 1978), p. 71; E. Wieschaus and W. J. Gehring, *Dev. Biol.* 50, 249 (1976).

23. J. H. Postlethwait, in *The Genetics and Biology of* Drosophila, M. Ashburner and T. R. F. Wright, Eds. (Academic Press, London, 1978), vol. 2C, p. 359.

24. P. J. Bryant and H. A. Schneiderman, *Dev. Biol.* 20, 263 (1969); A. Garcia-Bellido, P. Ripoll, G. Morata, *Nature New Biol.* 245, 251 (1973).

25. A. Garcia-Bellido and P. Santamaria, *Genetics* 72, 87 (1972); G. Morata and P. A. Lawrence, *Nature* 255, 614 (1975).

26. T. S. Painter, *Science* 78, 585 (1933); *Genetics* 19, 175 (1934); C. B. Bridges, *J. Hered.* 26, 60 (1935).

27. M. L. Pardue, in (*14*), p. 111.

28. A. Coulson, J. Sulston, S. Brenner, J. Karn, *Proc. Natl. Acad. Sci. U.S.A.* **83**, 7821 (1986).

29. A complete physical map of the *Drosophila* genome is being assembled in Europe, with funding from the Schlumberger Foundation "Les Treilles" and the European Economic Community in a consortium of European laboratories, including the Institute of Molecular Biology and Biotechnology in Crete (F. C. Kafatos, C. Louis, and C. Savakis), the Department of Genetics in Cambridge (M. Ashburner), and the Friedrich Miescher Laboratorium in Tübingen (H. Jäckle). The work is modeled on the fingerprinting method pioneered by J. Sulston for the nematode *Caenorhabditis elegans* (*28*), with important modifications that take into account the unique advantages of *Drosophila*. Researchers use minilibraries from microdissected polytene chromosomal segments (each corresponding to a numbered division—~1% of the euchromatic genome) as probes to select (rather than randomly pick) cosmids from a master library, and an ordered array of cosmid clones is being assembled sequentially.

30. W. Bender, P. Spierer, D. S. Hogness, *J. Mol. Biol.* **168**, 17 (1983).

31. F. Scalenghe, E. Tuvco, J. E. Edstroem, V. Pirrotta, M. Melli, *Chromosoma* **82**, 205 (1981).

32. P. M. Bingham, R. Levis, G. M. Rubin, *Cell* **25**, 693 (1981).

33. Transposable elements have been found in bacteria, fungi, plants, and animals. The frequency at which these elements move depends on a variety of poorly understood factors including element structure, genetic background, and environmental influences. More than 20 different transposable element families have been identified in *Drosophila melanogaster*, and these fall into four structural classes [G. M. Rubin, in *Mobile Genetic Elements*, J. A. Shapiro, Ed. (Academic Press, Orlando, FL, 1983), p. 329].

34. For a review on P elements see W. R. Engels, *Annu. Rev. Genet.* **17**, 315 (1983).

35. For a more detailed discussion of transposon tagging using P elements see M. G. Kidwell, in (*14*), p. 59.

36. Since P strains contain about 50 P elements in their genomes, many extraneous P elements will be present in addition to the one responsible for the mutation. Thus, only 1 of 50 recombinant phages that show P element homology will contain the gene of interest. Only that phage will contain DNA sequences that will label the cytogenetic location of the mutant when hybridized in situ to polytene chromosomes (*27*). Two tactics can be used to avoid having to isolate and screen large numbers of phages by hybridization in situ. First, the mutant strain can be back-crossed for several generations to M strain flies to remove most of the unwanted P elements. Second, the chromosomal region containing the mutation can be microdissected and a DNA library can be made from just this small fraction of the genome (*31*).

37. L. Cooley, R. Kelley, A. Spradling, *Science* **239**, 1121 (1988).

38. M. J. Simmons and J. K. Lim, *Proc. Natl. Acad. Sci. U.S.A.* **77**, 6042 (1980).

39. Reviewed in W. J. Gehring, *Science* **236** 1245 (1987).

40. Reviewed in B.-Z. Shilo, *Trends Genet.* **3**, 69 (1987).

41. E. Livneh *et al.*, *Cell* **40**, 599 (1985); also see (*96*).

42. C. Montell, K. Jones, E. Hafen, G. Rubin, *Science* **230**, 1040 (1985).

43. P. D. Boyer, P. A. Mahoney, J. A. Lengyel, *Nucleic Acids Res.* **15**, 2309 (1987).

44. S. L. Zipursky, T. R. Venkatesh, S. Benzer, *Proc. Natl. Acad. Sci. U.S.A.* **82**, 1855 (1985).

45. It has been possible to detect partial phenotypic rescue for genes affecting embryonic development by directly injecting the cloned gene [A. Preiss, U. B. Rosenberg, A. Kienlin, E. Seifert, H. Jäckle, *Nature* **313**, 27 (1985)] or an RNA transcript of the gene [C. Hashimoto, K. L. Hudson, K. V. Anderson *Cell* **52**, 269 (1988)] into early embryos.

46. S. H. Clark, M. McCarron, C. Love, A. Chovnick, *Genetics* **112**, 755 (1986).

47. D. Van Vactor, Jr., D. E. Krantz, R. Reinke, S. L. Zipursky, *Cell* **52**, 281 (1988).

48. A. C. Spradling and G. M. Rubin, *Science* **218**, 341 (1982); G. M. Rubin and A. C. Spradling, *ibid.* p. 348. For a recent review see A. C. Spradling, in (*14*), p. 175.

49. S. B. Scholnick, B. A. Morgan, J. Hirsh, *Cell* **34**, 37 (1983); A. C. Spradling and G. M. Rubin, *ibid.*, p. 47; D. Goldberg, J. Posakony, T. Maniatis, *ibid.*, p. 59.

50. S. J. Tojo, S. Germeraad, D. S. King, J. W. Fristrom, *EMBO J.* **6**, 2249 (1987); C. Zuker, D. Mismer, R. Hardy, G. M. Rubin, *Cell* **53**, 475 (1988).

51. S. Schneuwly, R. Klemenz, W. J. Gehring, *Nature* **325**, 816 (1987); G. Struhl, *ibid.* **318**, 677 (1985); D. Ish-Horowicz and S. M. Pinchin, *Cell* **51**, 405 (1987). Unfortunately,

even the heat-inducible *hsp*70 promoter, the best currently known for this purpose in *Drosophila* [J. T. Lis, J. A. Simon, C. A. Sutton, *Cell* 35, 403 (1983)] shows some transcription during the life cycle to the fly even when not intentionally induced.

52. J. A. Fischer, E. Giniger, T. Maniatis, M. Ptashne, *Nature* 332, 853 (1988).
53. S. Scherer and R. W. Davis, *Proc. Natl. Acad. Sci. U.S.A.* 76, 4951 (1979).
54. J. G. Izant and H. Weintraub, *Cell* 36, 1007 (1984).
55. Phenocopies of mutations affecting embryonic development have been generated by injection of antisense RNA into early embryos. For examples see: U. B. Rosenberg, A. Preiss, E. Seifert, H. Jäckle, D. C. Knipple, *Nature* 313, 703 (1985); C. V. Cabrera, M. C. Alonso, P. Johnston, R. G. Philips, P. A. Lawrence, *Cell* 50, 659 (1987).
56. E. Wieschaus and C. Nüsslein-Volhard, in (*14*), p. 199; J. A. Campos-Ortega and V. Hartenstein, *The Embryonic Development of* Drosophila melanogaster (Springer-Verlag, Berlin, 1985).
57. K. Illmensee, *Wilhelm Roux Arch. Entwicklungsmech. Org.* 170, 267 (1972); in *Genetic Mosaics and Cell Differentiation*, W. J. Gehring, Ed. (Springer-Verlag, Berlin, 1978).
58. P. Bryant, *Dev. Biol.* 26, 637 (1971).
59. M. Lohs-Schardin *et al.*, *Dev. Biol.* 68, 533 (1979).
60. Reviewed in J. H. Sang, *Adv. Cell Cult.* 1, 125 (1981); P. Martin and I. Schneider, in *The Genetics and Biology of* Drosophila, M. Ashburner and T. R. F. Wright, Eds. (Academic Press, London, 1978), vol. 2a, p. 219; I. Schneider and A. B. Blumenthal, *ibid.*, p. 266; K. Ui, R. Ueda, T. Miyake, *In Vitro Cell Dev. Biol.* 23, 707 (1987); P. M. Salvaterra, N. Bournias-Vardiabasis, T. Nair, G. Hou, C. Lieu, *J. Neurosci.* 7, 10 (1987).
61. G. M. Technau, *Development* 100, 1, (1987).
62. M. Wilcox, in (*14*), p. 243.
63. E. Hafen and M. Levine, in (*14*), p. 139.
64. C. S. Parker and J. Topol, *Cell* 36, 357 (1984); U. Heberlein and R. Tjian, *Nature* 331, 410 (1988).
65. D. Rio, *Proc. Natl. Acad. Sci. U.S.A.* 85, 2904 (1988).
66. E. C. Johnson and W. L. Pak, *J. Gen. Physiol.* 88, 651 (1986).
67. C. K. Solc, W. N. Zagotta, R. W. Aldrich, *Science* 236, 1094 (1987).
68. J. B. Thomas and R. J. Wyman, *Trends Neurosci.* 6, 214 (1983).
69. Reviewed in B. Ganetzky and C. F. Wu, *Annu. Rev. Genet.* 20, 13 (1986); M. A. Tanouye *et al.*, *Annu. Rev. Neurosci.* 9, 255 (1986).
70. D. M. Papazian, T. L. Schwarz, B. L. Tempel, Y. N. Jan, L. Y. Jan, *Science* 237, 749 (1987); A. Kamb, L. E. Iverson, M. A. Tanouye, *Cell* 50, 405 (1987).
71. Y. T. Kim and C. F. Wu, *J. Neurosci.* 10, 3245 (1987); C. K. Sok, W. N. Zagotta, R. W. Adrich, *Science* 236, 1094 (1987).
72. L. C. Timpe *et al.*, *Nature* 331, 143 (1988).
73. H. Steller, K.-F. Fishbach, G. M. Rubin, *Cell* 50, 1139 (1987); J. R. Jacobs, N. H. Patel, T. Elkins, C. S. Goodman, *Soc. Neurosci. Abstr.* 13, 1222 (1987).
74. J. Hall, *Trends Neurosci.* 9, 414 (1986).
75. R. Fernandez-Almunacid and O. M. Rosen, *Mol. Cell. Biol.* 7, 2718 (1987).
76. E. A. Fyrberg, B. J. Bond, N. D. Hershey, K. S. Mixter, N. Davidson, *Cell* 24, 107 (1981).
77. C. P. Emerson and S. I. Bernstein, *Annu. Rev. Biochem.* 56, 695 (1987); J. Toffenetti, D. Mischke, M. L. Pardue, *J. Cell Biol.* 104, 19 (1987).
78. C. C. Karlick and E. Fyrberg, *Mol. Cell. Biol.* 6, 1965 (1986); T. Tansey, M. D. Mikus, M. Dumoulin, R. V. Storti, *EMBO J.* 6, 1781 (1987).
79. W. E. Theurkauf, H. Baum, J. Bo, P. C. Wensink, *Proc. Natl. Acad. Sci. U.S.A.* 83, 8477 (1986); J. E. Rudolf, M. Kimble, H. D. Hoyle, M. A. Subler, E. C. Raff, *Mol. Cell. Biol.* 7, 2231 (1987).
80. T. J. Byers, R. Dubreuil, D. Branton, D. P. Kiehart, L. S. Goldstein, *J. Cell Biol.* 105, 2103 (1987).
81. L. M. Hall and P. Spierer, *EMBO J.* 5, 2949 (1986).
82. N. Itoh *et al.*, *Proc. Natl. Acad. Sci. U.S.A.* 83, 4081 (1986).
83. I. Hermans-Borgmeyer *et al.*, *EMBO J.* 5, 1503 (1986); B. Bossy, M. Ballivet, P. Spierer, *ibid.* 7, 611 (1988).
84. L. Salkoff *et al.*, *Science* 237, 744 (1987).
85. W. J. Gehring, *ibid.* 236, 1245 (1987).
86. U. B. Rosenberg *et al.*, *Nature* 319, 336 (1986); R. Schuh *et al.*, *Cell* 47, 1025 (1986).
87. V. L. Smith, K. E. Doyle, J. F. Maune, R. P. Munjaal, K. Beckingham, *J. Mol. Biol.* 196, 471 (1987); M. K. Yamanaka, J. A. Saugstad, O. Hanson-Painton, B. J. McCarthy, S. L. Tobin, *Nucleic Acids Res.* 15, 3317 (1987).
88. A. Rosenthal *et al.*, *EMBO J.* 6, 433 (1987).
89. The catalytic subunit is 82% identical [J. L. Foster, G. C. Higgins, F. R. Jackson, *J. Biol. Chem.* 263, 1676 (1988); D. Kalderon and

G. M. Rubin, unpublished data] and the R1 regulatory subunit is 71% identical (D. Kalderon and G. M. Rubin, unpublished data) to the corresponding mouse genes.

90. M. A. Simon, B. Drees, T. Kornberg, J. M. Bishop, *Cell* **42**, 831 (1985).
91. M. J. Henkemeyer, F. B. Gertler, W. Goodman, F. M. Hoffman, *ibid.* **51**, 821 (1987).
92. A. L. Katzen, T. B. Kornberg, J. M. Bishop, *ibid.* **41**, 449 (1985).
93. R. Steward, *Science* **238**, 692 (1987).
94. F. Rijsewijk *et al.*, *Cell* **50**, 649 (1987).
95. R. W. Padgett, R. D. St. Johnston, W. M. Gelbart, *Nature* **325**, 81 (1987).
96. L. Petruzzelli *et al.*, *Proc. Natl. Acad. Sci. U.S.A.* **83**, 4710 (1986); Y. Nishida, M. Hata, Y. Nishizuka, W. J. Rutter, Y. Ebina, *Biochem.*

Biophys. Res. Commun. **141**, 474 (1986).
97. E. D. Schejter, D. Segal, L. Glazer, B.-Z. Shilo, *Cell* **46**, 1091 (1986).
98. M. Leptin, R. Aebersold, M. Wilcox, *EMBO J.* **6**, 1037 (1987); T. Bogaert, N. Brown, M. Wilcox, *Cell* **51**, 929 (1987).
99. D. J. Montell and C. S. Goodman, *Cell* **53**, 463 (1988).
100. M. Seeger and T. Kaufman, unpublished data, as cited in T. Jessel, *Neuron* **1**, 3 (1988).
101. I thank the members of my laboratory, as well as K. Anderson, J. Fristrom, C. Goodman, J. Hall, and M. Siegler for useful comments on this manuscript. I also thank B. Baker for the mitotic chromosome panel of Fig. 2 and T. Laverty for the polytene chromosome panel.

10

Plants: Novel Developmental Processes

Robert B. Goldberg

The use of plants as objects of study has yielded important new biological principles (*1*). Hooke's experiments in 1665 with cork, an epidermis-like tissue on the exterior of woody plant stems and roots, first demonstrated the existence of cells. Schleiden contributed to the formulation of the cell theory in 1838 by demonstrating that every plant tissue is composed of cells and that cells are the fundamental units of living organisms. Studies with pea plants enabled Mendel to discover the laws of heredity in 1866, and genetic crosses with corn by McClintock in the 1940s established the concept of transposons and a fluid eukaryotic genome.

From a developmental perspective, Steward demonstrated in the late 1950s that differentiated carrot root cells retain the potential to undergo embryogenesis and develop into mature fertile plants; that is, plant cells are totipotent. At the biochemical level, Fraenkel-Conrat showed that tobacco mosaic virus particles can self-assemble in vitro and that RNA can function as the genetic material in the absence of DNA. Finally, experiments by Calvin, Hill, and others in the 1950s showed how carbon and light energy can enter the living world by photosynthesis in green plants.

During the last 5 years there has been a major refocusing on plants as a biological sys-
tem. The renewed interest in plants arose partly from the realization that gene transfer technology could be used to introduce novel genetic traits into crop plants (*2*). There are now many examples of plants that have been genetically engineered for resistance to herbicides, insects, and viruses. Advances in plant biotechnology also stimulated new strategies to regenerate plants from single cells in culture, provided new approaches to transfer genes from one plant to another, and intensified research on physiological and biochemical phenomena at the molecular level (*3*). Current excitement about plants as experimental organisms also derives from the perception that plants represent a new biological frontier, ready for experimentation on novel aspects of plant biology such as photosynthesis, seed development, reproduction, nitrogen fixation, fruit ripening, pollination, and light control. Although these phenomena have been studied extensively for many years, new technology and the development of accessible plant systems now permit plant-specific problems to be explored with a degree of sophistication not possible a few years ago.

In this chapter I outline many biological processes that are specific to plants and highlight approaches that can be used to solve important problems of plant biology. I have

R. B. Goldberg is at the Department of Biology, University of California, Los Angeles, CA 90024-1606. This chapter is reprinted from *Science* **240**, 1460 (1988).

focused on plant development because this area is under intense scrutiny and because an understanding of the molecular processes controlling gene expression during plant development should emerge in the not too distant future.

Plants Are a Diverse Group of Organisms

Plants are generally thought of as flowering species such as ornamentals, trees, shrubs, and vegetable crops. Although flowering plants constitute more than 90% of the 275,000 known plant species and are the focus of most plant research, the plant kingdom contains diverse groups of nonflowering plants with unique biological characteristics (4). In many cases, nonflowering plants offer experimental advantages over flowering plants for the study of common plant-specific processes.

The plant kingdom contains both nonvascular and vascular species

There are two broad categories of plants (Table 1). Nonvascular plants, such as mosses, do not have specialized water-conducting and food-conducting tissues and, therefore, lack true leaves, stems, and roots. This severely limits the size and habitat of nonvascular plants. By contrast, vascular seedless and seed-producing plants have highly differentiated xylem and phloem cells that conduct water and food over great distances. Both plant groups are thought to have evolved from a common green multicellular alga more than 450 million years ago (4). Support for this idea derives from the high degree of similarity in chloroplast genome organization in a flowering plant (tobacco) and a bryophytic nonvascular plant (*Marchantia*) (5).

What characteristics do nonvascular and vascular plants share that place them in the same kingdom? First, both are photosynthetic multicellular eukaryotic organisms that are highly adapted for growth and reproduction on land. Second, each has an alternation of a diploid spore-producing generation (sporophytic) and a haploid gamete-producing generation (gametophytic) in its life cycle (Fig. 1). Meiosis in all plant species, unlike that in animals, yields haploid spores rather than gametes (4). In vascular plants, the sporophytic phase dominates and the gametophytic phase is dependent on the sporophyte for nutrition and support. By contrast, the gametophyte is free-living in nonvascular plants and dominates the life cycle (Table 1). This characteristic makes nonvascular plants ideally suited for studying processes that are dominant in the gametophyte; for example, gamete production. Third, cellulose is the major polysaccharide found in nonvascular and vascular plant cell walls, and both groups form a cell plate during cell division. Finally, vascular and nonvascular plants store glucose as starch in their chloro-

Table 1. Plant diversity. Taxonomic groupings and time of first appearance are taken from (4); mya, million years ago.

Category	Division	Examples	Dominant generation	Time of appearance (mya)
Nonvascular	Bryophyta	Mosses, liverworts	Gametophyte	430
Vascular				430
Seedless	Psilophyta	Psilatum	Sporophyte	
	Lycophyta	Club mosses	Sporophyte	
	Sphenophyta	Horsetails	Sporophyte	
	Pterophyta	Ferns	Sporophyte	
Seeds				360
Nonflowering	Cycadophyta	Cycads	Sporophyte	
	Ginkgophyta	Maidenhair tree	Sporophyte	
	Coniferophyta	Pines, firs	Sporophyte	
	Gnetophyta	Genetum, welwitschia	Sporophyte	
Flowering	Anthophyta	Soybean, corn	Sporophyte	125

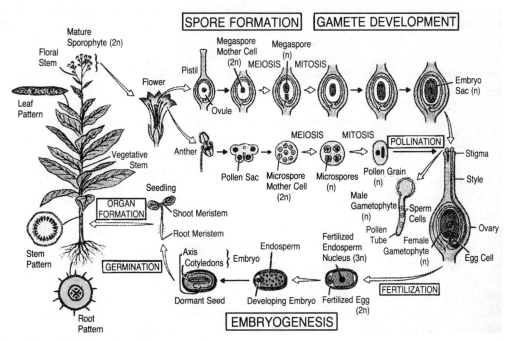

Fig. 1. Life cycle of a flowering plant.

plasts and utilize both chlorophyll a and chlorophyll b as light receptors.

Mosses and other nonvascular plants can provide insight into many plant-specific problems

Nonvascular plants offer opportunities to study plant processes that have not yet been fully realized. Because these plants are haploid for most of their life cycle, large numbers of mutants can be obtained for the genetic dissection of physiological and developmental events (6). One notable example is the moss *Physcomitrella patens* (7). This organism has been developed extensively as a model plant genetic system. *Physcomitrella* plants can be regenerated from protoplasts, and protoplast fusion permits complementation analysis to be carried out. Chemical mutagenesis has been used to induce auxotrophic, morphogenetic, hormonal, phototropic, and gravitropic mutants. Although transformation of *Physcomitrella* has not yet been reported, it should be possible to combine genetic and molecular approaches to

identify and study genes controlling important plant processes.

Algae are being used to study specialized plant processes

Although not plants, many protistans and bacteria with plant-like processes are being used for the study of specific plant problems. Blue-green algae (*Cyanobacteria*) are providing new insight into photosynthesis and nitrogen fixation because these processes are analogous to those that occur in plants and, in contrast to the situation with plants, gene replacement studies can be easily carried out (8). For example, deletion strains of *Cyanobacteria* have been constructed that lack critical photosynthesis genes and that permit functional dissection of both plant and blue-green algal genes important in photosynthesis (9). Similarly, the unicellular green alga *Chlamydomonas* has been used to dissect genetically photosynthetic reaction centers and chloroplast biogenesis (10). Uniparental genetics have permitted many mutants that are defective in important photosynthetic proteins to be induced (*11, 12*). For example, mutants containing defective photosystem II proteins have been used to localize amino acids required for interaction with electron acceptors in the photosystem II complex (*11, 12*). Because *Chlamydomonas* contains only one chloroplast and because numerous chloroplast mutants exist, the recent development of a chloroplast transformation system in this organism (*13*) should permit a functional dissection of genes and proteins critical for light reception and harvesting.

Finally, the multicellular brown alga *Fucus* has been used extensively for studying egg cell development and the establishment of embryonic form (*14*). Unlike higher plant egg cells that reside within the sporophyte and cannot be readily isolated (Fig. 1), *Fucus* eggs can be purified in large numbers, fertilized in vitro, and radioactively labeled (*14*). Although few molecular studies have been carried out and DNA transformation has not yet been reported, *Fucus* is a valuable model for studying how plant eggs become polarized after fertilization and how the embryonic axis is established in the absence of cell movement.

Higher Plants Have Novel Developmental Processes

Flowering plants are the most complex and intensely investigated organisms within the plant kingdom and have all of the features specific to plants as a biological system. These plants evolved from nonflowering seed-producing relatives at least 120 million years ago, and within a short time interval (60 million years) became the pre-eminent plant group (*4*). Flowering plants undergo many unique developmental events during their life cycle (Fig. 1). Events such as spore and gamete formation, seed development, and organogenesis from meristems are shared with one or more other plant groups. However, the presence of a flower, which functions exclusively in reproduction, adds additional developmental complexity and can be studied only by using flowering plants as experimental organisms. Developmental processes unique to higher

plants include regeneration potential of single cells, reproduction, seed development, indeterminate growth patterns, morphogenesis in the absence of cell movement, and environmental induction of specific developmental states.

Plants are morphologically simple organisms

A striking feature of higher plants is their apparent morphological simplicity. This contrasts greatly with the situation in higher animal forms such as vertebrates. During the life cycle of the flowering plant, only three vegetative organ systems (leaf, stem, and root) and three reproductive organ systems (petal, stamen, and pistil) are formed (Fig. 1). Additional organs, such as the fruit (a mature ovary) and embryonic cotyledon (seed leaf), form during specific developmental periods and are equivalent to vegetative and floral organ systems in structural complexity.

Each plant organ system is highly specialized to carry out specific biological functions such as photosynthesis (leaf), support and water conduction (stem), water absorption (root), spore formation (pistil and stamen), and food reserve synthesis (cotyledons). Irrespective of their specialized nature, plant organ systems are composed of anatomically similar cells and tissues. Collectively, only ten basic tissue types containing approximately 15 structurally diverse cells are represented in the organ systems of higher plants (*4*). For example, five prominent tissues can be visualized in the cross section of a tobacco floral stem (Fig. 2A): epidermis, cortex, xylem, phloem, and pith. With the exception of the cortex and pith, each tissue is structurally unique and contains specialized cell types. The pith and cortex both contain storage parenchyma cells and are similar tissues that differ only by their position relative to the vascular cylinder. In situ hybridization studies (*15–17*) indicate that many mRNAs are nonrandomly distributed within the stem and are preferentially localized within specific tissues (*18–20*). A stem mRNA, designated TS13 (*19*), is highly concentrated within the inner and outer phloem cells, as

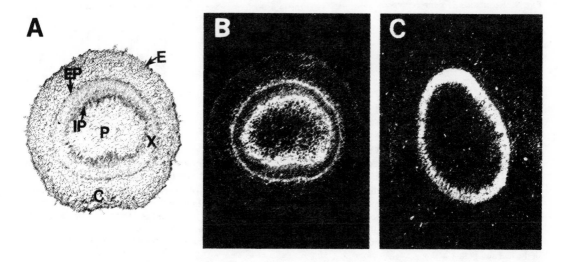

Fig. 2. Localization of mRNAs in the tobacco stem. Floral stems were fixed, embedded in paraffin, and hybridized in situ with single-stranded [32]S-labeled RNA probes (15–17). **(A)** Bright-field photograph. E, epidermis; C, cortex; P, pith; X, xylem; EP, external phloem; and IP, inner phloem. **(B)** Hybridization with the TS13 probe (18). TS13 represents a 0.7-kb mRNA that is present at high levels in the stem but is detectable at lower levels in heterologous organ systems (19). Photograph was taken by dark-field microscopy. The white grains represent regions containing RNA/RNA hybrids. **(C)** Hybridization with the TP7 probe (18). TP7 represents a 1.4-kb mRNA that is present at high levels in the stem and petal but is also detectable at lower concentrations in other organ systems (20). Dark-field micrograph as in (B). RNA/RNA hybrids are indicated by the white grains.

shown in Fig. 2B. TS13 mRNA sequences are not detectable in other floral stem tissues, although they may be present at concentrations below those observable with the in situ hybridization procedure. By contrast, Fig. 2C shows that a nonhomologous stem mRNA, designated TP7 (20), is localized preferentially within the xylem and is not detectable in phloem or other stem tissues. These findings indicate that plant tissues with distinct functions can be distinguished at the molecular level.

From an anatomical point of view, many tissues and cells in plant organ systems appear to be identical. Major differences in plant organ system morphology depend primarily on how the various tissues are organized into spatial patterns rather than apparent differences in tissues and cell types (Fig. 1). Thus, the leaf and root have tissues analogous to those shown in Fig. 2A for the floral stem. Because each organ system carries out unique functions, the morphological similarity must mask inherent differences at the biochemical level. Support for this notion derives from a comparison of the nuclear RNA and mRNA sequence sets in all tobacco organ systems (21, 22) (Fig. 3). First, each floral and vegetative organ system has a nuclear RNA complexity (1.2×10^5 kb) and an mRNA complexity (3.5×10^4 kb) equivalent to those observed for complex animal organ systems such as the liver and kidney (23). Thus, despite their apparent morphological simplicity, a very complex genetic program is expressed in each organ system. Second, each organ system has a large gene set that is not detectably expressed at the mRNA level in heterologous organs. For example, the stem has approximately 6000 diverse mRNA species—representing 25% of the stem mRNA complexity—that are not detectable on the polysomes of other vegetative and floral organ systems (21, 22). Finally, in situ hybridization studies with the TS13 stem mRNA (Fig. 2B) did not detect this sequence in the phloem of the root and petal, suggesting that there is a molecular diversification of similar tissues in distinct plant organ systems (18, 19). A major unsolved problem of plant biology is the identity of proteins and processes that confer

Fig. 3. Regulation of gene expression in flowering plants. Data represent the results of solution RNA–excess hybridization experiments with tobacco single-copy DNA and were taken from Kamalay and Goldberg (*21, 22*). Complexity values were calculated by using a tobacco single-copy DNA complexity of 6.4×10^5kb (*21*). **(A)** Hybridization with nuclear RNA populations. Total nuclear RNA [poly(A)$^+$ and poly(A)$^-$] was used for

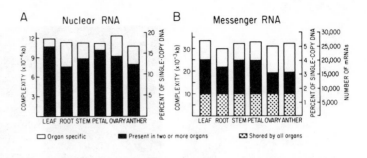

these experiments (*22*). In general, the nuclear RNA/mRNA complexity ratio for any given organ system is approximately 3.6. **(B)** Hybridization with mRNA populations. Experiments were performed with total polysomal RNA [poly(A)$^+$ and poly(A)$^-$] that was released from polysomes with EDTA (*21*). The number of diverse mRNA species in each mRNA population was calculated by assuming an average transcript size of 1.2 kb (*21*). Leaf and petal mRNA sets are qualitatively indistinguishable from each other in population hybridization experiments (*21*). Leaf and petal white bar areas represent mRNA species that are present in both the leaf and petal but are undetectable in heterologous organ system mRNAs. Recent experiments with petal cDNA clones indicate that there are quantitative differences with respect to specific mRNA species in the leaf and petal (*20*).

uniqueness on plant cells; that is, what is the nature of gene products expressed in specific plant cells and tissues, and how are these products integrated into processes that are unique to each organ system?

Plant cells are totipotent

A remarkable property that distinguishes plant cells from their animal counterparts is the capacity to regenerate into mature fertile plants (*24, 25*). Although not all plant cells retain their regenerative capacity, this property is the basis for the success of gene transfer technology in producing genetically engineered higher plant species (*2, 3*). Many plant cells have the capacity to reorganize into embryos and undergo somatic embryogenesis analogous to that which occurs after fertilization (*24, 25*). Alternatively, under the proper conditions, somatic cells can establish meristematic foci and are able to regenerate into mature plants through an organogenesis pathway (*24*). The regenerative capacity is not confined to diploid sporophytic cells, because microspores also have the potential to regenerate into haploid plants. The totipotency property of plant cells implies that they have a complete unrearranged set of genes and that this gene set retains the potential to establish and maintain all differentiated higher plant states.

The underlying cellular and molecular processes that enable plants to regenerate via embryogenic or meristematic pathways are not yet understood. In some cases, the inability to regenerate routinely from single cells commercially important varieties of many plants (for example, corn and soybean) has been an impediment to producing agriculturally superior crops by gene engineering technology (*2, 25*). The fact that differentiated plant cells with restricted functions (for example, microspore and parenchyma storage cells) can regenerate into fully mature plants raises two important questions regarding plant development. First, what is the meaning of determination in higher plants? Although a root cell or a microspore is determined within the context of the normal plant life cycle, these cells retain a flexible capacity to serve as foci for the differentiation of an entire multicellular plant. Clearly, this is very different from the situation in animals. Second, because many plant cells are capable of reorganizing into embryos, the role of the maternal or egg gene expression program on subsequent embryogenesis is unclear. That is, are maternal agents or factors important in establishing embryo-specific gene expression programs early in development? Either the egg cell plays a minor role in embryonic morphogenetic events, or somatic plant cells can be induced to express a maternal gene expression program before embarking on an embryogenetic pathway. Whatever the answer, the molecular basis of plant cell totipotency and the identification of cellular reorganization processes that must occur to establish a regenerative potential remain major unresolved problems.

Plants reproduce using both spores and gametes

As pointed out earlier, plants undergo an alternation of spore- and gamete-forming generations (Fig. 1). Sporogenous cells within the anther and ovary of the flower undergo meiosis and give rise to haploid microspores and megaspores, respectively. The haploid spores divide mitotically and differentiate into a three-celled male gametophyte (pollen grain) and a seven-celled female gametophyte (embryo sac). The pollen grain contains two sperm cells, whereas the embryo sac contains one egg cell. By contrast with gametogenesis in animals, the gametes are produced mitotically from cells that are not present in the embryo. Thus, plants do not have a specialized germ line that is set aside during embryogenesis.

The molecular events leading to gamete formation in higher plants are largely unexplored. Male gametophyte development has been investigated to a limited extent because developing pollen grains can be isolated and studied (26). Expression of both the diploid anther genome and the haploid microspore genome is required to produce a mature pollen grain with functional sperm cells (26). Gene expression studies have shown that there are approximately 25,000 diverse polyadenylated [poly(A)] RNAs present in the mature pollen grain and that many of these RNAs are represented in the sporophyte as well (26). The distribution of these RNAs within the three male gametophytic cells is unknown. Sperm cell isolation procedures (27) coupled with in situ hybridization studies (Fig. 2) should permit the cellular processes leading to sperm cell differentiation to be studied in more detail.

By contrast, egg cell development has been studied only with cytological procedures (24), and nothing is known about the gene expres-

sion programs that occur during egg cell formation. This differs greatly from the situation in animals for which there is a large body of information on the molecular biology of egg development (28). Because the higher plant egg cell is one of seven cells in the female gametophyte, and because the female gametophyte is buried within several layers of ovary and ovule cells (Fig. 1), it has not yet been possible to isolate egg cells in order to carry out molecular studies. Thus, as already pointed out, the contribution of maternal gene products to early embryonic events is largely unknown in plants. In the absence of techniques that permit easy access to large numbers of egg cells, plant egg cell development may be better studied by using a nonplant model such as *Fucus* (14). A conceptual understanding of plant development will require a functional characterization of the gene products and regulatory events required to differentiate haploid spores into male and female gametophytes. This information may suggest novel ways to create male-sterile and female-sterile plants useful for hybrid crop production (2).

Dual fertilization processes occur in higher plants

In addition to an alternation of spore- and gamete-forming generations, flowering plant reproduction requires two independent fertilization events to trigger sporophytic development (Fig. 1). Because the male and female gametophytes reside in different sporophytic organ systems, a pollination process is required to transfer sperm cells within the male gametophyte to the pistil where the egg cell resides (Fig. 1). Pollination and fertilization represent major examples of the small number of developmental processes in higher plants that require specific cell-cell recognition events to occur, in contrast to the situation in animals.

Specific pollen grain–pistil interactions must occur to ensure pollination of homologous plant species. Molecular studies suggest that specific proteins must be deposited on the pollen wall and on the surface of pollen receptor cells on the pistil (the stigma) (29). Similarly, the sperm cells must recognize two of the seven female gametophyte cells. One sperm cell fertilizes the egg to form a diploid zygote that will develop into the new sporophyte. The other unites with two previously fused nuclei (central cell nucleus) to form a triploid endosperm (Fig. 1). The endosperm develops into a terminally differentiated, nonembryonic tissue that is responsible for supplying nutrients to the embryo during embryogenesis or to the developing sporophyte after seed germination (Fig. 1). Sperm cells appear to be differentiated prior to fertilization and unite with specific female gametophytic cells (*30*)—that is, the double fertilization process is directed and not random. The molecular basis of sperm cell specification and the determinants that enable specific sperm cell–female gametophyte interactions to take place are not yet understood and require critical investigation.

Plant embryogenesis terminates with the formation of a dormant seed

In contrast to the situation in animal development (*28*), most major ontogenetic events in higher plants occur postembryonically. Plant embryogenesis terminates with the formation of a dormant embryo that is packaged within a seed (Fig. 1). Plant embryogenesis, unlike the development of many animal species (*28*), does not lead to the production of an organism that resembles a mature plant (Fig. 1). Rather, after seed dormancy ends, the sporophyte develops from meristematic tissues that are specified during embryogenesis.

The seed permits the embryo to persist for long periods of dormancy and is present within a fruit that is responsible for dispersing the embryo over large distances (*4*). Genetically controlled events cause the ovary to differentiate into a fruit with an elaborate structure designed for a particular dispersal process (*4*). What cellular signals trigger the differentiation of an ovule into a seed (Fig. 1), and what molecular processes program the development of an ovary into a fruit, remain unresolved plant-specific problems.

Higher plant embryos form two organ systems with different developmental fates

The embryo differentiates during embryogenesis into two organ systems—the axis and the cotyledon—that have different developmental fates. The cotyledon is a terminally differentiated organ system that senesces after germination and synthesizes highly specialized food reserves that are used by the germinating seedling. By contrast, the axis contains the root and shoot meristems that will give rise to sporophytic organ systems through the life cycle. Figure 4 shows the development of a soybean embryo and a tobacco embryo (*31*). In both cases an embryo can be visualized that is contained within the developing seed and that is embedded within nonembryonic endosperm tissue. A striking event in plant embryogenesis is the change from a globular embryo with radial symmetry (Fig. 4, A, B, and H) to a heart-shaped embryo with bilateral symmetry (Fig. 4, C, I, and J). At this embryogenic stage the cotyledons begin to differentiate, the plant body becomes polarized, and the root-shoot axis forms (Fig. 4, C and J). In analogy to the situation with the egg cell, the plant embryo is embedded deep within the developing fruit (Fig. 1), making it difficult to isolate large numbers of globular and heart-shaped embryos for molecular studies. As a result, the gene expression programs that occur during the very early stages of plant embryogenesis, as well as the molecular and cellular processes that direct embryo polarization and specify the cotyledon and axis cell lineages, are not known. The use of cultured somatic embryos (*32*) and embryolethal mutants (*33*) may provide a novel opportunity to investigate critical genetic events that take place during early embryogenesis.

Approximately 20,000 diverse genes are expressed by both the axis and the cotyledon during embryogenesis (*34*). Most of these genes encode rare class mRNAs of unknown function. A small gene set, however, directs the synthesis of highly prevalent mRNAs that encode seed proteins that are packaged preferentially into cotyledon cell storage bodies (*35*).

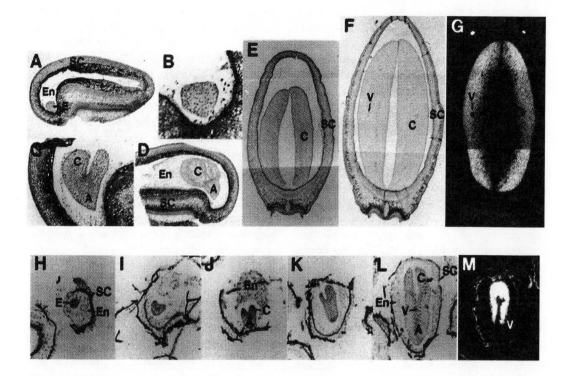

Fig. 4. Localization of seed protein mRNA in soybean and transformed tobacco seeds. (A–F) Bright-field photographs of soybean seed development (*31*). (A and B) Globular embryo, (C) heart stage embryo, (D) cotyledon stage embryo, (E and F) maturation-stage embryos. (G) In situ hybridization of a labeled single-stranded lectin RNA probe with a longitudinal section of a maturation stage soybean embryo (*16, 17*). The photograph was taken by dark-field microscopy, and the white grains represent regions with RNA/RNA hybrids. (H–L) Bright-field photographs of tobacco seed development. Tobacco seed development was characterized by Barker and Goldberg (*16*). (H) Globular stage embryo, (I and J) heart-stage embryos, (K and L) torpedo or maturation-stage embryos. (M) In situ hybridization of a labeled lectin mRNA probe with a longitudinal section of a transformed tobacco seed containing a soybean lectin gene (*37*). The photograph was taken by dark-field microscopy (*16, 17*). Abbreviations: E, embryo; En, endosperm; SC, seed coat; C, cotyledon; A, axis; and V, vascular tissue.

Seed protein genes are highly regulated during the plant life cycle. These genes encode mRNAs that accumulate and decay during embryogenesis and are either absent from or present at low concentrations in mature plant organ systems (*34–38*). One example of a cell-specific seed protein gene expression program is shown in Fig. 4G (*31*). In situ hybridization shows that soybean lectin mRNA (*36*) is highly localized within cotyledon parenchyma cells and is not detectable within the vascular tissue (*31*). Soybean lectin mRNA is localized within analogous cells in transformed tobacco plants (*31, 37*) (Fig. 4M), indicating that the regulatory machinery controlling seed protein gene expression is highly conserved in diver-

gent plant species. Recent studies have begun to identify the cis elements and trans factors that control seed protein gene expression (*38*). The precise regulatory network responsible for controlling seed protein gene expression is not yet known.

Plants have indeterminate developmental programs

As pointed out above, most higher plant developmental events occur after seed germination from meristems contained within the embryonic axis (Fig. 1). Both the shoot and root meristems are determined by the end of embryogenesis and are committed to pursue a given developmental pathway. Plant meristems consist of continuously dividing cells that regenerate themselves and are committed to produce specific organ systems. Meristematic cells are analogous to animal stem cells because they yield a new meristem upon division. However, in contrast with animal stem cells, plant meristems lead to the formation of complex organ systems. In addition, each meristem differentiates into three primary meristems: protoderm, ground meristem, and provascular cambium (4). In the case of the stem (Fig. 2), the primary meristems give rise to the epidermis, cortex and pith, and xylem and phloem tissues, respectively. The cellular processes responsible for specifying the meristems during embryogenesis are not yet understood.

The conversion of a vegetative shoot meristem into a flower-producing, floral meristem breaks the continuous development cycle. A floral meristem does not regenerate itself and leads to determinate growth and development of the flower. Diffusible substances, designated as florigens, trigger the conversion of a vegetative meristem to a floral meristem (39). What these flower-inducing agents are and how they mediate a pattern reorganization within the shoot meristem are not known.

Floral organ systems express highly regulated gene sets, and each organ of the flower has a unique mRNA collection that is not detectable in heterologous vegetative and floral organ system (21, 22) (Fig. 3). Experiments with individual floral mRNAs indicate that gene expression is regulated both temporally and spatially during flower development (40–42). Figure 5 shows the development of a tobacco flower, and the differentiation of the pistil and anther (40). The developmental dot blots (Fig. 5) indicate that two anther-specific mRNAs, designated TA13 and TA29, accumulate early in anther development and then decay prior to anther dehiscence and pollen grain release. By contrast, two different anther mRNAs, TA20 and TA25, persist throughout anther development. Localization of the TA20 and TA29 mRNAs within floral organ systems is shown in Fig. 6 (41). The TA20 mRNA is highly concentrated in the cell layer that con-

nects the ovule to the ovary, as well as in specific ovary wall regions (41) (Fig. 6, A and B). Within the anther the TA20 mRNA is localized in all tissues except the filament and the tapetum (41). By contrast, the TA29 mRNA is present within the anther tapetum cell layer, and is not detectable in other anther tissues (Fig. 6, C and D). The decay of the TA29 mRNA parallels the degeneration of the tapetal cell layer late in anther development (26, 41). These data indicate that different cell-specific and time-specific gene expression programs occur during flower development. The molecular processes responsible for establishing flower-specific gene expression programs are not yet understood.

Morphogenesis in plants occurs in the absence of cell movement

A major feature that distinguishes plant and animal cells is the presence of a plant cell wall. Because plant cells are glued together after division, morphogenesis must occur in the absence of cell movement. Distinct morphological patterns are produced by a combination of asymmetric cell division, division in distinct planes, differential growth after cell division, and different division rates (4).

As pointed out in the previous section, the morphological pattern that gives rise to each plant organ system is an intrinsic property of the meristem. How patterns are established within the root, shoot, and floral meristems to give rise to organs with specific morphologies is not yet known. Clonal analysis has provided some insight into the fate of specific meristematic cells (43). However, what causes the meristem to become organized and committed to a specific organ system pattern remains a central problem of plant development (44). As in *Drosophila*, homeotic mutants exist in plants that convert one floral organ system into another (for example, anther to pistil) (45). In addition, mutants exist that alter flower morphology (46). These mutants may prove useful for dissecting the genetic and cellular events required to establish unique plant morphological forms.

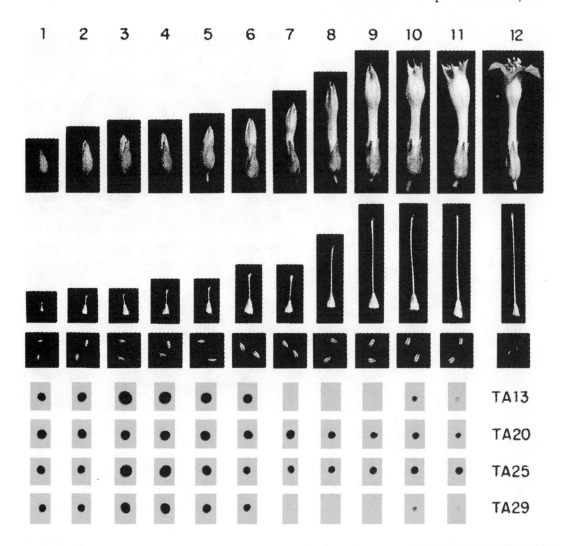

Fig. 5. Regulation of gene expression during tobacco flower development. Tobacco flower development was divided into 12 stages, with floral bud and petal lengths as developmental markers (*40*). In general, the transition from stage 1 to stage 12 occurred over a 1- to 2-week period depending on the time of year. Stage 1 and stage 12 flowers averaged 0.8 cm and 4.6 cm in length, respectively. Anther mRNAs were isolated from 12 stages of flower development, spotted onto Nytran membranes, and hybridized with labeled anther cDNA clones (*40*). TA13 and TA29 cDNA clones represent divergent mRNAs of the same gene family (*40*). By contrast, TA20 and TA25 cDNA clones represent unique anther mRNAs. At their peak levels, TA13, TA29, TA20, and TA25 mRNAs represent 0.42%, 0.26%, 0.48%, and 0.23% of the anther mRNA mass, respectively (*40*). TA13, TA25, and TA29 mRNAs are not detectable in heterologous vegetative and floral organ system mRNA populations. TA20 mRNA is represented at lower prevalences in pistil and petal mRNAs.

Environmental factors play a major role in plant development

Finally, in contrast with animals, plants are highly dependent on their environment and use extrinsic cues such as light, water, and temperature to trigger specific developmental events (*4*). For example, seed germination depends on temperature to break embryonic dormancy and on water to reinitiate cell division and metabolic processes (*4*). Physical forces such as gravity also play major roles in plant development. Much research has been carried out in these areas (*47*), particularly with light control (*48*). Photoreceptor proteins such as phytochrome have been identified and shown to par-

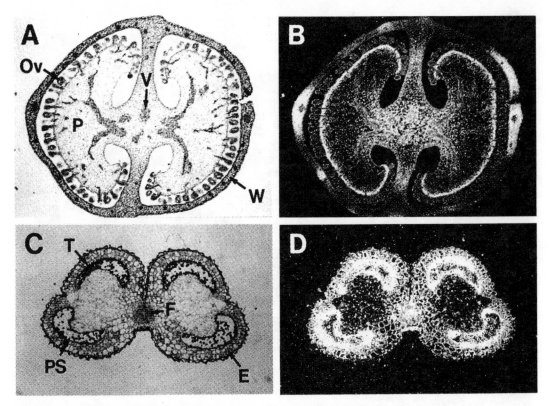

Fig. 6. Localization of mRNAs in the tobacco anther and ovary. Stage 6 ovaries and stage 4 anthers (Fig. 5) were fixed, embedded in paraffin, and hybridized in situ with single-stranded ^{35}S-labeled RNA probes (15–17). (**A** and **B**) Hybridization of an ovary cross section with the TA20 probe (Fig. 5). The TA20 mRNA is approximately 20% less prevalent in the pistil than in the anther (40, 41). (A) Bright-field photograph. (B) Dark-field photograph. White grains represent regions containing RNA/RNA hybrids. (**C** and **D**) Hybridization of an anther cross section with the TA29 probe (Fig. 5). (C) Bright-field photograph. (D) Dark-field photograph. White grains represent regions with RNA/RNA hybrids. Abbreviations: Ov, ovule; P, placenta; V, vascular tissue; W, ovary wall; PS, pollen sac; T, tapetum; and E, endothecium.

ticipate in light-regulated gene expression programs (47, 48). However, the exact mechanisms by which light and other environmental signals are perceived by plant cells and converted into molecular genetic information are not yet understood.

Higher Plants Have Complex Genetic Processes

As outlined above, plant developmental events are different from those found in the animal kingdom and are simpler in many cases. Paradoxically, gene expression processes that ultimately control and guide plant development are similar to those found in animal cells and are equally complicated. Plant genomes are as large and complex as those observed in the animal kingdom (49). The corn plant, for example, has a genetic potential equal to that of humans despite much less apparent biological complexity (50). In contrast with simple eukaryotes such as fungi (51), plants have complex nuclear RNA populations (22). The nuclear RNA complexity of each organ system is approximately four times that observed in the cytoplasm (Fig. 3), and reflects, in part, transcription of introns contained within plant genes (52). DNA sequencing studies and functional analysis of plant genes in transformed plants indicate that they have developmental control elements, splice junctions, promoters, and poly(A) addition sequences analogous to those found in the animal kingdom (3, 52). A large number of genes are expressed in plant organ systems, and plant gene expression is highly regulated (Fig. 3) (*21, 22*). Finally, both transcriptional and post-transcriptional events program plant gene expression (*21, 22, 53*). Experiments designed to unravel the molecular basis of plant gene expression will have to deal with relatively large genomes that express complex gene sets. Even the small *Arabidopsis* genome has 70,000 kb of diverse single-copy sequence (*45*). Ultimately, the major differences in plant and animal gene expression programs will involve the nature of signals that activate and repress specific genes—that is, how do unique plant physiological processes interface with gene expression events to trigger novel plant developmental states?

Higher Plants Have Many Advantages as an Experimental System

In the preceding sections I outlined many problems specific to plants. How difficult is it to study these problems at the molecular level? First, plant gene transfer technology is highly developed, easy to use, and applicable to a large number of plant species (*2, 3*). Plant cells can be transformed by microinjection, by use of *Agrobacterium* Ti plasmid vectors, or by direct uptake of exogenous DNA (*2, 3*); for example, it is not uncommon to obtain a 20% plant cell transformation frequency with Ti plasmid vectors (*54*). In addition, a single transformed tobacco cell can produce at least 250,000 seeds (*55*). Thus, investigating a large number of transformed plants is relatively easy. There are now many examples of development-specific and constitutively expressed genes that have been investigated in either transformed plants or protoplast cultures (*2, 3*). Second, because plant cells are totipotent, expression of modified genes can be studied throughout the plant life cycle and in any cell type that can be visualized by in situ hybridization or marker gene localization procedures (Fig. 4) (*2, 3, 15*). Transformation experiments have been carried out with genes expressed during embryogenesis (Fig. 4), as well as with those active in the mature sporophyte (*3*). Third, genetically

defined transposon systems in corn, snapdragon, and petunia are available for the identification of specific plant genes by transposon tagging (*56–58*). Several corn regulatory loci have been isolated by this procedure. These include the *cI* gene that regulates anthocyanin biosynthesis (*57*) and the *opaque-2* gene that controls seed protein deposition (*58*). Transformation of heterologous plant species by transposons indicates that these elements can move in foreign cell environments (*59*). For example, the corn Ac element is transposable in tobacco, carrot, and *Arabidopsis* plants (*59*). Thus, insertional mutagenesis schemes can be devised to isolate any gene that produces a scorable phenotype. Fourth, the *Arabidopsis* plant provides an accessible organism with which to study plant-specific processes (*45*). Because *Arabidopsis* has a small genome (70,000 kb), a short life cycle (6 weeks), a small size (30 cm), and very little repetitive DNA, large transposon and chemical mutagenesis programs can be carried out for the selection of genes controlling specific plant processes (*45*). Genes with chemically induced defects can be isolated, in principle, by complementation with wild-type DNA segments that flank restriction fragment length polymorphism (RFLP) markers linked to mutant genes (*45*). Finally, general advances in molecular biology, such as protein microsequencing (*60*), DNA binding protein purification procedures (*61*), and antisense gene technology (*62*) are being used to identify and study plant genes by a biochemical approach. The relative ease by which both molecular and genetic approaches can be applied in plants should facilitate the solution to many plant-specific problems.

Conclusions

Plants offer a large number of exciting and unexplored biological questions that cannot be examined in other organisms. Insight into plant-specific processes may suggest novel ways to produce superior crops by gene engineering technology. Despite the difficulties in studying certain aspects of the higher plant life cycle of higher plants and the long generation time in many plant species, there is an unprecedented opportunity to study plants at the molecular and cellular level.

References and Notes

1. G. Karp, *Cell Biology* (McGraw-Hill, New York, 1984); G. Mendel, *J. Heredity* **42**, (1951); B. McClintock, *Science* **226**, 792 (1984); D. S. Bendall and R. Hill, *Annu. Rev. Plant Physiol.* **19**, 167 (1968); H. Fraenkel-Conrat and R. C. Williams, *Proc. Natl. Acad. Sci. U.S.A.* **41**, 690 (1955); F. C. Steward, M. D. Mapes, K. Mears, *Am. J. Bot.* **45**, 705 (1958); F. C. Steward, *Sci. Am.* **209**, 104 (March 1963); F. C. Stewart, M. O. Mapes, A. E. Kent, R. D. Holsten, *Science* **143**, 20 (1964).

2. R. M. Goodman, H. Hauptli, A. Crossway, V. C. Knauf, *Science* **236**, 48 (1987); E. C. Cocking and M. R. Davey, *ibid.*, p. 1259; M. Vaeck *et al.*, *Nature* **328**, 33 (1987); L. Comai *et al.*, *ibid.* **317**, 741 (1984); D. M. Shah *et al.*, *Science* **233**, 478 (1986); P. P. Abel *et al.*, *ibid.* **232**, 738 (1986); R. S. Nelson *et al.*, *Biotechnology* **6**, 403 (1988).

3. J. St. Schell, *Science* **237**, 1176 (1987); H. Klee, R. Horsch, S. Rogers, *Annu. Rev. Plant Physiol.* **38**, 467 (1987).

4. P. H. Raven, R. F. Evert, S. E. Eichhorn, *Biology of Plants* (Worth, New York, 1986).

5. K. Umesono and H. Ozeki, *Trends Genet.* **3**, 382 (1987).

6. A. F. Dyer and J. G. Duckett, *The Experimental Biology of Bryophytes* (Academic Press, New York, 1984).

7. D. J. Cove, *Heredity* **43**, 295 (1979); I. Jenkins, G. R. M. Courtice, D. J. Cove, *Plant Cell Environ.* **9**, 637 (1986); N. W. Ashton and D. J. Cove, *Mol. Gen. Genet.* **154**, 87 (1977); N. H. Grimsley, N. W. Ashton, D. J. Cove, *ibid.*, p. 97; G. I. Jenkins, D. J. Cove, *Planta* **158**, 357 (1983); G. I. Jenkins and D. J. Cove, *ibid.* **159**, 432 (1983); T. S. Futers, T. L. Wong, D. J. Cove, *Mol. Gen. Genet.* **203**, 529 (1986).

8. S. E. Stevens and R. D. Porter, *Proc. Natl. Acad. Sci. U.S.A.* **77**, 6052 (1980); P. J. Lammers, J. W. Golden, R. Haselkorn, *Cell* **44**, 905 (1986).

9. W. F. J. Vermaas, J. K. Williams, A. W. Rutherford, P. Mathis, C. J. Arntzen, *Proc. Natl. Acad. Sci. U.S.A.* **83**, 9474 (1986).

10. E. Harris, *The Chlamydomonas Source Book* (Academic Press, New York, in press); J. D. Rochaix, *Microbiol. Rev.* **46**, 13 (1987).

11. J. D. Rochaix and J. Erickson, *Trends Biochem. Sci.* **13**, 56 (1988).

12. J. M. Erickson, M. Rahire, J.-D. Rochaix, L. Mets, *Science* **228**, 204 (1985).

13. J. E. Boynton *et al.* [*ibid.* **240**, 1534 (1988)] used high velocity microprojectiles (particle gun) to rescue a chloroplast photosynthesis deficiency mutant with wild-type DNA.

14. D. L. Kropf, B. Kloareg, R. S. Quatrano, *ibid.* **239**, 187 (1988).

15. K. H. Cox, D. V. DeLeon, L. M. Angerer, R. C. Angerer, *Dev. Biol.* **101**, 485 (1984).

16. S. J. Barker, J. J. Harada, R. B. Goldberg, *Proc. Natl. Acad. Sci. U.S.A.* **85**, 458 (1988).

17. K. H. Cox and R. B. Goldberg, in *Plant Molecular Biology: A Practical Approach*, C. H. Shaw, Ed. (IRL Press, Oxford, in press).

18. R. Yadegari, K. L. Cox, G. N. Drews, R. B. Goldberg, unpublished observations.

19. K. L. Cox and R. B. Goldberg, unpublished results. A stem cDNA library was constructed in pBR329 according to the procedure of U. Gubler and B. J. Hoffman [*Gene* **25**, 263 (1983)]. The TS13 cDNA clone was identified as representing a stem-specific mRNA by screening the stem library with stem, leaf, root, anther, pistil, and petal ^{32}P-labeled cDNAs.

20. G. N. Drews and R. B. Goldberg, unpublished results. A petal cDNA library was constructed by producers similar to that used for the stem (*19*). The TP7 cDNA clone was identified by screening the petal cDNA library with ^{32}P-labeled cDNAs prepared from mRNAs isolated from pink and white petal regions (Fig. 1). The TP7 mRNA is preferentially concentrated in pigmented petal regions.

21. J. C. Kamalay and R. B. Goldberg, *Cell* **19**, 935 (1980).

22. _____, *Proc. Natl. Acad. Sci. U.S.A.* **81**, 2801 (1984).

23. R. Axel, P. Feigelson, G. Schutz, *Cell* **7**, 247 (1976); N. J. Savage, J. M. Sala-Trepat, J. Bonner, *Biochemistry* **17**, 462 (1978); N. Chaudhari and W. E. Hahn, *Science* **220**, 924 (1983).

24. V. Rhaghavan, *Experimental Embryogenesis in Angiosperms* (Cambridge Univ. Press, Cambridge, MA, 1986).

25. T. K. Hodges, K. K. Kano, C. W. Imbrie, M. R. Becwar, *Biotechnology* **4**, 219 (1986); I. K. Vasil, *ibid.* **6**, 397 (1988).

26. R. P. Willing and J. P. Mascarenhas, *Plant Physiol.* **75**, 865 (1984); J. R. Stinson *et al.*, *ibid.* **83**, 442 (1987); J. P. Mascarenhas, in *Temporal and Spatial Regulation of Plant Genes*, D. P. S. Verma and R. B. Goldberg, Eds. (Springer-Verlag, Vienna, in press).

27. S. D. Russell, *Plant Physiol.* **81**, 317 (1986); E. Mattys-Rochon, P. Vergue, S. Detchepare, C. Dumas, *ibid.* **83**, 464 (1987).

28. E. H. Davidson, *Gene Activity in Early Develop-*

ment (Academic Press, New York, 1986).

29. J. B. Nasrallah, T. H. Kao, C. H. Chen, M. L. Goldberg, M. E. Nasrallah, *Nature* **326**, 617 (1987); E. C. Cornish, J. M. Pettitt, I. Bonig, A. E. Clarke, *ibid.*, p. 99; E. C. Cornish, J. M. Pettitt, A. E. Clarke, in *Temporal and Spatial Regulation of Plant Genes*, D. P. S. Verma and R. B. Goldberg, Eds. (Springer-Verlag, Vienna, in press).

30. S. D. Russell, *Proc. Natl. Acad. Sci. U.S.A.* **82**, 6129 (1985).

31. L. Perez-Grau and R. B. Goldberg, unpublished results. Seeds were fixed, embedded and hybridized in situ as in (*16, 17*). *Glycine max* cv 'Dare' (soybean) and *Nicotiana tabacum* cv 'SR1' (tobacco) were used for these experiments. Soybean and tobacco seed development takes place over approximately 110 and 30 days, respectively (*16*).

32. A. M. Breton and Z. R. Sung, *Dev. Biol.* **90**, 58 (1982); Z. R. Sung and R. Okimoto, *Proc. Natl. Acad. Sci. U.S.A.* **80**, 2661 (1983); J. H. Choi, L. S. Liu, C. Borkird, Z. R. Sung, *ibid.* **84**, 2906 (1987).

33. D. W. Meinke, *Theor. Appl. Genet.* **69**, 543 (1985).

34. R. B. Goldberg, G. Hoschek, S. H. Tam, G. S. Ditta, R. W. Breidenbach, *Dev. Biol.* **83**, 201 (1981); G. A. Galau and L. Dure, *Biochemistry* **20**, 4169 (1981); H. Morton, I. M. Evans, J. A. Gatehouse, D. Boulter, *Phytochemistry* **22**, 807 (1983).

35. R. B. Goldberg, G. Hoschek, G. S. Ditta, R. W. Breidenbach, *Dev. Biol.* **83**, 218 (1981); L. Dure, J. B. Pyle, C. A. Chan, J. C. Baker, G. A. Galau, *Plant Mol. Biol.* **2**, 199 (1983).

36. L. O. Vodkin, *Plant Physiol.* **68**, 766 (1981); R. B. Goldberg, G. Hoschek, L. O. Vodkin, *Cell* **33**, 465 (1983); L. O. Vodkin, P. R. Rhodes, R. B. Goldberg, *ibid.* **34**, 1023 (1983).

37. J. K. Okamuro, K. D. Jofuku, R. B. Goldberg, *Proc. Natl. Acad. Sci. U.S.A.* **83**, 8240 (1986); B. B. Mathews, L. Perez-Grau, R. B. Goldberg, unpublished results. Lectin mRNA represents approximately 0.75% of the embryo mRNA mass at mid-maturation (*36*). Lectin mRNA is reduced 200-fold, 20,000-fold, and 1,000,000-fold in the axis, root, and leaf, respectively.

38. Z. L. Chen, M. A. Schuler, R. N. Beachy, *Proc. Natl. Acad. Sci. U.S.A.* **83**, 8560 (1986); K. D. Jofuku, J. K. Okamuro, R. B. Goldberg, *Nature* **328**, 734 (1987).

39. A. Lang, in *Plant Gene System and Their Biology*, J. L. Key and L. McIntosh, Eds. (Liss, New York, 1987); S. R. Singer and C. N. McDaniel, *Dev. Biol.* **118**, 587 (1986); *Proc. Natl. Acad. Sci. U.S.A.* **84**, 2790 (1987).

40. J. Truettner and R. B. Goldberg, unpublished results. The common tobacco, *Nicotiana tabacum* cv 'Samsun' was used for these experiments. Anther mRNAs were isolated as described in (*21*) and by K. H. Cox and R. B. Goldberg in *Plant Molecular Biology: A Practical Approach*, C. H. Shaw, Ed. (IRL Press, Oxford, in press). An anther cDNA library was constructed in pBR329 according to the procedure outlined in (*19*), with stage 6 flowers. Complementary DNA clones representing mRNAs present exclusively, or at increased levels, in the anther were identified by screening the anther cDNA library separately with anther, pistil, petal, leaf, root, and stem ^{32}P-labeled cDNAs. The TA13 and TA29 cDNA clones have approximately 85% sequence identity with each other (J. Botterman, J. Leemans, R. B. Goldberg, unpublished results) and are complementary to 1.1- and 1.2-kb mRNAs. By contrast, the TA20 and TA25 cDNA clones do not detectably cross-react with TA13, TA29, or with each other, and are complementary to 0.6-kb mRNAs.

41. M. Wallroth and R. B. Goldberg, unpublished results. In situ hybridization was carried out as described (*16, 17*). Ovaries and anthers were fixed as described by G. S. Avery [*Am. J. Bot.* **20**, 309 (1937)]. The TA20 anther localization experiment is not shown.

42. C. S. Gasser *et al.*, in *Temporal and Spatial Regulation of Plant Genes*, D. P. S. Verma and R. B. Goldberg, Eds. (Springer-Verlag, Vienna, in press).

43. S. Satina, A. F. Blakeslee, A. G. Avery, *Am. J. Bot.* **27**, 895 (1940); S. Satina and A. F. Blakeslee, *ibid.* **28**, 862 (1941); R. S. Poethig, *ibid.* **74**, 581 (1987).

44. P. B. Green, *Am. Zool.* **27**, 657 (1987).

45. E. M. Meyerowitz and R. E. Pruitt, *Science* **229**, 1214 (1985); E. M. Meyerowitz, *Annu. Rev. Genet.* **21**, 93 (1987); P. S. Pang and E. M. Meyerowitz, *Biotechnology* **5**, 1177 (1987).

46. E. S. Coen, J. Almeida, T. P. Robbins, A. Hudson, R. Carpenter, in *Temporal and Spatial Regulation of Plant Genes*, D. P. S. Verma and R. B. Goldberg, Eds. (Springer-Verlag, Vienna, in press); M. Freeling, D. K. Bongard-Pierce, N. Hanlead, B. Lana, S. Hake, in *ibid.*

47. B. G. Pickard, *Annu. Rev. Plant Physiol.* **36**, 55 (1985); J. M. Morgan, *ibid.* **35**, 199 (1984); P. L. Steponkus, *ibid.*, p. 543; R. Malkin, *ibid.* **34**, 455 (1983); L. Pruitt, *ibid.*, p. 557.

48. R. Fluhr, C. Kuhlemeier, E. Nagy, N.-H. Chua, *Science* **232**, 1106 (1986); E. M. Tobin and J. Silverthorne, *Annu. Rev. Plant Physiol.* **36**, 569 (1985).

49. M. D. Bennett and J. B. Smith, *Proc. R. Soc.*

London Ser. B **274**, 227 (1976).

50. S. Hake and V. Walbot, *Chromosoma* **79**, 251 (1980).
51. W. E. Timberlake, D. S. Shumard, R. B. Goldberg, *Cell* **10**, 623 (1977).
52. G. Heidecker and J. Messing, *Annu. Rev. Plant Physiol.* **37**, 439 (1986); J. W. S. Brown, *Nucleic Acids Res.* **14**, 9549 (1986).
53. L. R. Beach, D. Spencer, P. J. Randall, T. J. V. Higgens, *Nucleic Acids Res.* **13**, 999 (1985); I. M. Evans *et al.*, *Planta* **160**, 559 (1984); L. Walling, G. N. Drews, R. B. Goldberg, *Proc. Natl. Acad. Sci. U.S.A.* **83**, 2123 (1986); C. Kuhlemeier, P. J. Green, N. H. Chua, *Annu. Rev. Plant Physiol.* **38**, 221 (1987).
54. S. R. Rogers, R. B. Horsch, R. T. Fraley, *Methods Enzymol.* **118**, 627 (1986); R. B. Horsch *et al.*, *Science* **227**, 1229 (1984).
55. T. S. Goodspeed, *The Genus Nicotiana* (Chronica Botanica, Waltham, MA, 1954).
56. N. Federoff, S. Wessler, M. Shure, *Cell* **35**, 235 (1983); H. Dooner, J. English, E. Ralston, E. Weck, *Science* **234**, 210 (1986); N. Theres, T. Scheele, P. Starlinger, *Mol. Gen. Genet.* **209**, 193 (1987); C. Martin, R. Carpenter, H. Sommer, H. Saedler, E. S. Coen, *EMBO J.* **4**, 1625 (1985).
57. J. Paz-Ares, D. Ghosal, W. U. Wienand, P. A. Anderson, H. Saedler, *EMBO J.* **6**, 3553 (1987); K. C. Cone, F. A. Burr, B. Burr, *Proc. Natl. Acad. Sci. U.S.A.* **83**, 9631 (1986).
58. R. J. Schmidt, F. A. Burr, B. Burr, *Science* **238**, 960 (1987).
59. B. Baker, J. Schell, H. Lorz, N. Federoff, *Proc. Natl. Acad. Sci. U.S.A.* **83**, 4844 (1986); B. Baker, G. Coupland, N. Federoff, P. Starlinger, J. Schell, *EMBO J.* **6**, 1547 (1987); M. A. Van Sluys, J. Tempé, N. Federoff, *ibid.*, p. 3881.
60. G. Bauw, M. DeLoose, D. Inzé, M. Van Montagu, J. Vandekerckhove, *Proc. Natl. Acad. Sci. U.S.A.* **84**, 4806 (1987).
61. U. Heberlein and R. Tjian, *Nature* **331**, 410 (1988); J. Kadonaga and R. Tjian, *Proc. Natl. Acad. Sci. U.S.A.* **83**, 5889 (1986); H. Singh, J. H. LeBowitz, A. S. Baldwin, P. A. Sharp, *Cell* **52**, 495 (1988).
62. S. J. Rothstein, J. DiMaio, M. Strand, D. Rice, *Proc. Natl. Acad. Sci. U.S.A.* **84**, 8439 (1987); D. A. Knecht and W. F. Loomis, *Science* **236**, 1081 (1987).

63. I thank G. Drews for suggestions regarding this manuscript. In addition, I thank my associates, J. Treuttner, R. Yadegari, L. Perez-Grau, K. Cox, M. Wallroth, and G. Drews, for letting me use their unpublished observations. The work presented in this paper was supported by grants from the National Science Foundation and the U.S. Department of Agriculture.

11

Genetically Engineering Plants for Crop Improvement

Charles S. Gasser and Robert T. Fraley

The stable introduction of foreign genes into plants represents one of the most significant developments in a continuum of advances in agricultural technology that includes modern plant breeding, hybrid seed production, farm mechanization, and the use of agrichemicals to provide nutrients and control pests. The first-generation applications of genetic engineering to crop agriculture are targeted at issues that are currently being addressed by traditional breeding and agrichemical discovery efforts: (i) improved production efficiency, (ii) increased market focus, and (iii) enhanced environmental conservation. Genetic engineering methods complement plant breeding efforts by increasing the diversity of genes and germplasm available for incorporation into crops and by shortening the time required for the production of new varieties and hybrids. Genetic engineering of plants also offers exciting opportunities for the agrichemical, food processing, specialty chemical, and pharmaceutical industries to develop new products and manufacturing processes.

The first transgenic plants expressing engineered foreign genes were tobacco plants produced by the use of *Agrobacterium tumefaciens* vectors (*1*). Transformation was confirmed by the presence of foreign DNA sequences in both primary transformants and their progeny and by an antibiotic resistance phenotype conferred by a chimeric neomycin phosphotransferase gene. These early transformation experiments often utilized plant protoplasts as the recipient cells; the subsequent development of transformation methods based on regenerable explants (*2*) such as leaves, stems, and roots contributed significantly to the facile and routine transformation methods that are used today for many dicotyledonous plant species. There are a variety of free DNA delivery methods, including microinjection, electroporation, and particle gun technology, that are being developed for the transformation of monocotyledonous plants such as corn, wheat, and rice. In view of the rapid progress that is being made, it is likely that all major dicotyledonous and monocotyledonous crop species will be amenable to improvement by genetic engineering within the next few years.

In this chapter, we describe transformation methods that have been developed for plants and discuss some of the applications of genetically engineered plants in agriculture. We also address some of the critical issues that will influence the commercialization of genetically engineered crops.

C. S. Gasser is at the Department of Biochemistry and Biophysics, University of California, Davis, CA 95616. R. T. Fraley is at Monsanto Company, 700 Chesterfield Village Parkway, St. Louis, MO 63198. This chapter is reprinted from *Science* **244**, 1293 (1989).

Methods for Introducing Genes into Plants

Agrobacterium tumefaciens–*mediated gene transfer*

Derivatives of the plant pathogen *Agrobacterium tumefaciens* have proved to be efficient, highly versatile vehicles for the introduction of genes into plants and plant cells. Most transgenic plants produced to date were created through the use of the *Agrobacterium* system. *Agrobacterium tumefaciens* is the etiological agent of crown gall disease and produces tumorous crown galls on infected species. The utility of this bacterium as a gene transfer system was first recognized when it was demonstrated that the crown galls were actually produced as a result of the transfer and integration of genes from the bacterium into the genome of the plant cells (*3*). Virulent strains of *Agrobacterium* contain large Ti (for tumor inducing) plasmids, which are responsible for the DNA transfer and subsequent disease symptoms. Genetic and molecular analyses showed that Ti plasmids contain two sets of sequences necessary for gene transfer to plants; one or more T-DNA (transferred DNA) regions that are transferred to the plant, and the *Vir* (virulence) genes which are not, themselves, transferred during infection. The T-DNA regions are flanked by border sequences that were shown to be responsible for the definition of the region that is to be transferred to the infected plant cell. The T-DNA contains 8 to 13 genes (*4*), including a set for production of phytohormones, which are responsible for formation of the characteristic tumors when transferred to infected plants. Several excellent reviews on the biology of this and other pathogenic species of *Agrobacterium* have been published for those who desire more detailed information (*4*).

Early experiments demonstrated that heterologous DNA inserted into the T-DNA could be transferred to plants along with the existing T-DNA genes (*5*). Efficient plant transformation systems were constructed by removing the phytohormone biosynthetic genes from the T-DNA region, thereby eliminating the ability of the bacteria to induce aberrant cell proliferation (*6*). Modern plant transformation vectors are capable of replication in *Escherichia coli* as well as *Agrobacterium*, allowing for convenient manipulations (*7*). The general features of these vectors and the process of transfer to plant cells are outlined in Fig. 1. Recent technological advances in vectors for *Agrobacterium*-mediated gene transfer have involved improvements in the arrangements of genes and restriction sites in the plasmids that facilitate construction of new expression vectors. Vectors in current use have convenient multilinker regions, which may be flanked by a promoter and a polyadenylate addition site for direct expression of inserted coding sequences (*8*).

Agrobacterium constitutes an excellent system for introducing genes into plant cells, since (i) DNA can be introduced into whole plant tissues, which bypasses the need for protoplasts, and (ii) the integration of T-DNA is a relatively precise process. The region of DNA to be transferred is defined by the border sequences; occasional rearrangements do occur, but in most cases an intact T-DNA region is inserted into the plant genome (*9*). This contrasts with free DNA delivery systems in which the plasmids routinely undergo rearrangment and concatenation reactions before insertion and can lead to chromosomal rearrangements during insertion in both animal (*10*) and plant (*11*) systems. Sequencing of insertion sites shows that only small duplications or other changes occur in flanking sequences during T-DNA integration (*12*). The stability of expression of most genes that are introduced by *Agrobacterium* appears to be excellent. Published studies have shown that integrated T-DNAs give consistent genetic maps and appropriate segregation ratios (*1, 13*). Introduced traits have been found to be stable over at least five generations during cross-breeding and seed increase on genetically engineered tomato and oilseed rape plants (*14*). This stability is critical to the commercialization of transgenic plants. The list of plant species that can be transformed by *Agrobacterium* has been greatly expanded and now includes several of the most important broadleaf crops (Table 1).

A

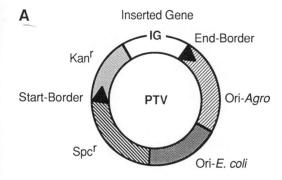

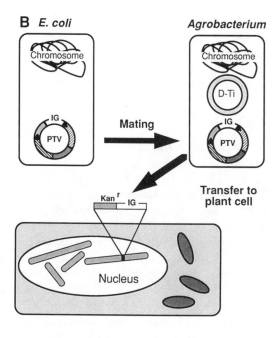

Fig. 1. *Agrobacterium* mediated plant transformation. (**A**) Generalized plant transformation vector (PTV). The plasmid contains an origin of replication that allows it to replicate in *Agrobacterium* (Ori-*Agro*), and a high copy number origin of replication functional in *E. coli* (Ori-*E. coli*). This allows for easy production and testing of engineered plasmids in *E. coli* prior to transfer to *Agrobacterium* for subsequent introduction into plants. Two resistance genes are usually carried on the plasmid, one for selection in bacteria, in this case for spectinomycin resistance (*Spc*[r]), and the other that will express in plants; in this example encoding kanamycin resistance (*Kan*[r]). Also present are sites for the addition of one or more inserted genes (IG) and directional T-DNA border sequences which, when recognized by the transfer functions of *Agrobacterium*, delimit the region that will be transferred to the plant. (**B**) Diagram of the plant transformation process. The PTV constructed in *E. coli* is transferred to an engineered *Agrobacterium* by a "triparental" mating procedure (*6*). The engineered *Agrobacterium* contains a "disarmed" Ti plasmid (D-Ti) from which the genes necessary for pathogenesis have been removed (*6*). Virulence functions on the D-Ti interact in trans with the border sequences on the PTV mobilizing the region between them into a plant cell and inserting it into one of the plant's chromosomes within the nucleus. The kanamycin-resistant phenotype conferred by the *Kan*[r] gene allows the selection of transformed plant cells during plant regeneration.

Advances in other transformation technologies

In those systems where *Agrobacterium*-mediated transformation is efficient, it is the method of choice because of the facile and defined nature of the gene transfer. Few monocotyledonous plants appear to be natural hosts for *Agrobacterium*, although transgenic plants have been produced in asparagus with *Agrobacterium* vectors (*15*) and transformed tumors have been observed in yam (*16*). Cereal grains such as rice, corn, and wheat have not been successfully transformed by *Agrobacterium*, despite encouraging evidence for T-DNA transfer in corn (*17*). Extensive efforts have consequently been directed toward the development of systems for the delivery of free DNA into these species. The first of these systems to give demonstrable transformation of plant cells relied on physical means similar to those used in the transformation of cultured animal cells. Transformation has been achieved in plant protoplasts through facilitation of DNA uptake by calcium phosphate precipitation, polyethylene glycol treatment, electroporation, or combinations of these treatments (*18*). These methods have allowed the production of transgenic cells for the study of gene expression in systems that cannot be transformed by other means (*19*).

The applicability of these systems to the production of transgenic plants is limited by the difficulties involved in regenerating plants

Table 1. Species for which the production of transgenic plants have been reported. Abbreviations: At, *Agrobacterium tumefaciens*; Ar, *Agrobacterium rhizogenes*; FP, free DNA introduction into protoplasts; PG, particle gun; MI, microinjection; IR, injection of reproductive organs.

Plant species	Method (reference)
Herbacious dicots	
Petunia	At (2)
Tomato	At (83)
Potato	At (84)
Tobacco	At (1), FP (85), PG (26)
Arabidopsis	At (86)
Lettuce	At (87)
Sunflower	At (88)
Oilseed rape	At (89), MI (31)
Flax	At (90)
Cotton	At (91)
Sugarbeet	At (92)
Celery	At (93)
Soybean	At (38), PG (27)
Alfalfa	At (94)
Medicago varia	At (95)
Lotus	At (96)
Vigna aconitifolia	FP (97)
Cucumber	Ar (98)
Carrot	Ar (99)
Cauliflower	Ar (100)
Horseradish	Ar (101)
Morning glory	Ar (102)
Woody dicots	
Poplar	At (103)
Walnut	At (104)
Apple	At (105)
Monocots	
Asparagus	At (15)
Rice	FP (21)
Corn	FP (23)
Orchard grass	FP (106)
Rye	IR (30)

from protoplasts. There have been significant advances in the regeneration of cereals (traditionally one of the most recalcitrant groups) from protoplasts. Several laboratories have succeeded in regenerating fertile rice plants from protoplasts (20). This advance was rapidly followed by the production of transgenic rice plants through the delivery of free DNA to protoplasts followed by regeneration (21). Progress in regeneration of corn has been more limited; one group demonstrated regeneration

of mature plants from protoplasts and succeeded in producing transgenic plants (22, 23). However, all plants were sterile, apparently as a result of the necessary period in culture or the regeneration procedure. While this progress is encouraging, limitations remain in the application of this technology to cereal crop improvement. In corn and rice, the ability to form regenerable protoplasts appears to be primarily confined to a small number of varieties. Even if the fertility problems are overcome, introduction of the transferred genes into the broad range of commercial varieties in use today would require a lengthy period of backcrossing.

In parallel with the work on protoplast transformation, efforts to find novel ways to introduce DNA into intact cells or tissues have been emphasized. Regeneration of cereals from immature embryos or from explants is relatively routine (24). One of the most significant developments in this area has been the introduction of "particle gun" or high-velocity microprojectile technology. In this system, DNA is carried through the cell wall and into the cytoplasm on the surface of small (0.5 to 5 μm) metal particles that have been accelerated to speeds of one to several hundred meters per second (25–27). The particles are capable of penetrating through several layers of cells and allow the transformation of cells within tissue explants. Production of transformed corn cells (28) and fertile, stably transformed tobacco (26) and soybean (27) plants with particle guns has already been demonstrated. By eliminating the need for passage through a protoplast stage, the particle gun method has the potential to allow direct transformation of commerical genotypes of cereal plants. Intensive efforts to produce transgenic cereals by the use of particle guns are currently under way in many laboratories around the world.

Other methods that have the potential to influence the production of transgenic cereals include gene transfer into pollen (29), direct injection into reproductive organs (30), microinjection into cells of immature embryos (31), and rehydration of desiccated embryos (32). There has been some demonstration of transient or stable gene expression through the use of each of these methods in some species, but the range of their applicability remains to be demonstrated.

Application of Genetic Engineering to Crop Improvement

The availability of efficient transformation systems for crop species is of intense interest to biotechnology, agrichemical, and seed companies for the application of this technology to crop improvement. Initial research has been focused on the engineering of traits that relate directly to the traditional roles of industry in farming, such as the control of insects, weeds, and plant diseases. Progress has been rapid, and genes conferring these traits have already been successfully introduced into several important crop species. Genetically engineered soybean, cotton, rice, corn, oilseed rape, sugarbeet, tomato, and alfalfa crops are expected to enter the marketplace between 1993 and 2000.

Weed control

Engineering herbicide tolerance into crops represents a new alternative for conferring selectivity and enhancing crop safety of herbicides. Research has largely concentrated on those herbicides with properties such as high unit activity, low toxicity, low soil mobility, and rapid biodegradation and with broad spectrum activity against various weeds. The development of crop plants that are tolerant to such herbicides would provide more effective, less costly, and more environmentally attractive weed control. The commercial strategy in engineering herbicide tolerance is to gain market share through a shift in herbicide use (*33*)—not to increase the overall use of herbicides, as is popularly held. Herbicide-resistant plants will have the positive impact of reducing overall herbicide use through substitution of more effective and environmentally acceptable products.

Two general approaches have been taken in engineering herbicide tolerance: (i) altering the level and sensitivity of the target enzyme for

the herbicide and (ii) incorporating a gene that will detoxify the herbicide. As an example of the first approach, glyphosate, the active ingredient of Roundup herbicide, acts by specifically inhibiting the enzyme 5-enolpyruvylshikimate-3-phosphate synthase (EPSPS) (*34*). Glyphosate is active against annual and perennial broadleaf and grassy weeds, has very low animal toxicity, and is rapidly inactivated and degraded in all soils (*35*). Tolerance to glyphosate has been engineered into various crops by introducing genetic constructions for the overproduction of EPSPS (*36*) or of glyphosate-tolerant variant EPSPS enzymes (*37, 38*). Similarly, resistance to sulfonylurea compounds, the active ingredients in Glean and Oust herbicides, has been produced by the introduction of mutant acetolactate synthase (ALS) genes (*39*). Glean and Oust are broadspectrum herbicides and are effective at low application rates. Since both EPSPS and ALS activities are present in wild-type plants, the possibility of deleterious effects on crop performance or product quality due to their reintroduction is unlikely. The use of these herbicides in new crop applications may require reexamination of residues of the herbicides; however, since the residue safety levels for these two compounds in food crops have already been established, this is not an issue unique to genetically engineered plants.

Resistance to gluphosinate (*40*) and bromoxynil (*41*) has been achieved by the alternative approach of introducing bacterial genes encoding enzymes that inactivate the herbicides by acetylation or nitryl hydrolysis, respectively. In field tests the gluphosinate-tolerant plants have shown excellent tolerance to the herbicide (*42*). Evaluation of the biological activity of the specific herbicide conjugates and metabolites that may be present in the transgenic plants will be carried out according to existing chemical residue regulations.

Current crop targets for engineered herbicide tolerance include soybean, cotton, corn, oilseed rape, and sugarbeet. Factors such as herbicide performance, crop and chemical registration costs, potential for out-crossing to weed species, proprietary rights issues, and competing herbicide technologies must all be considered before final decisions on commercialization of specific herbicide-tolerant crops can be made.

Insect resistance

The production of insect-resistant plants is another application of genetic engineering with important implications for crop improvement and for both the seed and agrichemical industries. Progress in engineering insect resistance in transgenic plants has been achieved through the use of the insect control protein genes of *Bacillus thuringiensis* (B.t.). *Bacillus thuringiensis* is an entomocidal bacterium that produces an insect control protein which is lethal to selected insect pests (*43*). Most strains of B.t. are toxic to lepidopteran (moth and butterfly) larvae, although some strains with toxicity to coleopteran (beetle) (*44*) or dipteran (fly) (*45*) larvae have been described. The insect toxicity of B.t. resides in a large protein; this protein has no toxicity to beneficial insects, other animals, or humans (*46*). The mode of action of the B.t. insect control protein is thought to be exerted at the level of disruption of ion transport across brush border membranes of susceptible insects (*47*).

Transgenic tomato, tobacco, and cotton plants containing the B.t. gene exhibited tolerance to caterpillar pests in laboratory tests (*48*). The level of insect control observed in the field tests with tobacco and tomato plants has been excellent; in one such test tomato plants containing the B.t. gene suffered no agronomic damage under conditions that led to total defoliation of control plants (*49*).

The excellent insect control observed under field conditions indicates that this technology may have commercial application in the near future. Early market opportunities for caterpillar resistance are leafy vegetable crops, cotton, and corn. Crop targets for beetle resistance are potato and cotton. Other types of insecticidal molecules are necessary to extend biotechnology approaches for controlling additional insect pests in these and other target crops. Plants genetically engineered to express a proteinase inhibitor gene are partially resistant to tobacco budworm in laboratory experiments (*50*); field tests will be necessary to determine the agronomic utility of this approach.

Disease resistance

Significant resistance to tobacco mosaic virus (TMV) infection, termed "coat protein–mediated protection," has been achieved by expressing only the coat protein gene of TMV in transgenic plants (*51*). This approach produced similar results in transgenic tomato, tobacco, and potato plants against a broad spectrum of plant viruses, including alfalfa mosaic virus, cucumber mosaic virus, potato virus X, and potato virus Y (*52*). One mechanism of coat protein–mediated cross protection appears to involve interference with the uncoating of virus particles in cells before translation and replication (*53*).

Transgenic tomatoes carrying the TMV coat protein gene have been evaluated in greenhouse and field tests and shown to be highly resistant to viral infection (Fig. 2) (*54*). The transgenic plants showed no yield loss after virus inoculation, whereas the yield was reduced 23% to 69% in control plants. The level of capsid protein in the engineered plants [typically 0.01% to 0.5% of the total protein (*52*)] is well below the levels found in plants infected with this endemic virus. This fact should facilitate registration and commercialization of virus-resistant plants. Virus resistance could provide significant yield protection in important crops such as vegetables, corn, wheat, rice, and soybean.

While limited success in engineering resistance to fungal diseases has been reported (*55*), genetically engineered resistance to fungal pathogens and to bacteria remains in the early research stages.

Key Advances in Expression and Gene Isolation Technology

Dramatic progress has been made in our understanding of and ability to alter the regulation of gene expression in plants and in techniques for the identification and isolation of genes of interest. In many cases, this progress has been facilitated by the availability of efficient gene transfer systems. The engineered plants discussed in the previous section generally depend on the use of continuously expressed promoters driving dominant single gene traits. Future plant genetic engineering will probably include alteration of traits that require subtle temporal and spatial regulation of gene expression and introduction or alteration of entire biosynthetic pathways.

Regulated gene expression

Genes that show precise temporal and spatial regulation in leaves (*56*), floral organs (*57*), seeds (*58*), and other plant organs have now been identified and isolated from a number of species of higher plants (*59*). Within the next few years, genetic engineers will have in hand a large battery of regulatory sequences that will

allow for accurate targeting of gene expression to specific tissues within transgenic plants. In addition, a number of genes that respond to external influences, such as heat shock, anaerobiosis, wounding, nutrients, and applied phytohormones, have been isolated and characterized (*60*). The control regions of these genes may also find utility in genetic engineering strategies.

The ability to decrease the expression of a gene in a transgenic plant also has potential utility in the study of plant gene expression and function as well as in crop improvement. Significant successes have already been achieved with genes that produce antisense RNAs to the messengers for polygalacturonase in tomato fruits (*61*) and chalcone synthase in petunia and tobacco plants (*62*). In all of these studies, substantial reductions (up to 90%) in the levels of the mRNA and protein products of the target genes were observed. Striking phenotypic alterations were observed in some of these transgenic plants (*62*). This method of con-

Fig. 2. Virus-resistant plants. (**A**) Greenhouse evaluation of tomato plants containing the TMV coat protein gene. Tomato cotyledons were transformed (*83*) with an *Agrobacterium* strain containing a TMV coat protein cDNA chimeric gene (*51*). Transgenic regenerates were screened for coat protein production by immunoblot analysis. The R₁ progeny of a representative plant that expressed high levels of coat protein were anlayzed for virus resistance after inoculation with TMV. The control plant on the left is a segregant that lacks the TMV coat protein gene; the plant on the right has inherited the gene.

(**B**) Field test (1988) of tomato plants containing TMV coat protein gene. Control and transgenic tomato seedlings were grown in a greenhouse and transplanted 6 weeks later to the field test site located in Jersey County, Illinois. The control (left) and engineered (right) plants were inoculated with the PV230 strain of TMV (10 g/ml) 2 weeks after planting, and the photo was taken 4 weeks later. The fruit yield on the control plants was 19.6 kg per plot compared to 62.4 kg per plot for the engineered plants; the yield of the inoculated engineered plants was equivalent to that of noninoculated plants.

structing mutant phenotypes will significantly enhance biochemical and physiological studies on protein and enzyme function. In an alternative approach to reducing expression of a gene, the enzymatic regions derived from self-splicing RNA molecules are used to design RNA enzymes capable of specific RNA cleavage (63). In vitro studies have demonstrated the potential of this method, but it has yet to be applied in plants (63). Preliminary work on insertion of donor DNA into plant chromosomes by homologous recombination (64) indicates that it may also be possible to use this approach for the selective inactivation of a gene.

Gene tagging

Advances in methods for the identification and isolation of new gene coding sequences are of great importance to the engineering of improved plants. The cloning of transposon sequences has allowed the isolation of genes from several species by transposon-mediated gene tagging (65). The demonstration that mobile elements isolated from maize are able to transpose when introduced into dicot species (66) indicates that this powerful technique is applicable to any plant species for which transformation is possible. It has also been shown that under appropriate transformation conditions, the T-DNA of a plant transformation vector can itself serve as an insertional mutagen (67).

Gene mapping

Major efforts have been mounted to obtain high-resolution restriction fragment length polymorphism (RFLP) genetic maps in a number of plant species (68). The availability of such a map in tomato has already led to the resolution of several loci affecting quantitative quality traits (69). The RFLP mapping technique will be especially powerful in *Arabidopsis*, where the small genome size and lack of significant repetitive sequences (70) will simplify the process of genome "walking" from an

RFLP marker to a closely linked gene. The availability of *Arabidopsis* genomic libraries in cosmids, which can also act as plant transformation plasmids (71), will allow direct testing of the isolated DNA for its ability to complement the mutation of interest at each step of the walking process. In addition, such libraries may be used in large-scale transformation experiments to directly rescue genes by complementing mutants with a selectable phenotype (71).

Key Issues Affecting Introduction of Genetically Engineered Plants

The advances in crop improvement by genetic engineering have occurred so rapidly that the initial introduction of these crops in the marketplace will be primarily influenced by nontechnical issues. These issues include regulatory approval, proprietary protection, and public perception.

Regulatory approval

In the United States, genetically engineered plants potentially come under the statutory jurisdiction of three federal agencies: the United States Department of Agriculture (USDA), Food and Drug Administration (FDA), and Environmental Protection Agency (EPA). The field testing of genetically engineered crops has been less controversial than the introduction of other recombinant organisms into the environment. In the last 3 years there have been over a dozen tests of engineered crops in diverse locations across the United States (72)—by year end there will be over 30 such tests. All of these tests have been reviewed in detail by the USDA, with input from the other government agencies. The key consideration in approval of these tests has been a scientific evaluation of the risk and environmental impact of a particular field test experiment. Several studies and discussions of the issues and perceptions that surround the release of genetically engineered crops have produced a consensus that such engineered crops present virtually no direct risk to human or animal health (73). The specific knowledge of the introduced DNA sequences, the detailed understanding of the known functions of the gene products, and the high level of biological or physical containment were cited as key reasons for the inherent low risk to human and animal health.

The "success" of such small field tests, while important, has overshadowed other needs in the regulatory process. For example, many unanswered questions remain regarding the cost and regulatory requirements for large-scale multisite field tests. It is important that an approval process be developed to accommodate the rapid transition that will occur as testing of engineered crops goes from small, isolated field plots to large-scale, multisite testing; the development of genetically engineered crop varieties and hybrids will ultimately occur in the fields around the world—not in the research laboratory. The mechanism for FDA or EPA approval or endorsement of genetically engineered plants and food products remains undefined. Issues such as regulatory requirements, registration costs, and commercialization timelines are already becoming significant issues for companies attempting to develop improved genetically engineered crops for the mid-1990s. Several groups (74), such as the International Food Biotechnology Council (IFBC) and the Federation of American Scientists for Experimental Biology (FASEB) expert panel on criteria for determining the regulatory status of food and food ingredients produced by new technologies, consisting of academic scientists and representatives of major food, chemical, biotechnology, and seed companies, are working with government agencies to develop appropriate registration guidelines. The regulation of transgenic plants must be based on scientific principles that (i) meet the general public's need for a safe and reasonably priced food supply and (ii) recognize the inherent low risk of gene transfer technology and the benefits afforded by genetically engineered crops to growers, food processors, and consumers.

Proprietary protection

Patent protection for genetically engineered plants is considered essential to offset the cost of developing crops with significant new traits. The Supreme Court decision in Diamond *v.* Chakrabarty (75) ruled that microorganisms were not unpatentable simply because they were living cells, and in 1985, the U.S. Board of Patent Appeals and Interferences ruled specifically that whole plants were patentable (76). Numerous companies have since filed patent applications that cover the genes, the processes of isolating genes, and making the genetically modified plants and seeds themselves. Patent protection provides a broader proprietary right than is provided under either the International Union for the Protection of New Varieties of Plants (UPOV) or the U.S. Plant Variety Protection Act (PVPA). The scope of the proprietary right of a patent on a plant is broadened by the absence of the "experimental use" exceptions found in protection afforded by plant varietal certification status. Although no one disputes that companies that have invested heavily in R&D to isolate, test, and commercialize genes are entitled to protection for their inventions, there is considerable debate within the seed industry concerning how much protection is deserved and what impact patents will have on the cooperative nature of the seed industry itself (77). The concern has been voiced that patents on plants will favor large seed companies and reduce the overall number of companies. In contrast, while there were three private soybean seed companies before PVPA, now there are more than 40; patenting plants will likely create further incentive to invest in the seed industry in order to position it to meet the technological challenges and supply needs of the future. Much of this debate results from confusion surrounding the restrictions imposed by patent rights versus the incentive they provide for the competitive research and product development that stimulates innovation. Many of the conciliatory proposals, including patenting of genes (but not plants) and compulsory licensing in the event that plant patenting is permitted, if implemented, could significantly reduce the incentive for private industry funding in this field.

Lack of proprietary protection for genetically engineered plants outside the United States remains a serious limitation; plant and animal varieties are largely excluded from patent protection by European countries that signed the 1973 European Patent Convention. At this time only specific processes can be patented. The European Patent Office (EPO) is currently readdressing the patenting of plants and animals, but this seems certain to be appealed. It may be several years before the situation is clear, and only then will begin the wave of oppositions, appeals, and infringement actions that have marked the early pharmaceutical patents in the biotechnology area (78). Enforceability of plant patents in other countries, including Japan, China, and Eastern Bloc countries, is questionable. While there are numerous initiatives to harmonize both registration and proprietary protection throughout the key trading countries in the world, the outcome is not imminent and will be unlikely to have an impact on first-generation products.

Public perception

Genetically engineered crops are being developed at a time when a lack of understanding regarding the importance of agricultural research exists. Current issues, including concerns about (i) periodic, temporary production surpluses, (ii) changing farm infrastructure, (iii) inconsistency in farm policies, and (iv) a general distrust for new technologies, have at times overshadowed the long-term need for the provision of economical, high-quality food products for a growing world population. Currently, at the beginning of the 1989 cropping season, world reserves of grain are at their lowest level since the years immediately following World War II; another drought in 1989 could create a world food emergency (79).

Despite this background, recent polls conducted by the Office of Technology Assessment indicate that most people believe that the benefits of agricultural biotechnology research

outweigh remote risks (72). In view of the initial public debate that has occurred over the last several years on field testing and environmental release of genetically engineered organisms, it would seem that agricultural biotechnology has indeed passed its first major public perception obstacle.

The next test of the public acceptance of this technology will come in several years when food products derived from genetically engineered crops enter the general food supply. The current focus on issues of risk and environmental release has heightened the need for increased science education and open discussion of issues. It is essential that the safety and benefits of agricultural biotechnology research and the critical role that it will play in providing for world food demand (80) be communicated and understood, so that informed decisions by the public are possible.

A Future Perspective on Genetically Engineered Plants

During the last 5 years, the availability of gene transfer systems has catalyzed a major refocusing on plants as a biological system; the use of genetically engineered plants as an analytical tool to explore unique aspects of gene regulation and development and the potential to produce novel commercial crop varieties has created a high level of scientific excitement and has driven research into many new areas. The breadth of information to be gained from the study of transgenic plants is serving as an important focus for unifying basic plant science research in plant breeding, pathology, biochemistry, and physiology with molecular biology. Regulation of gene expression is the fundamental basis for manipulating cellular metabolism, and this new research tool offers the possibility of extending physiological and genetic observations to a mechanistic level. In the next few years we can expect to see major advances in our understanding of basic plant processes.

These advances, in turn, will accelerate the application of genetically engineered plants in the seed production and agrichemical industries. The major crops that can currently be improved with genetic techniques are soybean, cotton, rice, and alfalfa (Table 1), and commercial introductions of genetically engineered varieties are likely in the mid-1990s. Rapid progress is being made in the genetic engineering of corn, and it is likely that genetically engineered corn hybrids carrying traits for resistance to herbicides, insects, and viral diseases will reach the marketplace by the year 2000. The timing of commercialization of genetically engineered crops is ultimately determined by the need to address each of the following issues: (i) evaluation of field performance, (ii) breeding and seed increase for commercial-scale release, (iii) establishment of optimal agronomic practices, and (iv) regulatory approval and crop certification.

The worldwide agrichemical industry has been and will continue to be a leading sponsor of agricultural biotechnology research. All major agrichemical companies have R&D efforts in the area of biotechnology for crop improvement. These companies see opportunities to develop new products and extend the use of existing products, as well as to be positioned at the leading edge of new technologies that may have a significant impact on existing agrichemical businesses.

Genetic engineering of plants also offers exciting opportunities for the food processing industry to develop new products and more cost-effective processes. While many of the early successful examples of genetically engineered plants have focused on agronomic genes, it is possible that the food processing and specialty chemical industries may represent the greatest commercial opportunity for biotechnology. Examples of such applications include production of (i) larger quantities of starch or specialized starches with various degrees of branching and chain length to improve texture and storage properties, (ii) higher quantities of specific oils or the elimination of particular fatty acids in seed crops, and (iii) proteins with nutritionally balanced amino acid composition. The ability to reduce processing costs by the elimination of anti-nutritive or off-flavor components in foods is quite feasible with antisense nucleic acid tech-

nology. The enzymes and genes involved in biosynthesis of coloring materials and flavors are important to the food industry and to the consumer. Studies on the biosynthesis of some of these compounds have been hampered by the low quantities of enzymes present in the producing cells, but new techniques based on gene tagging may overcome these difficulties.

Enormous opportunity lies in the successful use of crops for both commodity and specialty chemical products. Plants have traditionally been a source of a wide variety of polymeric materials. These range from starch and celluloses, which are carbohydrate-based, to polyhydrocarbons such as rubber and waxes. Many of these polymers have been replaced in the last two to three decades by synthetic materials derived from petroleum-based products. However, the cost, supply, and waste-stream problems often associated with petroleum-based products are issues that are focusing renewed attention on the use of biological polymers. Genetic engineering will significantly enlarge the spectrum and composition of available plant polymers.

Plants also offer the potential for production of foreign proteins with various applications to health care. Proteins such as neuropeptides, blood factors, and growth hormones could be produced in plant seeds, and this may ultimately prove to be an economical means of production. Several mammalian proteins have been produced in genetically engineered plants (*81*), and expression of pharmaceutical peptides in oilseed rape plants has been reported (*82*).

References and Notes

1. R. B. Horsch *et al.*, *Science* **223**, 496 (1984); M. De Block, L. Herrera- Estrella, M. Van Montagu, J. Schell, P. Zambryski, *EMBO J.* **31**, 681 (1984).
2. R. B. Horsch *et al.*, *Science* **227**, 1229 (1985).
3. M.-D. Chilton *et al.*, *Cell* **11**, 263 (1977).
4. E. W. Nester, M. P. Gordon, R. M. Amasino, M. F. Yanofsky, *Annu. Rev. Plant Physiol.* **35**, 387 (1984); G. Gheysen, P. Dhaese, M. Van Montagu, J. Schell, in *Genetic Flux in Plants*, B. Hohn and E. S. Dennis, Eds. (Springer-

Verlag, New York, 1985), pp. 11-47; P. Zambryski, J. Tempe, J. Schell, *Cell* 56, 193 (1989).

5. K. A. Barton *et al.*, *Cell* 32, 1033 (1983).

6. R. T. Fraley *et al.*, *Bio/Technology* 3, 629 (1985).

7. H. Klee and S. G. Rogers, in *Plant DNA Infectious Agents*, T. Hohn and J. Schell, Eds. (Springer-Verlag, New York, 1985), pp. 179–203.

8. S. G. Rogers *et al.*, *Methods Enzymol.* 153, 253 (1987).

9. A. Spielmann and R. B. Simpson, *Mol. Gen. Genet.* 205, 34 (1986); R. Jorgensen, C. Snyder, J. D. G. Jones, *ibid.* 207, 471 (1987).

10. W. Mark, K. Signorelli, E. Lacy, *Cold Spring Harbor Symp. Quant. Biol.* 50, 453 (1985); L. Covarrubias, Y. Nishida, B. Mintz, *Proc. Natl. Acad. Sci. U.S.A.* 83, 6020 (1986).

11. I. Potrykus *et al.*, in *Plant DNA Infectious Agents*, T. Hohn and J. Schell, Eds. (Springer-Verlag, New York, 1985), pp. 229–247.

12. G. Gheysen *et al.*, *Proc. Natl. Acad. Sci. U.S.A.* 84, 61 (1987).

13. M. Wallroth *et al.*, *Mol. Gen. Genet.* 202, 6 (1986); S. C. Deroles and R. C. Gardner, *Plant Mol. Biol.* 11, 355 (1988).

14. X. Delannay, private communication.

15. B. Bytebier *et al.*, *Proc. Natl. Acad. Sci. U.S.A.* 84, 5345 (1987).

16. W. Schäfer, A. Görz, G. Kahl, *Nature* 327, 529 (1987).

17. N. Grimsley, T. Hohn, J. W. Davies, B. Hohn, *ibid.* 325, 177 (1987).

18. I. Potrykus *et al.*, *Mol. Gen. Genet.* 199, 183 (1985); H. Lorz, B. Baker, J. Schell, *ibid.*, p. 178; M. Fromm, L. P. Taylor, V. Walbot, *Nature* 319, 791 (1986); H. Uchimiya *et al.*, *Mol. Gen. Genet.* 204, 204 (1986).

19. J. Callis, M. Fromm, V. Walbot, *Genes Dev.* 1, 1183 (1987); W. R. Marcotte, Jr., C. C. Bayley, R. S. Quatrano, *Nature* 335, 454 (1988).

20. T. Fujimura, M. Sakurai, H. Akagi, T. Negishi, A. Hirose, *Plant Tissue Culture Lett.* 2, 74 (1985); K. Toriyama, K. Hinata, T. Sasaki, *Theor. Appl. Genet.* 73, 16 (1986); Y. Yamada, Z. Q. Yang, D. T. Tang, *Plant Cell Rep.* 4, 85 (1986); R. Abdullah, E. C. Cocking, J. A Thompson, *Bio/Technology* 4, 1087 (1986).

21. K. Toriyama *et al.*, *Bio/Technology* 6, 1072 (1988).

22. C. Rhodes, K. Lowe, K. Ruby, *ibid.*, p. 56.

23. C. A. Rhodes *et al.*, *Science* 240, 204 (1988).

24. I. K. Vasil, *Bio/Technology* 6, 397 (1988).

25. T. M. Klein, E. D. Wolf, R. Wu, J. C. Sanford,

Nature 327, 70 (1987).

26. T. M. Klein *et al.*, *Proc. Natl. Acad. Sci. U.S.A.* 85, 8502 (1988).

27. D. E. McCabe *et al.*, *Bio/Technology* 6, 923 (1988).

28. T. M. Klein, B. A. Roth, M. E. Fromm, in *Genetic Engineering*, vol. 11, J. K. Setlow, Ed. (Academic Press, New York, in press).

29. G. Zhou *et al.*, *Methods Enzymol.* 101, 433 (1983); D. Hess, *Intern Rev. Cytol.* 107, 367 (1987); Z.-x. Luo and R. Wu, *Plant Mol. Biol. Reporter* 6, 165 (1988).

30. A. de la Peña, H. Lörz, J. Schell, *Nature* 325, 274 (1987).

31. G. Neuhaus, G. Spangenberg, O. Mittelsten Scheid, H.-G. Schweiger, *Theor. Appl. Genet.* 75, 30 (1987).

32. R. Töpfer, B. Gronenborn, J. Schell, H.-H. Steinbiss, *Plant Cell* 1, 133 (1989).

33. C. Benbrook and P. Moses, in *Proceedings: BioExpo 1986* (Butterworth, Stoneham, MA, 1986), pp. 27–54.

34. H. Steinrücken and N. Amrhein, *Biochem. Biophys. Res. Commun.* 94, 1207 (1980); D. M. Mousdale and J. R. Coggins, *Planta* 160, 78 (1984).

35. D. Atkinson, in *Glyphosate*, E. Grossbard and D. Atkinson, Eds. (Butterworth, London, 1985), pp. 127-133; L. Torstensson, *ibid.*, pp. 137–150.

36. D. M. Shah *et al.*, *Science* 233, 478 (1986).

37. L. Comai *et al.*, *Nature* 317, 741 (1985).

38. M. A. Hinchee *et al.*, *Bio/Technology* 6, 915 (1988).

39. G. W. Haughn *et al.*, *Mol. Gen. Genet.* 211, 266 (1988).

40. M. De Block *et al.*, *EMBO J.* 6, 2513 (1987).

41. D. M. Stalker, K. E. McBride, L. D. Malyj, *Science* 242, 419 (1988).

42. W. De Greef *et al.*, *Bio/Technology* 7, 61 (1989).

43. H. T. Dulmage, in *Microbial Control of Pests*, H. D. Burgess, Ed. (Academic Press, New York, 1981), pp. 193–222; A. I. Aronson, W. Beckman, P. Dunn, *Microbiol. Rev.* 50, 1 (1986).

44. S. A. McPherson *et al.*, *Bio/Technology* 6, 61 (1988).

45. T. Yamamoto and R. E. McLaughlin, *Biochem. Biophys. Res. Commun.* 103, 414 (1981); L. J. Goldberg and J. Margalit, *Mosquito News* 37, 355 (1977).

46. A. Llausner, *Bio/Technology* 2, 408 (1984); D. R. Wilcox *et al.*, in *Protein Engineering*, M. Inouye and R. Sarma, Eds. (Academic Press, New York, 1986), p. 395.

47. V. F. Sacchi *et al.*, *FEBS Lett.* **204**, 213 (1986).
48. D. A. Fischhoff *et al.*, *Bio/Technology* **5**, 807 (1987); M. Vaeck *et al.*, *Nature* **238**, 33 (1987); D. A. Fischhoff, private communication.
49. X. Delannay *et al.*, *Bio/Technology*, in press.
50. V. A. Hilder *et al.*, *Nature* **330**, 160 (1988).
51. P. Powell-Abel *et al.*, *Science* **232**, 738 (1986).
52. N. E. Tumer *et al.*, *EMBO J.* **6**, 1181 (1987); C. Hemenway, R.-X. Fang, W. K. Kaniewski, N.-H. Chua, N. E. Tumer, *ibid.* **7**, 1273 (1988); M. Cuozzo *et al.*, *Bio/Technology* **6**, 549 (1988); A. Hoekema, M. J. Huisman, L. Molendijk, P. J. M. van den Elzen, B. J. C. Cornelissen, *ibid.* **7**, 273 (1989); C. M. P. Van Dun, J. F. Bol, L. Van Vloten-Doting, *Virology* **159**, 299 (1987); E. C. Lawson and P. R. Sanders, private communication.
53. J. C. Register and R. N. Beachy, *Virology* **166**, 524 (1988).
54. R. S. Nelson *et al.*, *Bio/Technology* **6**, 403 (1988).
55. J. D. G. Jones *et al.*, *Mol. Gen. Genet.* **212**, 536 (1988).
56. F. Nagy, G. Morelli, R. T. Fraley, S. G. Rogers, N.-H. Chua, *EMBO J.* **4**, 3063 (1985); R. Fluhr and N.-H. Chua, *Proc. Natl. Acad. Sci. U.S.A.* **83**, 2358 (1986).
57. M. A. Anderson *et al.*, *Nature* **321**, 38 (1986); C. S. Gasser, K. A. Budelier, A. G. Smith, D. M. Shah, R. T. Fraley, *Plant Cell* **1**, 15 (1989).
58. C. Sengupta-Gopalan *et al.*, *Proc. Natl. Acad. Sci. U.S.A.* **82**, 3320 (1985); R. N. Beachy *et al.*, *EMBO J.* **4**, 3047 (1985).
59. D. P. S. Verma and R. B. Goldberg, Eds., *Temporal and Spatial Regulation of Plant Genes* (Springer-Verlag, New York); R. B. Goldberg, *Science* **240**, 1460 (1988); P. N. Benfey and N.-H. Chua, *ibid.* **244**, 174 (1989).
60. R. Broglie *et al.*, *ibid.* **224**, 838 (1984); D. E. Rochester, J. A. Winter, D. M. Shah, *EMBO J.* **5**, 451 (1986); A. Theologis, *Annu. Rev. Plant Physiol.* **37**, 407 (1986); M. M. Sachs and T.-H. D. Ho, *ibid.*, p. 363 (1986); R. W. Thornberg, G. An, T. E. Cleveland, R. Johnson, C. A. Ryan, *Proc. Natl. Acad. Sci. U.S.A.* **84**, 744 (1987); J. Gómez *et al.*, *Nature* **334**, 262 (1988); N. M. Crawford, M. Smith, D. Bellissimo, R. W. Davis, *Proc. Natl. Acad. Sci. U.S.A.* **85**, 5006 (1988); C. J. Lamb, M. A. Lawton, M. Dron, R. A. Dixon, *Cell* **56**, 215 (1989).
61. C. J. S. Smith *et al.*, *Nature* **334**, 724 (1988); R. E. Sheehy, M. K. Kramer, W. R. Hiatt, *Proc. Natl. Acad. Sci. U.S.A.* **85**, 8805 (1988).
62. A. R. van der Krol *et al.*, *Nature* **333**, 866 (1988).
63. J. Haseloff and W. L. Gerlach, *ibid.* **334**, 585 (1988).
64. J. Paszkowski, M. Baur, A. Bogucki, I. Potrykus, *EMBO J.* **7**, 4021 (1988).
65. N. V. Federoff, D. B. Furtek, O. E. Nelson, *Proc. Natl. Acad. Sci. U.S.A.* **81**, 3825 (1984), K. C. Cone, F. A. Burr, B. Burr, *ibid.* **83**, 9631 (1986); E. M. Rinchik, L. B. Russell, N. G. Copeland, N. A. Jenkins, *Genetics* **112**, 321 (1986); R. J. Schmidt, F. A. Burr, B. Burr, *Science* **238**, 960 (1987).
66. B. Baker, G. Coupland, N. Federoff, P. Starlinger, J. Schell, *EMBO J.* **6**, 1547 (1987); M. A. Van Sluys, J. Tempe, N. Federoff, *ibid.*, p. 3881.
67. K. A. Feldmann *et al.*, *Science* **243**, 1351 (1989).
68. R. Bernatzky and S. D. Tanksley, *Genetics* **112**, 887 (1986); T. Helentjaris, *Trends Genet.* **3**, 217 (1987); B. S. Landry, R. Kesseli, H. Leung, R. W. Michelmore *Theor. Appl. Genet.* **74**, 646 (1987); B. Burr, F. A. Burr, K. H. Thompson, M. C. Albertson, C. W. Stuber, *Genetics* **118**, 519 (1988); C. Chang *et al.*, *Proc. Natl. Acad. Sci. U.S.A.* **85**, 6856 (1988).
69. A. H. Paterson *et al.*, *Nature* **335**, 721 (1988); S. D. Tanksley and J. Hewit, *Theor. Appl. Genet.* **75**, 811 (1988).
70. R. E. Pruitt and E. M. Meyerowitz, *J. Mol. Biol.* **187**, 169 (1986).
71. H. J. Klee, M. B. Hayford, S. G. Rogers, *Mol. Gen. Genet.* **210**, 282 (1987).
72. U. S. Congress, Office of Technology Assessment, *New Developments in Biotechnology—Field Testing Engineered Organisms: Genetic and Ecological Issues*, OTA-BA-350 (U.S. Government Printing Office, Washington, DC, May 1988).
73. Council of the National Academy of Sciences, *Introduction of Recombinant DNA–Engineered Organisms into the Environment: Key Issues* (National Academy Press, Washington, DC, 1987); Boyce Thompson Institute for Plant Research, *Regulatory Considerations: Genetically Engineered Plants* (Center for Science Information, San Francisco, CA, 1987).
74. International Food Biotechnology Council, *Food and Chemical News*, 6 February 1989, pp. 10-12; Expert Panel on Criteria for Determining the Regulatory Status of Food and Food Ingredients Produced by New Technologies, *Federal Register*, vol. 53, no. 168, 30 August 1988.
75. Diamond *v.* Chakrabarty, 447 U.S. 303

(1980).

76. Ex parte Hibberd, *United States Patents Quarterly* **227**, 303 (1985).

77. J. Johnson, *Seedsmen's Digest*, 6 to 8 October 1987; Executive Committee of the American Seed Traders Association, *Diversity* **14**, 27 (1988).

78. B. J. Fowlston, *Bio/Technology* **6**, 911 (1988).

79. L. R. Brown, *Worldwatch Paper 85* (Worldwatch Institute, Washington, DC, October 1988).

80. Council for Agricultural Science and Technology, *Report No. 114* (University of Iowa, Ames, IA, 1988).

81. D. A. Eichholtz *et al.*, *Somat. Cell. Mol. Genet.* **13**, 67 (1987); D. D. Lefebvre, B. L. Miki, J.-F. Laliberté, *Bio/Technology* **5**, 1053 (1987); A. Barte *et al.*, *Plant Mol. Biol.* **6**, 347 (1986).

82. J. Vandekerckhove *et al.*, *Bio/Technology*, in press.

83. S. McCormick *et al.*, *Plant Cell Rep.* **5**, 81 (1986).

84. G. Ooms, M. M. Burrell, A. Karp, M. Bevan, J. Hille, *Theor. Appl. Genet.* **73**, 744 (1987); S. Sheerman and M. W. Bevan, *Plant Cell Rep.* **7**, 47 (1988).

85. J. Paszkowski *et al.*, *EMBO J.* **3**, 2717 (1984).

86. A. M. Lloyd *et al.*, *Science* **234**, 464 (1986).

87. R. Michelmore, E. Marsh, S. Seely, B. Landry, *Plant Cell Rep.* **6**, 439 (1987).

88. N. P. Evrette *et al.*, *Bio/Technology* **5**, 1201 (1987).

89. J. E. Fry, A. Barnason, R. B. Horsch, *Plant Cell Rep.* **6**, 321 (1987). E.-C. Pua, A. Mehra-Palta, F. Nagy, N.-H. Chua, *Bio/Technology* **5**, 815 (1987).

90. N. Basiran, P. Armitage, R. J. Scott, J. Draper, *Plant Cell Rep.* **6**, 396 (1987); M. C. Jordan and A. McHughen, *ibid.* **7**, 281 (1988).

91. P. Umbeck, G. Johnson, K. Barton, W. Swain, *Bio/Technology* **5**, 263 (1987); E. Firoozabady *et al.*, *Plant Mol. Biol.* **10**, 105 (1987).

92. A. R. Barnason and J. E. Fry, private communication.

93. D. Catlin *et al.*, *Plant Cell Rep.* **7**, 100 (1988).

94. E. A. Shahin *et al.*, *Crop Sci.* **26**, 1235 (1986).

95. M. Deak, G. B. Kiss, C. Koncz, D. Dudits, *Plant Cell Rep.* **5**, 97 (1986).

96. H. P. Hernalsteens *et al.*, *EMBO J.* **3**, 3039 (1984).

97. F. Köhler, C. Golz, S. Eapen, H. Kohn, O. Schieder, *ibid.* **6**, 313 (1987).

98. A. J. Trulson, R. B. Simpson, E. A. Shahin, *Theor. Appl. Genet.* **73**, 11 (1986).

99. C. David and M.-D. Chilton, J. Tempé, *Bio/Technology* **2**, 73 (1984).

100. C. David and J. Tempé, *Plant Cell Rep.* **7**, 88 (1988).

101. T. Noda *et al.*, *ibid.* **6**, 283 (1987).

102. D. Tepfer, *Cell* **37**, 959 (1984).

103. J. J. Fillatti *et al.*, *Mol. Gen. Genet.* **206**, 192 (1987); F. Pythoud, V. P. Sinkar, E. W. Nester, M. P. Gordon, *Bio/Technology* **5**, 1323 (1987).

104. G. H. McGranahan *et al.*, *Bio/Technology* **6**, 800 (1988).

105. D. J. James *et al.*, *Plant Cell Rep.* **7**, 658 (1989).

106. M. E. Horn, R. D. Shillito, B. V. Conger, C. T. Harms, *ibid.*, p. 469.

107. We thank X. Delannay, D. Fischhoff, P. Sanders, C. Lawson, A. Barnason, J. Fry, and J. Leemans for giving us permision to cite their work prior to publication. We thank G. Kishore, S. Rogers, H. Klee, J. Callis, D. Gunning, N. Tumer and S. Brown for help with the content, readability, and references in the manuscript. Special thanks to all other members of the Plant Molecular Biology Group at Monsanto.

12

Fish as Model Systems

Dennis A. Powers

Darwin's *The Origin of Species* emphasized the importance of systematic and zoogeographic studies and, as a result, ichthyologists like David Starr Jordan became major figures in American evolutionary thought. At the turn of the century, Jacques Loeb, Thomas Hunt Morgan, and others felt that Louis Agassiz's dictum to "study nature, not books" should include experimental manipulation as well as natural history (*1*). In the next decades, as fish were used to probe the secrets of nature, a few species emerged as particularly useful models.

Fish models have been exploited by essentially every biological discipline (*2–4*). Because there are more than 20 disciplines and thousands of fish species, I have chosen a few representatives to communicate the flavor of this research. I will also point out some instances where fish have played particularly important roles in specific disciplines (for example, neurobiology) and enumerate some advantages of these model systems.

Advantages of Fish as Models

Fish are the oldest and most diverse vertebrates. They evolved around 500 million years ago, and today there are more fish species than all other vertebrates combined. Research with fishes provides a conceptual framework and evolutionary reference point for other ver-

tebrate studies. They live in a wide variety of habitats that range from fresh to salt water, from cold polar seas to warm tropical reefs, and from shallow surface waters to the intense pressures of the ocean depths. Elucidating the evolutionary strategies and mechanisms that fish use to adapt to these diverse environments is one of the exciting challenges for modern biologists.

Many fish species are amenable to both field and laboratory experiments and are easily raised and bred under laboratory conditions. There is extensive animal husbandry information available from hundreds of years of practical experience by fish farmers, hobbyists, and aquaculturists. Many fish are much less expensive to buy and raise than their mammalian, avian, reptilian, or amphibian counterparts. They are generally the most fecund, some producing hundreds of eggs on a periodic basis, whereas others produce thousands. These eggs are usually large and externally fertilized, and, because some are transparent, embryonic development can be easily followed. Historically, these advantages and the economic importance of some fish have made them favored models for such studies and, as a consequence, the detailed embryology of many species has been carefully documented.

Fish are useful models for genetic manipulations. There are several highly homozygous strains, and general methods for

The author is the director of the Hopkins Marine Station, and Harold A. Miller Professor of the Department of Biological Sciences, Stanford University, Pacific Grove, CA 93950. This chapter is reprinted from *Science* 246, 352 (1989).

obtaining new strains and mutants have been established. For example, inbred strains of medaka (*Oryzias latipes*), top minnows (*Poeciliopsis lucida*), and others have been produced by classical repetitive inbreeding. In addition, naturally occurring hermaphroditic (*Rivulus marmoratus*) and gynogenetically reproducing fish are available. Some scientists have imitated nature, successfully producing gynogenetic diploids in the laboratory. Streisinger and his colleagues (5, 6), for example, introduced methods for large-scale production of homozygous diploid zebrafish (*Brachydanio rerio*). With this simple technique, eggs were activated by sperm having DNA that had been inactivated by ultraviolet irradiation, then the maternal haploid genome was duplicated. The first cell division is then prevented by heat stress or hydrostatic pressure; however, subsequent cell divisions are allowed to proceed without intervention. Thus, the offspring is a diploid homozygote with the maternal genome. Because of the unusual sex-determination characteristics, the offspring are both males and females, so that normal breeding can continue in the next generation. This general approach has been applied to other fish, but some require hormones to produce males. A novel variation on Streisinger's theme used active sperm to initiate development of trout eggs with DNA that had been photoinactivated (7).

Mammalian models range from small species (for example, mice, rats, and guinea pigs) to large ones (for example, cows, sheep, and primates), each of which may be used to answer different types of scientific questions. Fish represent an even more diverse morphological group than mammals and, thus, the choice of a particular model depends on the question being addressed. Large fish, such as dogfish sharks (*Squalus acanthus*), the electric ray (*Torpedo californica*), winter flounder (*Pseudopleuronectes americanus*), rainbow trout (*Onchorhynchus mykiss*), and carp (*Cyprinus carpio*), tend to be used in studies where experimental manipulations are significantly facilitated by the larger size or distinct adaptation of the organ systems of the fish, for example, the historic advances with the fish kidney model. Although a number of these large fish have also been the focus of genetic analysis, the time required for sexual maturation and the cost of maintaining a large number of genetic stocks have made them less useful than for other studies. On the other hand, a number of small fish species with shorter life cycles, such as zebrafish (*Brachydanio rerio*), medaka (*Oryzias latipes*), killifish (*Fundulus heteroclitus*), guppies (*Lebistes reticulatus*), mollies (*Poecilia formosa*), platyfish (*Platypoecilus maculatus*), and swordtails (*Xiphophorus helleri*), may be used as models for even the most sophisticated genetic analyses.

Recently, a number of researchers have successfully microinjected cloned DNA into fertilized fish eggs and, in some cases, the transferred gene has been integrated into genomic DNA, the protein has been expressed, and the transferred gene inherited in a Mendelian manner (8). For example, studies with growth hormone constructs indicate significant enhancement of growth in transgenic fish. Because the first generation offspring of these transgenic fish are usually genetic mosaics, some of their F_1 offspring may carry and express the foreign gene, whereas others may not. This phenomenon is dramatically illustrated in Fig. 1.

Neurobiology

For decades, neurobiologists have found fish to be excellent models. In fact, most biologists and physicians over 40 years of age had their first exposure to vertebrate neuroanatomy when they dissected the brain and cranial nerves of the dogfish shark. Comparative neurobiology can provide insight about the human nervous system and its role in health. Fish models can lead to understanding of vertebrate neurology in general and provide perspective by their fundamental evolutionary relationship with other vertebrates. Our present concept of vertebrate color vision, for example, was significantly influenced by a series of classical studies on fish retina (9). Certain progressive neurological diseases are best understood in the context of the evolutionary states of the nervous system in which evolutionarily "higher"

Fig. 1. The carp in this figure are the F_1 offspring of mosiac transgenic parents. The larger individual is transgenic and carries and expresses a mammalian GH gene whose expression is driven by a metallothionein promoter. The smaller individuals in the figure are the larger individual's nontransgenic siblings. This photograph was provided by Z. Zhu of the Institute of Hydrobiology, Academica Sinica in the People's Republic of China, and senior research scientist of the Center of Marine Biotechnology, University of Maryland.

central nervous system functions are lost first, then sequentially "lower" evolutionary states, with the order being reversed during recovery (*10*). Bullock has provided several examples that are consistent with that hypothesis (*10*).

Prosser described neurobiology as the neuronal basis of animal behavior, determined by neural circuits that are controlled by cell-to-cell communications, including chemical coupling through neurotransmitters (*3*). Most neurotransmitters are amino acids, their derivatives, or peptides. Acetylcholine (ACh) is a combination of choline (from serine) and acetyl coenzyme A. Acetylcholine is perhaps the most universally recognized neurotransmitter among nonneurobiologists. It is widespread among most taxa and performs a variety of important functions. Research on the ACh receptor and the Na^+ channel of fish has played a critical role in the development of neurobiology.

Approximately one-third of vertebrate neurons respond to ACh. In the brain, there are at least two types of ACh or cholinergic receptors, muscarinic and nicotinic. The most important studies on the nicotinic receptor have used fish models. The electric organ of the ray, *Torpedo californica*, contains cholinergic neurons that innervate electrocytes that develop from myotubes but have 1000 times more ACh receptors than muscle cells (*11*). The abundant ACh receptor has been extensively characterized, including the cloning and sequencing of all four of its protein subunits (*12*). Moreover, this receptor has been reconstituted in artificial membranes, and its role has been elucidated in the disease myasthenia gravis (*13*). These studies on the ACh

receptor are an important success in molecular neurobiology and a model approach.

Another example in which research on fish has paved the way for molecular approaches to neurobiology is provided by the studies on the voltage-controlled Na^+ channel. This integral membrane protein is responsible for the extremely rapid depolarization associated with propagated action potentials in nerve and muscle cells of animals from many phyla. Studies of Na^+ channels have guided research into other voltage-controlled channels, such as K^+ and Ca^{2+} channels, that are even more widespread and functionally diverse.

The Amazonian electric eel, *Electrophorus electricus*, like the electric ray, *Torpedo californica*, is well endowed with electroplax tissue for delivering strong electric shocks to both predators and prey. Extensive research during the 1950s and 1960s on the electric organ of this fish resolved a conflict between bioelectric and biochemical interpretations of the nervous system. Since that time, this fish has been a favorite model for other neurochemical studies. For example, because the electric organ is such a rich source of Na^+ channels, it was used as a tissue source for the purification of the first Na^+ channel (*14*). This Na^+ channel was also the first from which essentially normal functional activity was successfully reconstituted (*15*). Those studies on fish Na^+ channels have become the model for studies on other channels (*16*).

As was the case for the ACh receptor, a fish Na^+ channel was the first of such molecules to be cloned (*17*). The primary structure revealed four repeating homologous units, each of which contains a unique sequence in which five to seven positively charged lysine or arginine residues occur at every third position, with most of the intervening positions occupied by polar residues (Fig. 2). Because of the similarity between this sequence and the hypothetical structure of the voltage-dependent gating machinery for the channel (*18*), it was proposed that this macromolecular machine was a transmembrane structure that underlies the intermembrane charge movement that triggers opening of the channel, referred to as gating current. This concept is central to all proposed structural models that attempt to account for Na^+ channel function (*19*).

The unusual S4 sequence identified in the eel Na^+ channel is not unique. Nearly identical structures have now been identified in essentially every other voltage-controlled channel that has been cloned and sequenced: two more Na^+ channels from rat brain (*20*) and one from *Drosophila* (*21*), the *Shaker* K^+ channel from *Drosophila* (*22*), and the dihydropyridine receptor from rabbit skeletal muscle, a putative Ca^{2+} channel (*23*). The presence of an S4-like sequence has been used to identify genes coding for voltage-controlled ion channels (Fig. 2). Work on the electric eel has directly shaped our present understanding of the molecular structure and function of these and perhaps all channels gated by membrane voltage.

Endocrinology

Approximately 400 million years ago, the ancestors of modern bony fishes (teleosts) invaded fresh water and diversified. A few hundred million years later, new species returned to the ocean and proliferated; afterward, new teleosts reinvaded fresh water. Today, some fish are restricted to either fresh water or salt water and others spend part of their life in each environment. Freshwater fish are in a Na^+-poor environment and have evolved mechanisms to retain salt. Although they do not drink water and their skin is relatively impermeable, a significant influx of water occurs across the gills, which is eliminated as a dilute urine through the glomerular kidney.

Salt lost in the feces and urine is replaced from food and by active Na^+ uptake at the gills. The opposite situation exists for saltwater fish, which must conserve water and exclude salt. They drink water and eliminate excess ions either at the gills or through feces; urine output is minimized. Fish that migrate between fresh and salt water must, therefore, regulate these mechanisms in order to survive (*24*).

Pickford and Phillips (*25*) showed that hypophysectomized killifish, *Fundulus heteroclitus*, failed to survive when transferred from salt water to fresh water but did survive when injected with prolactin (Prl). The rapid loss of Na^+ at the gills was essentially halted by Prl injections. This and other papers by Pickford and her colleagues set the stage for delineating of the role of Prl in the complex process of osmoregulation. Although all the details are not yet known, in at least a few fish species (*24*), Prl appears to stimulate active Na^+ uptake, inhibit Cl^- excretion, retain Ca^{2+}, enhance production of dilute urine by inhibiting water reabsorption, and facilitate Na^+ retention by suppression of a Na^+,K^+-ATPase (*24, 26–29*). In fact, when fish migrate from fresh water to salt water, the Na^+,K^+-ATPase can increase by as much as an order of magnitude (*29*).

Pickford also studied the growth-promoting effect of growth hormone in hypophysectomized killifish (*30*). Since growth hormone (GH) and Prl arose from a common evolutionary precursor (*31*), it should not be surprising that there were reports of overlapping biological activities (*32, 33*). However, it was eventually shown that Prl and GH from the same species had mutually exclusive effects (*33*). In trout (*33*), GH functions in a manner opposite to Prl in osmoregulation; it enhances Na^+,K^+-ATPase activity and Na^+ exclusion at the gills. When some fish migrate from salt water to fresh water, Prl increases, GH decreases, and Na^+ is retained by the reduction of Na^+,K^+-ATPase activity. When they migrate from fresh water to salt water, the opposite is true and growth is accelerated because of increased GH concentrations (*24*).

The detailed mechanism of this control and its general applicability to other species (*24*) is

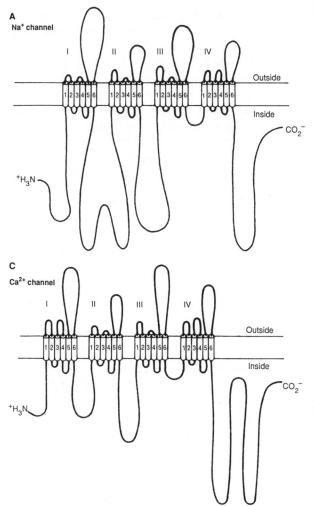

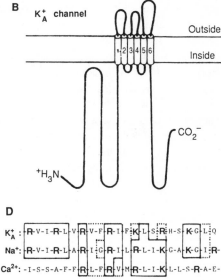

Fig. 2. (A to C) Models for the membrane-spanning regions of three voltage-controlled ion channels. The S4 region of each domain is segment 4. Both the Na^+ channel from *Electrophorous* electric organ (Fig. 2A) and the dihydropyridine receptor (a putative Ca^{2+} channel) from mammalian skeletal muscle (Fig. 2C) have four pseudo-subunit domains, each of which has six transmembrane segments. A gene product from the *Shaker* locus of *Drosophila* (Fig. 2B) is composed of only one such domain, which is homologous to domain IV of Na^+ and Ca^{2+} channels. This protein codes for a K^+ channel, but it is not yet known how many subunits are necessary to form a functional channel. (D) Conservation of the S4 region among voltage-controlled channels described in text and pictured in (A to C). Positively charged amino acids (arginine = R, lysine = K) regularly spaced with two (mostly nonpolar or hydrophobic) intervening residues. Solid lines encircle identical amino acids; dotted lines denote conservative replacements. M, methionine; V, valine; I, isoleucine; A, alanine; G, glycine; S, serine; T, threonine; Q, glutamine; N, asparagine; D, aspartate; E, glutamate. Adapted from (*19*).

not yet known, but the availability of large quantities of biosynthetic GH and Prl from cloned cDNAs (*35*) should provide adequate material to study these mechanisms. Now that the fish GH gene structure is known (*31*) and the Prl gene structure is being elucidated, the mechanisms controlling gene regulation should follow. These molecular studies as well as similar efforts on other pituitary hormones (*36*), hormones from the hypothalamus (*37*), and other endocrine hormones, signal that a new phase in fish endocrinology has begun. Moreover, these tools can be used on other fish species to explore evolutionary variations on this theme of osmoregulation.

Developmental Biology

Modern developmental biology tends to focus on (i) embryonic pattern formation, including the movement and eventual fate of specific cells (*38*); (ii) the mechanisms responsible for developmental stability; (iii) the expression of specific genes during development, including their regulatory mechanisms; (iv) agents responsible for initiating new developmental programs and shifting the timing of developmental events; (v) sex determination; (vi) the mechanisms of cellular and tissue differentiation; and (vii) the mechanisms that control organ system development (*38*). Although many of these topics can be addressed in a variety of organisms, a vertebrate is, clearly, needed for questions relating to typical vertebrate development. For previously mentioned reasons, fish have been favorite models of embryogenesis for over a century. For example, the killifish, *Fundulus heteroclitus*, and the medaka, *Oryzias latipes*, played central roles in embryology in the United States and Japan, respectively. In a recent review of fish developmental genetics (*39*), the authors emphasize that gene regulation is intrinsically tied to evolutionary adaptation.

In the absence of an extensive array of laboratory mutants, researchers have taken advantage of interspecific fish hybrids, unisexual fish, and species derived from polyploid ancestors. The use of interspecific fish hybrids to study the fate of alleles and gene regulation is an in vivo analog to the in vitro somatic cell hybrid technique commonly used to study gene regulation in mammalian cell culture. The application of this approach has been reviewed by Whitt (*40*). He has shown that the greater the evolutionary distance between the parental stocks of hybrids, the greater the frequency of expression of abnormal characters; also maternal alleles are generally expressed at their normal time, whereas paternal alleles are delayed (*40*). With the recent advances in genetic techniques to manipulate fish genomes, the mechanisms responsible for these interesting observations may be examined.

Shifts in the timing of developmental events (heterochrony) have also been studied with fish models. These shifts can sometimes be traced to a single locus. For example, Kallman has shown that differences in the time required to reach sexual maturity in the swordtail, *Xiphophorus maculatus*, is a function of a single locus that regulates luteinizing hormone-releasing hormone (*41*).

Some investigators have shown that developmental rate and time required for hatching are correlated with specific enzyme-encoding loci that presumably play a role in the timing of developmental events. For example, DiMichele and Powers (*42*) showed that developmental rate and hatching in the killifish, *Fundulus heteroclitus*, was highly correlated with genetic variation of the "heart" locus of lactate dehydrogenase (*Ldh-B*). The homozygote for one allele, Ldh-B^a, consumed oxygen faster and hatched earlier than the homozygote with the other allele, Ldh-B^b. Recently, DiMichele (*43*) showed that oxygen consumption was altered in a predictable way by the type of lactate dehydrogenase microinjected into fertilized eggs, indicating that the enzyme had a direct effect on development. Development and hatching differences have been observed for other loci in *Fundulus* (*44*) and trout (*45, 46*).

One of the major drawbacks to the use of vertebrates for developmental studies has been the paucity of mutants. For example, because of the nature of mammalian development, the problems associated with identifying and isolating mutants have been both costly and difficult (*47*). On the other hand, the zebrafish has been an unusually successful model for generating and analyzing developmental mutants (*48*). The success of the zebrafish system is largely the result of methods that (i) allow developmental mutants to be identified in a single generation, (ii) engender completely isogenic stocks in a single generation, (iii) permit the cryopreservation of gametes, (iv) make possible the artificial creation of mutants, and (v) enable the formation of transgenic fish (*8, 49*).

The movement and eventual fate of cells during development can be studied by direct observation of unmarked or marked cells (*38*). The latter is usually accomplished by introducing one of several nontoxic tags that facilitate observation. Kimmel and his colleagues (*50*) have used direct observation of unmarked cells, of cells tagged with fluorescent dyes, and of genetic mosaics to study the lineages and eventual fates of embryonic cells. They have shown that blastomere lineages are indeterminate but lineage restriction exists after gastrulation. However, it is not yet clear that postgastrula cells are irrevocably committed.

The studies of Grunwald and his collaborators illustrate the power of Streisinger's method for isolating and analyzing developmental mutants (51). One of Grunwald's mutants causes degeneration of later developing central nervous system components but does not affect early primary neural tissue. This and other studies on zebrafish mutants are providing insights about vertebrate development that could not have been as easily perceived with other vertebrate models. Zebrafish are becoming a powerful tool to study vertebrate development, and recent molecular studies on fish neuropeptides (36), hormones (31, 35, 37), and homeo box genes (52), coupled with the ability to make transgenic zebrafish (8), promise an expanded role for zebrafish in developmental biology.

Aquatic Toxicology and Carcinogenesis Research

Fish models cannot replace mammals for research into mammalian physiology, but they can offer an inexpensive and, in some instances, more acceptable alternative for chemical carcinogen testing. Fish are particularly useful for the assessment of water-borne and sediment-deposited toxins where they may provide advanced warning of the potential danger of new chemicals and the possibility of environmental pollution. Recent public awareness of increasing contamination of the oceans and the potential associated health risks should encourage further research in aquatic toxicology.

Russel and Kotin (53) were the first to suggest a correlation between pollution and incidence of fish tumors. Those correlations were emphasized when an outbreak of liver cancers in cultured rainbow trout was traced to the presence of an aflatoxin in their food (54). Recently, hepatocellular carcinomas have been found to be significantly elevated in the winter flounder (*Pseudopleuronectes americanus*) from Boston Harbor (55) and in the English sole (*Pleuronectes vetulus*) from Puget Sound (56). The latter study suggested that the tumors might be the result of excess polycyclic aromatic hydrocarbons in the sediments. The

elegant studies of Hendricks *et al.* (57) demonstrated that, indeed, the polycyclic aromatic hydrocarbon, benzo[*a*]pyrene, was capable of inducing hepatomas (58).

The ability to induce neoplastic lesions in fish tissues has inspired innovative methods for detecting chemical carcinogens. One particularly novel test involves the use of fish embryos. In this test, fertilized eggs from rainbow trout (59) or medaka (60) are exposed for a short period to a defined concentration of a toxin. After removal, the embryos are allowed to develop, and evidence of tumors is recorded. This is a highly sensitive test that, in some ways, is superior to the analogous test in rodents. The main advantages and disadvantages are discussed elsewhere (59, 60).

In addition to chemical carcinogens, viruses have been suggested as potential causes of fish neoplasms. For example, Papas and his colleagues suggested that lymphosarcomas in the northern pike were the result of a C-type virus (61). Kimura *et al.* (62) and Sano *et al.* (63) indicated that lesions or tumors in salmon could be induced by exposure to some herpes viruses.

The discovery that melanomas could be generated in hybrid crosses between swordtails (*Xiphophorus helleri*) and platyfish (*Platypoecilus maculatus*) opened a new area of melanoma research (64). After this discovery, a number of scientists analyzed the genetic basis of this interesting phenomenon (65). Platyfish have melanin pigment patterns that are coded by specific color genes. For example, one strain has a black pigmented spot or spots on the dorsal fin, the expression of which is controlled by an allele of a sex-linked locus. If a female is crossed with a melanin-lacking male swordtail, the offspring will have abnormal melanization that will often lead to melanomas. Anders *et al.* (66) described the color gene as an oncogene and suggested that it was correlated with c-*src*. This "color oncogene" is controlled by at least one regulatory locus and a host of environmental and physiological factors.

There have been other oncogenes studied in fish and some have been analyzed at the molecular level. The *ras* gene was cloned from goldfish liver DNA (67), and the c-*myc* gene was isolated and characterized from rainbow trout (*68*). Each of these showed a remarkable similarity to their mammalian and avian counterparts (*68*).

Schartl and Peter (*69*) demonstrated progressive growth of fish tumors when malignant melanotic melanoma tissue, from the swordtail fish, *Xiphophorous*, was transplanted into thymus-aplastic nude mice. Moreover, while the tumor adapted to the physiological conditions of the mammalian host, it retained its fish-specific morphology and biochemical specificity. In a recent study, winter flounder tissues that contained histopathological lesions were assayed for oncogenes by transfection of the DNA into mouse fibroblasts. The transfected fibroblasts induced subcutaneous sarcomas when transferred into nude mice, and there is some evidence that these sarcomas may contain fish c-K-*ras* oncogenes (*70*).

Biochemical and Genetic Adaptation

Comparative physiologists and biochemists are interested in the mechanisms that organisms use to adapt to environmental stress. Fish are particularly good models because they live in a variety of habitats and must adapt to environmental parameters, like temperature, pressure, oxygen, *p*H, and salinity, which are easily measured and controlled under laboratory conditions. Hochachka and Somero (*4*) pointed out many of the adaptive mechanisms and strategies of aquatic animals.

Temperature is a useful parameter to illustrate biochemical adaptation. Fish are cold-blooded organisms, and their success involves adaptations to changes in environmental temperature. This is accomplished by a host of metabolic, physiological, and behavioral changes (*3, 4*). Like many organisms, fish use heat shock genes in response to elevated temperature (*71*), but for extreme cold, some have evolved a novel set of antifreeze genes to encode proteins that keep their blood from freezing. DeVries (*72*), who discovered antifreeze proteins (AFP), found that polar fish express these genes all year, whereas temperate

species, like winter flounder, express AFPs only in winter (*73*). This seasonal variation is an excellent system to study the role of environmentally regulated gene expression.

AFPs of Antarctic fish are composed of repeating units of Ala-Ala-Thr with a disaccharide, galactosyl-*n*-acetylgalactose amine, glycosidically linked to the threonine (*74*). Winter flounder AFPs differ in that they do not have disaccharides. The flounder express an alanine-rich helical protein whose primary, secondary, and tertiary structures have been determined (*74–76*), and the elegant mechanism by which these proteins bind micro–ice crystals and lower the freezing point of the blood has been formulated (*75*). Recently, the DNA sequences encoding these genes have been elucidated (*76*), and microinjection of the gene into other species has been successful (*8*). Now the tasks of understanding the regulation of this gene family and delineating the molecular mechanisms involved in the transfer of information from the environment to the target tissue remain.

The potential adaptive role of genetic variation at enzyme-synthesizing loci has been the subject of intense investigation for almost three decades. Indeed, few subjects in biology have been more debated than the evolutionary significance of protein polymorphisms (*77*). Most of the debate centered around two contrasting views: the "selectionist" and "neutralist." Proponents of the first idea assert that natural selection maintains protein polymorphisms, whereas those of the second school argue that the vast majority of such variation is selectively neutral. In 1974, Lewontin (*77*) summarized the failure of evolutionary biologists to resolve this important issue. Since that time, a number of researchers have addressed the controversy with renewed vigor and with the sophisticated tools of molecular biology (*78*).

A few in vitro studies of enzymes have suggested selection, however, there is little corroborating evidence at other levels of biological organization (*78*). My colleagues and I are addressing the significance of genetic variation in the killifish, *Fundulus heteroclitus*. Using temperature, *p*H, oxygen, and salinity as model environmental variables, we are investigating a series of allelic isozymes for evidence of natural selection at several levels of biological organization. We have analyzed enzyme kinetics, protein structure, gene sequences and their regulation, cell metabolism, oxygen transport, developmental heterochrony, swimming performance, population biology, zoogeography, and have carried out laboratory and field selection experiments (*44, 78, 79*). Based on the results of some of those studies, DiMichele predicted heterochrony and mortality differences between phenotypes as a function of temperature and salinity, and some of those predictions were realized as he found hatching time and mortality differences for both single and multilocus phenotypes at high temperatures (*43*). Moreover, those that survived the highest temperature regime were also the most common phenotypes at the warm southern extreme of the natural distribution of the species. Although this multidiscipline approach is just beginning to bear fruit, it already appears that some enzyme loci are maintained by natural selection while others are not.

Another example of this approach is the work of Vrijenhoek and his colleagues, who have shown dramatic differences in survival among populations of the fish genus *Poeciliopsis* (*80, 81*). In addition, laboratory and field selection experiments reveal the action of natural selection on genetic variants marked by four enzyme- encoding loci. Using acute cold, heat, and hypoxia, variables that mimic seasonal environmental stress, they demonstrated that allozyme diversity and survival were intrinsically linked (*81*). It will be exciting to explore the biochemical and molecular basis for this phenomenon in the future.

Elucidating the array of mechanisms that animals use to adapt to diverse environments is one of the exciting challenges for modern biology; fish provide excellent models to meet that challenge.

Conclusions

Fish represent the largest and most diverse group of vertebrates. Their evolutionary position relative to other vertebrates and their ability to adapt to a wide variety of environments make them ideal for studying both organismic and molecular evolution. A number of other characteristics make them excellent experimental models for studies in embryology, neurobiology, endocrinology, environmental biology, and other areas. In fact, they have played a critical role in the development of several of these disciplines. Research techniques that enable scientists to make isogenic lines in a single generation, create and maintain mutants, culture cells, and transfer cloned genes into embryos signal an increasing role for fish as experimental models.

References and Notes

1. Loeb worked extensively with fish and was first author on 70 publications on *Fundulus heteroclitus* alone. Using the same species, Morgan published seven papers on the regeneration and identification of embryo-forming regions [see, for example, T. H. Morgan, *Science* **28**, 287 (1908)]. They both worked on marine organisms in Woods Hole, MA, and Pacific Grove, CA.
2. W. S. Hoar and D. J. Randall, *Fish Physiology* (Academic Press, New York, 1969–1984), vols. 1–10.
3. C. L. Prosser, *Comparative Animal Physiology* (Saunders, Philadelphia, PA, 1973); *Adaptational Biology: Molecules to Organisms* (Wiley, New York, 1986).
4. P. W. Hockachka and G. S. Somero, *Biochemical Adaptation* (Princeton Univ. Press, Princeton, NJ, 1984).
5. G. Streisinger, C. Walker, N. Dower, D. Knauber, F. Singer, *Nature* **291**, 293 (1981).
6. G. Streisinger, *Natl. Cancer Inst. Monogr.* **65**, 53 (1984).
7. G. H. Thorgaard, P. D. Scheerer, J. E. Parsons, *Theor. Appl. Genet.* **71**, 119 (1985).
8. Z. Zhu, G. Li, L. He, S. Chen, Z. *Angew. Ichthyol.* **1**, 31 (1985); *Kexue Tongbao* **31**, 988 (1986); D. Chourrout, R. Guyomard, L. M. Houdebine, *Aquaculture* **51**, 143 (1986); R. A. Dunham, J. Eash, J. Askins, T. M. Townes, *Trans. Am. Fish. Soc.* **116**, 87 (1987); G. L. Fletcher, M. A. Shears, P. L. King, M. J. Davies, C. L. Hew, *Can. J. Fish. Aquat. Sci.* **45**, 352 (1988); N. D. Maclean, D. Penman, Z. Zhu, *BioTechnology* **5**, 257 (1987); G. W. Stuart, J. V. McMurry, M. Westerfield, *Development* **103**,

403 (1988); T. McEvoy et al., Aquaculture 68, 27 (1988); K. Ozato et al., Cell Differ. 19, 237 (1986); T. T. Chen et al., 1989 UCLA Symposium on Transgenic Animals, in press; T. T. Chen et al., Aquaculture, in press; P. Zhang et al., Mol. Reprod. Dev., in press; D. A. Powers, L. I. Gonzales-Villasenor, P. Zhang, T. T. Chen, R. A. Dunham, NIH Symposia on Transgenic Animals, in press.

9. N. Daw, J. Physiol. 197, 567 (1968); J. T. Schmidt, Trends Neurosci. 4, 111 (1982); M. K. Powers and S. S. Easter, in Fish Neurobiology, R. G. Northcutt and R. E. Davis, Eds. (Univ. of Michigan Press, Ann Arbor, 1983), vol. 1, pp. 377–404.

10. T. H. Bullock, in Fish Neurobiology, R. G. Northcutt and R. E. Davis, Eds. (Univ. of Michigan Press, Ann Arbor, 1983), vol. 2, pp. 362–368. Bullock emphasized that Hughlings Jackson pointed this out almost 100 years ago. See A. M. Lassek, The Unique Legacy of Dr. Hughlings Jackson (Thomas, Springfield, IL, 1970).

11. V. P. Whittaker, Neurochem. Res. 12, 121 (1987).

12. M. Noda et al., Nature 302, 528 (1983); A. Devillers-Thiery et al., Proc. Natl. Acad. Sci. U.S.A. 80, 2067 (1983); M. Noda et al., Nature 301, 251 (1983); M. Mishima et al., ibid. 307, 604 (1984); M. A. Raftery, M. W. Hunkapiller, C. D. Strader, L. E. Hood, Science 208, 1454 (1980).

13. E. Heibronn, in Handbook of Neurochemistry, A. Lajtha, Ed. (Plenum, New York, 1985), vol. 10, pp. 241–248; M. Schumacker et al., Nature 319, 6025 (1986).

14. W. S. Agnew, A. C. Moore, S. R. Levinson, M. A. Raftery, Proc. Natl. Acad. Sci. U.S.A. 75, 2606 (1978).

15. R. L. Rosenberg, S. A. Tomiko, W. S. Agnew, ibid. 81, 5594 (1984).

16. V. Flockerzi et al., Nature 323, 66 (1986).

17. M. Noda et al., ibid. 312, 121 (1984).

18. C. M. Armstrong, Physiol. Rev. 61, 644 (1981); W. F. Gilly and C. M. Armstrong, J. Gen. Physiol. 79, 965 (1982).

19. W. A. Catterall, Science 242, 50 (1988).

20. M. Noda et al., Nature 312, 188 (1986).

21. L. Salkoff et al., Science 237, 744 (1987).

22. B. L. Tempel, D. M. Papazian, T. L. Schwarz, Y. N. Han, L. Y. Jan, ibid., p. 770.

23. S. B. Ellis et al., ibid., 241, p. 1661.

24. While this is an accepted generality, it is built on a very few species. There is evidence that other hormones are also involved and, in some species, alternate mechanisms may be used. See W. S. Hoar and D. J. Randall, Eds., Fish Physiology, vol. 10, part B, Gills: Ion and Water Transfer

(Academic Press, New York, 1984); H. A. Bern, Am. Zool. 23, 663 (1986).

25. G. E. Pickford and J. G. Phillips, Science 130, 454 (1959).

26. G. E. Pickford and J. W. Atz, The Physiology of the Pituitary Gland of Fishes (New York Zoological Society, New York, 1957).

27. K. Fiedler, Zool. Jahrb. Abt. Allg. Zool. Physiol. Tiere 69, 609 (1962).

28. V. Blum and K. Fiedler, Gen. Comp. Endocrinol. 5, 186 (1965); O. Riddle, J. Natl. Cancer Inst. 31, 1039 (1963); Animal Behav. 11, 419 (1963), and references therein.

29. G. De Renzis and M. Bornancin, in Fish Physiology, vol. 10, part B, Gills: Ion and Water Transfer, W. S. Hoar and D. J. Randall Eds. (Academic Press, New York, 1984), pp. 65–104.

30. G. E. Pickford, Bull. Bingham Oceanogr. Coll. 14 (no. 2), 5 (1953); ibid., p. 46.

31. Growth hormone gene sequence and discussion of evolution are in L. B. Agellon, S. L. Davies, T. T. Chen, D. A. Powers, Proc. Natl. Acad. Sci. U.S.A. 85, 5136 (1988); other articles on evolution are W. L. Miller and N. L. Everhardt, Endocr. Rev. 4, 97 (1983); E. P. Slater, J. D. Baxter, N. L. Eberhardt, Am. Zool. 26, 939 (1986); and references in both.

32. D. C. W. Smith, Mem. Soc. Endocrinol. 5, 83 (1956); W. S. Hoar, in The Pituitary Gland, G. W. Harris and B. T. Donovan, Eds. (Butterworths, London, 1966), vol. 1, pp. 242–294.

33. W. C. Clarke, S. W. Farmer, K. M. Hartwell, Gen. Comp. Endocrinol. 33, 174 (1977).

34. W. C. Clarke, H. A. Bern, C. H. Li, D. C. Cohen, Endocrinology 93, 960 (1973); B. A. Doneen, Gen. Comp. Endocrinol. 30, 34 (1976); C. S. Nicoll and P. Light, ibid. 17, 490 (1971).

35. S. Sekine et al., Proc. Natl. Acad. Sci. U.S.A. 82, 4306 (1985); L. B. Agellon and T. T. Chen, DNA 5, 463 (1986); L. L. Gonzalez, P. J. Zhang, T. T. Chen, D. A. Powers, Gene 65, 239 (1988); C. S. Nicoll et al., Gen. Comp. Endocrinol. 68, 387 (1987); Y. Kuwana et al., Agric. Biol. Chem. 52, 1033 (1988); S. Song et al., Eur. J. Biochem. 172 (no. 2), 279 (1988); L. B. Agellon, S. L. Davies, C. Lin, T. T. Chen, D. A. Powers, Mol. Reprod. Dev. 1, 11 (1988).

36. N. Kitahara et al., Comp. Biochem. Physiol. B 91 (no. 3), 551 (1988).

37. Y. Okawara et al., Proc. Natl. Acad. Sci. U.S.A. 85, 8439 (1988).

38. J. P. Trinhaus, Cells into Organs: The Forces that Shape the Embryo (Prentice-Hall, Englewood Cliffs, NJ, 1984), p. 1; Am. Zool. 24, 673 (1984); J. Exp. Zool. 118, 269 (1951); T. Betchaku and J. P. Trinkaus, Am. Zool. 26, 193 (1986).

39. G. H. Thorgaard and F. W. Allendorf, in *Developmental Genetics of Higher Organisms*, G. M. Malacinski, Ed. (Macmillan, New York, 1981), pp. 363–391.

40. G. S. Whitt, *Am. Zool.* 21, 549 (1981); *Isozymes: Curr. Top. Biol. Med. Res.* 10, 1 (1983).

41. K. D. Kallman, *Copeia* 1983, 755 (1983).

42. L. DiMichele and D. A. Powers, *Nature* 260, 563 (1982); *Physiol. Zool.* 57, 46 (1984); *ibid.*, p. 52.

43. L. DiMichele, personal communication.

44. _____, D. A. Powers, J. DiMichele, *Am. Zool.* 26, 201 (1986).

45. F. W. Allendorf, K. L. Knudsen, R. F. Leary, *Proc. Natl. Acad. Sci. U.S.A.* 80, 1397 (1983).

46. J. E. Wright, J. B. Heckman, L. M. Atherton, in *Isozymes*, vol. 3, *Developmental Biology*, C. L. Markert, Ed., (Academic Press, New York, 1975), pp. 375–401; R. G. Danzmann, M. M. Ferguson, F. W. Allendorf, *Dev. Genet.* 5, 117 (1985); J. H. Knudsen, R. F. Leary, M. Talluri, *Genetics* 107, 57 (1984).

47. There have been a number of important developmental mouse mutants isolated over the last 20 years, but they cannot be easily maintained, and lethal genes are almost impossible to isolate and maintain.

48. Reviewed by C. B. Kimmel and R. W. Warga, *Trends Genet.* 4 (no. 3), 68 (1988).

49. C. Walker and G. Streisinger, *Genetics* 103, 125 (1983); S. Chakrabarti, G. Streisinger, F. Singer, C. Walker, *ibid.*, p. 109; G. Streisinger, F. Singer, C. Walker, D. Knauber, N. Dower, *ibid.* 112, 311 (1986).

50. C. B. Kimmel and R. D. Law, *Dev. Biol.* 108, 78 (1985); *ibid.*, p. 94; C. B. Kimmel and R. W. Warga, *Science* 231, 365 (1986); *Nature* 327, 234 (1987); *Dev. Biol.* 124, 269 (1987).

51. D. Grunwald, C. B. Kimmel, M. Westerfield, C. Walker, G. Streisinger, *Dev. Biol.* 126, 115 (1988).

52. H. G. Eiken, P. R. Njolstad, A. Molven, A. Fjose, *Biochem. Biophys. Res. Commun.* 149, 1165 (1987); P. R. Njolstad and A. Fjose, *ibid.* 152, 426 (1988).

53. F. E. Russel and P. Kotin, *J. Natl. Cancer Inst.* 6, 857 (1957).

54. Reviewed by M. C. Mix, *Marine Environ. Res.* 20, 1 (1986).

55. R. A. Murchelano and R. E. Wolke, *Science* 228, 587 (1985).

56. D. C. Malins et al., *J. Natl. Cancer Inst.* 74, 487 (1985).

57. J. D. Hendricks, T. R. Meyers, D. W. Shelton, J. L. Casteel, G. S. Bailey, *ibid.*, p. 839.

58. Induction of fish tumors has been demonstrated in a variety of fish and with many carcinogens. For example, R. O. Sinnhuber, J. D. Hendricks, J. H. Wales, G. B. Putman, *Ann. N.Y. Acad. Sci.* 298, 389 (1977); M. E. Schultz and R. J. Schultz, *J. Hered.* 73, 43 (1982); *Exp. Mol. Pathol.* 42, 320 (1985); M. F. Stanton, *J. Natl. Cancer Inst.* 34, 117 (1965); K. Aoki and H. Matsudaira, *ibid.* 59, 1747 (1977); Y. Hyodo-Taguchi and H. Matsudaira, *ibid.* 73, 1219 (1984).

59. J. D. Hendricks et al., *Natl. Cancer Inst. Monogr.* 65, 129 (1984).

60. J. E. Klaunig, B. A. Barut, R. J. Goldblatt, *ibid.*, p. 155.

61. T. S. Papas, J. E. Dahlberg, R. A. Sonstegard, *Nature* 261, 506 (1977).

62. T. Kimura, M. Yosimizu, M. Tanaka, *Fish Pathol.* 15, 149 (1981).

63. T. Sano, H. Fukuda, N. Okamoto, F. D. Kaneko, *Bull. Japan Soc. Sci. Fish.* 48, 1159 (1983).

64. C. Kosswig, *Z. Indukt. Abstammungs. Vererbungsl.* 47, 150 (1928); G. Haussler, *Klin. Wochenschr.* 7, 1561 (1928).

65. There are hundreds of papers that should be cited here; these represent only a few examples. H. D. Reed and M. Gordon, *Am. J. Cancer* 15, 1524 (1931); M. Levine and M. Gordon, *Cancer Res.* 6, 197 (1962); F. Anders, *Experientia* 23, 1 (1967); A. Anders, F. Anders, K. Klinke, in *Genetics and Mutagenesis of Fish*, J. H. Schroder, Ed. (Springer-Verlag, Berlin, 1973), pp. 33–52; A. Anders and F. Anders, *Biochim. Biophys. Acta* 516, 61 (1978); K. Ozato and Y. Wakamatsu, *Differentiation* 24, 181 (1983); A. Permutter and H. Potter, *J. Cancer Res. Clin. Oncol.* 114, 359 (1988).

66. R. Anders, M. Schartl, A. Barnekow, A. Anders, *Adv. Cancer Res.* 42, 191 (1984).

67. N. Nemoto, K. Kodama, A. Tazawa, P. Masahito, T. Ishikawa, *Differentiation* 32, 17 (1986).

68. R. J. Van Beneden, D. K. Watson, T. Chen, J. A. Lautenberg, T. S. Papas, *Proc. Natl. Acad. Sci. U.S.A.* 83, 3698 (1986); *Murine Environ. Res.* 24, 339 (1988).

69. M. Schartl and R. U. Peter, *Cancer Res.* 48, 741 (1988).

70. J. Stegeman, personal communication.

71. M. Koban, personal communication.

72. A. L. DeVries, thesis, Stanford University, Stanford, CA, (1969); *Science* 172, 1152 (1971).

73. Reviewed by: A. L. DeVries, in *Fish Physiology*, vol. 6, *Environmental Relations and Behavior*, W. S. Hoar and D. J. Randall, Eds. (Academic Press, New York, 1971), pp. 157–190; *Annu. Rev. Physiol.* 73A (no. 4), 627 (1983); R. E. Feeney, *Am. Sci.* 62, 712 (1974); C. L. Hew and G. L. Fletcher, in *Circulation, Respiration*

and Metabolism, R. Gilles, Ed. (Springer-Verlag, Berlin, 1985), pp. 553–690.

74. J. A. Raymond, P. Wilson, A. L. DeVries, *Proc. Natl. Acad. Sci. U.S.A.* **86**, 881 (1989).

75. For example, A. L. DeVries, J. Vandenheede, R. E. Feeney, *J. Biol. Chem.* **246**, 305 (1971); Y. Lin, J. G. Duman, A. L. DeVries, *Biochem. Biophys. Res. Commun.* **46**, 87 (1972); D. S. C. Yang, M. Sax, A. Chakrabartty, C. L. Hew, *Nature* **333**, 232 (1988); A. Chakrabartty and C. L. Hew, *Marine Can. J. Zool.* **66**, 403 (1988).

76. For example, Y. Lin and J. K. Gross, *Proc. Natl. Acad. Sci. U.S.A.* **78**, 2825 (1981); P. L. Davies, A. H. Roach, C. L. Hew, *ibid.* **79**, 335 (1982); B. Gourlie *et al.*, *J. Biol. Chem.* **259**, 14960 (1984); P. L. Davies *et al.*, *ibid.*, p. 9241; G. K. Scott, G. L. Fletcher, P. L. Davies, *Can. J. Fish. Aquat. Sci.* **43**, 1028 (1986); G. K. Scott, P. L. Davies, M. H. Kao, G. L. Fletcher, *J. Mol. Evol.* **27**, 29 (1988).

77. R. C. Lewontin, *The Genetic Basis of Evolutionary Change* (Columbia Univ. Press, New York, 1974).

78. Reviewed by D. A. Powers, in *New Directions in Physiological Ecology*, M. Feder, A. Bennet, W. Burggren, R. Huey, Eds. (Cambridge Univ. Press, Cambridge, 1987), pp. 102–134.

79. A. R. Place and D. A. Powers, *J. Biol. Chem.* **259**, 1299 (1984); *ibid.*, p. 1309; *Biochem. Genet.* **16**, 577 (1978); *Proc. Natl. Acad. Sci. U.S.A.* **76**, 2354 (1979); R. J. Van Beneden and D. A. Powers, *J. Biol. Chem.* **260**, 14596 (1985); *Mol. Biol. Evol.* (no. 2), 155 (1989); I. J. Ropson and D. A. Powers, *ibid.*, p. 171; D. L. Crawford, H. R. Costantino, D. A. Powers, *ibid.* (no. 4), p. 369; D. Brown, I. Ropson, D. A. Powers, *Heredity* **79** (no. 5), 359 (1988); D. A. Powers *et al.*, *Am. Zool.* **26**, 131 (1986); D. A. Powers, P. Dalessio, E. Lee, L. DiMichele, *ibid.*, p. 235; D. A. Powers, L. DiMichele, A. R. Place, *Isozymes*, **10**, 147 (1983); R. J. Van Beneden, R. E. Cashon, D. A. Powers, *Biochem. Genet.* **19**, 701 (1981); R. E. Cashon, R. J. Van Beneden, D. A. Powers, *ibid.*, p. 715; D. A. Powers and A. R. Place, *ibid.* **16**, 593 (1978); L. DiMichele and D. A. Powers, *Science* **216**, 1014 (1982); L. I. Gonzalez-Villasenor and D. A. Powers, *Evolution*, in press; I. Ropson, D. Brown, D. A. Powers, *ibid.*, in press.

80. J. Quattro and R. Vrijenhoek, *Science* **245**, 976 (1989).

81. R. Vrijenhoek, personal communication.

82. I thank D. Mazia, D. Epel, W. Gilly, D. Crawford, and M. Powell for their helpful comments on this manuscript. In addition, I would like to thank R. Vrijenhoek, J. Stegeman, Z. Zhu, M. Koban, and L. DiMichele, who provided unpublished information. Finally, I would like to apologize to the hundreds of outstanding researchers whose excellent work on fish models could not be accommodated. I could have just as easily written the entire paper using a completely different set of examples, without mentioning all the exciting work on fish models.

13

Contributions of
Bird Studies to Biology

Masakazu Konishi, Stephen T. Emlen,
Robert E. Ricklefs, John C. Wingfield

Each taxon presents two faces: one of the uniformity that embodies the common features shared by all members of the group and the other of diversity that expresses the ecological and evolutionary responses of individuals and populations to the array of environments on Earth. When compared to vertebrate classes and invertebrate taxa of similar rank, birds are remarkable for their uniformity of anatomy, physiology, and life cycle. Penguins, flycatchers, geese, sparrows, and hawks resemble each other much more closely than their mammalian ecological counterparts (for example, seals, bats, sheep, mice, and cats). All birds lay eggs, most have extended parental care, and most enjoy relatively long lives, even compared to mammals. Yet, despite their uniformity, birds inhabit all regions of the earth and assume a wide variety of ecological roles.

We know the species-level taxonomy and geographical distribution of birds better than that of any other taxon of comparable diversity. Birds also show tremendous diversity in their social systems, making them valuable models for understanding the population biology and social behavior of other animal groups such as

primates. Most birds explore their environments and communicate with one another by the same senses of sight and hearing that we use. These sense organs, and the brain areas devoted to these senses, are highly developed in birds, making avian subjects ideal for the study of neural design and function.

Birds are active by day and generally conspicuous. One can attach markers or leg bands for individual recognition and follow individuals on their daily rounds of activity and over their lifetimes. Like butterflies and a few other groups noted for their beauty, birds have won the devotion of dedicated amateurs and thereby stimulated the amassing of a formidable database on life-history attributes (birth-rate, age at first reproduction, life span, and so forth) and populations from all parts of the world. In a volume on the lifetime reproductive success of animals (*1*), 13 of 25 nontheoretical chapters were drawn from studies of birds. The *Zoological Record* for 1985 indexed over 9,300 articles on birds (excluding work on domesticated forms), slightly more articles than there are species. In the journal *Ecology*, 11% of articles published in 1987 and

M. Konishi is the Bing Professor of Behavioral Biology, Division of Biology, California Institute of Technology, Pasadena, CA 91125. S. T. Emlen is Professor of Animal Behavior, Section of Neurobiology and Behavior, Cornell University, Ithaca, NY 14853. R. E. Ricklefs is Professor of Biology, Department of Biology, University of Pennsylvania, Philadelphia, PA 19104. J. C. Wingfield is Professor of Zoology, Department of Zoology, University of Washington, Seattle, WA 98195. This chapter is reprinted from *Science* **246**, 465 (1989).

1988 concerned avian studies, while in the journal *Animal Behaviour*, 37% of articles published in the same 2 years dealt with birds.

All these attributes make birds ideal subjects for investigation in various fields of biology. In this article, we shall discuss how the study of birds has contributed to important discoveries in biology and to the formation and testing of significant new ideas in selected fields of biology.

Evolution, Ecology, and Sociobiology

Speciation. Ornithologists pioneered the study of the formation of species. Details of morphological variation and geographical distribution gleaned from field studies and museum collections led Mayr to conclude that the splitting of one species into two is usually preceded by a period of geographical isolation during which genetic differences accumulated (2). This idea, known as the theory of allopatric speciation, has stimulated many theoretical, empirical, and experimental studies of speciation. The ease of observing and capturing birds in the wild has also helped investigators to collect the best information on the heritability (3) and responses to selection of morphological traits in natural settings (4). Although birds have proved favorable subjects for the study of speciation, they have not figured so prominently as other groups in the study of evolution at higher phyletic levels. Nevertheless, birds have been subjected to the most ambitious molecular analysis of phylogeny of any taxonomic group, which, along with more traditional data, has allowed researchers to obtain a better understanding of avian historical biogeography and diversification (5).

Population and community ecology. Bird studies have had their greatest impact in the area of population and community ecology. For three decades beginning in the 1940s, the strongest and most influential voice for these new ideas was that of Lack (6). Lack visualized connections between the theory of interacting populations, developed by Lotka and Gause in the 1920s and 1930s (7), and patterns of morphological (8) and ecological (9) diversification among species; he also championed the position that populations are regulated by factors that have effects which intensify at higher density (10). He also initiated the field of evolutionary ecology (8), which seeks to interpret behavior and life-history attributes in the context of evolutionary optimization. Among Lack's most important insights was that ecological and geographical distributions of birds could be interpreted as the outcome of interspecific interactions (11). This insight expressed itself early in Lack's analysis of evolutionary diversification of Darwin's finches on the Galapagos Archipelago (8).

The principle of ecological divergence or resource partitioning mediated by competition became the basis of Hutchinson's (12) concept of the multidimensional niche, in which species compete and diversify with respect to many environmental factors, and that of limiting similarity set by interspecific competition. This concept quickly emerged, largely through a series of papers on birds by MacArthur (13), into a paradigm of community structure regulated by competitive interaction. Birds were also the subjects of 17 of 50 studies, summarized by Schoener (14), that quantified resource partitioning in terrestrial animal communities. Although the momentum of this paradigm sent much of community ecology onto an unproductive sidetrack during the 1970s (15), it has provided a rich empirical base of data on community structure, and the controversy it generated greatly clarified thinking in the 1980s.

An important paradigm to emerge from the principles of population and community ecology was MacArthur and Wilson's (16) "equilibrium" theory of island biogeography, which related the number of species on islands to diversity-specific rates of immigration and extinction. Because their distributions on islands are so well known, birds have figured prominently in testing the theory (17), as well as applying it to the optimal design of nature reserves, which are, all too often, islands of habitat (18). Somewhat ironically, articles on avian biogeography affirming the connection between competition and coexistence raised questions in many minds about the statistical validity of community patterns and caused substantial rethinking of the competition paradigm (*19*).

Another of Lack's inspirations was the idea that natural selection adjusts life history traits in order to maximize fitness of the individual. Lack (*8*) initially applied this thinking to clutch size—the number of eggs laid per nest—in birds, but later (*20*) broadened his scope to include other attributes, including mating system and social behavior. Wynne-Edwards, drawing on his lifelong experiences with seabirds, disagreed with Lack by concluding that most life-history traits, including social behavior, had evolved to regulate population numbers at levels that the environment could sustain (*21*). This debate over the units of selection (individuals versus populations) sparked a widespread reexamination of "species level" thinking in the 1950s and early 1960s, and led to the realization that selection at the level of groups could only occur under very restrictive conditions (*22*). Experiments involving the artificial alteration of clutch sizes and the measurement of resulting fledgling success provided strong support for the theory of individual selection (*23*).

The early ideas and studies of Lack and others stimulated many serious long-term studies of life history traits and the adaptive significance of behavior. These investigations resulted in a considerable elaboration of optimization theory in studies of life histories (*24*) and foraging tactics (*25*). Birds have provided most of the empirical and experimental tests of optimal foraging theory, whether it addresses the choice of prey within foraging locality, the choice of switching to a new location as prey are locally depleted, or the avoidance of unpredictable supplies of food (*25*). Birds have also proven to be suitable experimental subjects for studies of life history evolution (*26*).

Behavioral ecology. Bird studies have also been important in forging new interdisciplinary bridges between animal behavior and ecology. In 1964, Crook (*27*) published a monograph on the comparative ecology and social organization of nearly 50 species of weaverbirds. He concluded that ecological factors are impor-

tant in shaping the form of their societies. The degree of gregariousness, the presence or absence of territorial defense, and even the basic form of the mating system are all profoundly influenced by the abundance and the spatial and temporal distribution of key resources, predators, and competitors.

Crook's ideas of the ecological shaping of social behavior gained unusually rapid and widespread acceptance. The reason, in part, was because of the extraordinary wealth of preexisting avian field data that had been accumulating in the literature since the turn of the century. Many ornithologists made immediate use of this abundance of information to verify Crook's correlational findings for other avian groups. Soon such analyses extended beyond ornithology, and parallel correlations between simple ecological predictors and social structure were uncovered in groups as diverse as coral reef fishes (*28*), anurans (*29*), bats (*30*), ungulates (*31*), and primates (*32*).

Comparative studies were soon supplemented by experimental ones, and descriptive explanations were replaced by analytical models. Brown (*33*) incorporated economic thinking into his models of adaptive behavior. In his theory of economic defendability, he argued that the distribution pattern of a resource determined the benefit-to-cost ratio of defending that resource, and he used this approach to model the evolution of territorial behavior (*33*). Studies of nectar-feeding birds, in which the caloric costs and benefits of territorial defense could be measured with accuracy, have provided strong support for the model (*34*).

Economic considerations are also pivotal to our understanding of why animals live in groups. Two ecological factors, predator avoidance and feeding efficiency, are believed to be primary determinants of grouping tendencies. Depending on the types of predators and abundance of predators, grouping may make individuals more conspicuous and thus more susceptible to predation. On the other hand, grouping may also provide the means for more efficient predator detection and deterrence (for example, through sentinel behavior or group mobbing). Similarly, when food resources are not renewable and are spatially stable, grouping can lead to over-exploitation of these resources. But, if food resources are ephemeral and unpredictable in location, group living can provide a means for enhanced food localization (for example, through the pooling of information gathered by multiple foragers). These types of cost-benefit trade-offs determine when grouping is advantageous, as well as the optimal group size. Avian examples have been fundamental in the development and testing of these ideas. Trained raptors have been used to quantify the benefit that prey species gain by living in flocks of different sizes (35). Colonial birds provide the best evidence to date for information pooling in the localization of unpredictable food sources (36).

Mating systems. Avian studies have been central to the development of mating system theory. Orians and his colleagues were the first to propose the adaptive significance of polygyny (one male pairing with more than one female) (37). They argued that polygynous mating occurs when resource distributions are sufficiently uneven that a female can maximize her chance of leaving offspring by mating bigamously with an already mated male on a high-quality territory rather than mating monogamously with an unmated male occupying a lower quality territory. This "polygyny threshold" model was based on years of research on blackbirds and wrens (38).

Emlen and Oring (39) later incorporated ideas of economic defendability with sexual selection theory to develop an ecological classification of avian mating systems. The reproductive output of males is limited by the availability of potential female partners, and males compete among themselves to gain access to receptive females. Ecological factors often determine the distribution, and hence the defendability, of female mates. Emlen and Oring suggested that polygyny occurs when individual males can control the access of other males to potential mates and thus can monopolize several females for themselves. Social and ecological variables that influence the mating options of individual organisms have proven to be strong predictors of mating systems, not only of birds, but also for many other animal groups including amphibians (29), insects (40),

and mammals (41).

Traditionally, biologists have classified mating systems according to the pair bonds that form between individuals. Recent evidence indicates, however, that individuals of both sexes commonly engage in reproductive activities outside of the pair bond. Such "mixed mating strategies" (42) take the form of extra-pair copulations by both sexes and, among birds, of intraspecific nest parasitism by females. Studies of birds previously assumed to be monogamous have demonstrated a surprisingly high level of mixed paternity and maternity of clutches (43). The availability of DNA fingerprinting is giving ornithologists and other field biologists a new tool with which to investigate the basis of mate choice and the conditions under which individuals elect to mate with individuals other than their nominal partners.

Altruism and kinship. Bird studies have a key role in testing recent ideas of the importance of kinship in the evolution and maintenance of altruistic behaviors. The existence of such behaviors as restraint from reproduction by honey bee workers, alarm calling by social rodents, and helping at the nest by birds have long posed paradoxes for evolutionary biologists because they appear to contradict the basic self-interest fostered by natural selection. How can behaviors evolve if their performance is detrimental to the performer? Hamilton's (24) concept of inclusive fitness showed that altruistic behaviors could be favored by natural selection provided that they were directed toward genetic relatives, which have a high probability of carrying the genes responsible for the behavior. Cooperatively breeding birds (birds in which extra adults assist the breeding pair in the rearing of young) have provided a test arena for Hamilton's theory, because the costs and benefits associated with helping behaviors can be accurately quantified (44).

Several long-term studies of cooperatively breeding birds provide data on the relative importance of kinship in helping (45-47). Helpers may benefit from their actions in two major (and additive) ways: they may reap kinship benefits by increasing the survival of nondescendant relatives; and, by helping in the present, they may increase their own prospects for becoming successful breeders in the future. Different species, depending on their ecology and demography, vary in the relative importance of the two types of benefits. White-fronted bee-eaters (*Merops bullockoides*) provide an example of kin favoritism and kin benefit (Fig. 1) (*46, 48*), whereas Florida scrub jays *Aphelocoma C. coerulescens*) llustrate a case where the future direct (personal) benefits of helping are large (*47*).

Ethology

Communication. The study of communication is an important component of animal sociology, and birds have been favorite subjects in this field. The realization that displays and vocalizations of animals are their means of communication was the key to the development of ethology as a discipline. It is not coincidental that two of the founders of ethology, Lorenz (*49*) and Tinbergen (*50*), and even their predecessors, Heinroth (*51*) and Whitman (*52*), all studied birds. Comparative and experimental studies of signaling have developed side by side in this field. Lorenz's study of behavior in ducks and geese (*53*), Kortlandt's study of cormorant display (*54*), and the studies of various gulls by Tinbergen and his associates (*55*) are examples in which an understanding of the evolution of signaling behavior

by comparing closely related species was attempted. Similarly, a comparative study of alarm calls among different bird species led Marler to the concept of evolutionary convergence in animal signals (*56*), whereas the diversity of bird songs led him to the concept of species specificity in animal signals that serve in species recognition (*57*).

Experimental studies of communication aim at elucidating the rules of encoding information and the function of signals. The methods for studying the coding rules involve the use of dummies and recorded and synthetic sounds. Consistent with the conclusion inferred from comparative studies, the results of experimental work show that birds use species-specific visual displays and vocalizations to recognize the members of their own species or geographical populations. Different bird species use different sets of acoustical cues to recognize the song of their own species. Any one of two or three cues appears to be sufficient for species recognition; there is redundancy in the encoding of species specificity in song (*58*). Birds can also use individual variations in song and calls for recognition of neighbors, mates, and family members (*59*). The generalizations drawn from comparative and experimental studies of acoustic communication in birds have not only contributed to our knowledge of animal communication, but also have guided the study of vocal communication in other animals, including primates (*60*).

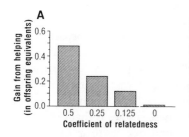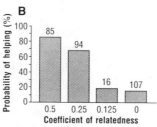

Fig. 1. The importance of kinship in explaining helping-at-the-nest behavior in a cooperatively breeding bird, the white-fronted bee-eater (*Merops bullockoides*). In this species, approximately 55% of all nonbreeding adults act as helpers; the remaining 45% do not. Birds that do help incubate eggs, and defend and provision nestlings. (**A**) The indirect fitness gain (kin benefit) realized by a helper is plotted as a function of the helper's relatedness to the nestlings it aids. This gain is measured in offspring equivalents (*48*) and is calculated as the product of two terms: the average number of additional offspring successfully fledged as a result of a helper's activities, and the coefficient of relatedness between the nestlings and the helper (sibling = 0.5, half-sibling = 0.25, and so on). (**B**) The probability that a potential helper (a nonbreeding bee-eater with a recipient nest available) becomes an actual helper is plotted as a function of its coefficient of relatedness to the recipient nestlings. Numbers above histograms are sample sizes of potential helpers in each kin category. White-fronted bee-eaters are most likely to become helpers when the recipients are close genetic relatives and, as a result, the kin benefits realized from helping are large.

Ontogeny and learning. Studies of bird song led to important advances in understanding the processes of behavioral ontogeny and the factors that control them (*61*). Early bird fanciers knew that young birds must have a good tutor to become a good singer. They also knew that birds could be bred for more elaborate song, as in the domestic canary. Thus, differences in song between species and between individuals contain both heritable and experiential components. The experiential component has been most extensively studied since the seminal work of Thorpe (*62*).

Learning of voice, whether it occurs in human or avian babies, involves translation of the auditory percept of the voice into motor coordinations that give rise to the same voice, that is, the sound that causes the same auditory percept. Song learning consists of perceptual and sensorimotor phases. Young birds commit a tutor song to memory in the first phase and reproduce the memorized song in the second. The perceptual phase may be restricted to a relatively short period in early life or it may be extended, the details varying according to such factors as the photoperiod and the conditions under which song tutoring occurs. Given a choice between the species' own song and alien songs, young birds learn the song of their own species. In the sensorimotor phase, birds must hear themselves sing in order to reproduce the memorized song. Birds deafened after the perceptual phase, but before or during the sensorimotor phase, cannot reproduce the memorized song (*63*).

Birds are the only known nonhuman animal that learns complex vocalizations (*64*). The characteristics of song learning have intriguing resemblances to those of speech acquisition in human infants (*65*). Speech sounds are particularly attractive to human babies. Infants can discriminate between different speech sound categories even in languages they have never heard (*66*). The babbling stage of speech development resembles an early stage of bird song development. The impressionable phase of speech acquisition and the need for auditory feedback in human speech development also parallel the major features of song development in birds. Because intrusive experiments cannot be performed on humans, avian song learning is a valuable animal model of speech acquisition.

Orientation and navigation. Orientation and navigation is another field in which the nature and function of sensory cues are the subject of investigation. Birds are the champions of long-distance travel. Hundreds of species make annual migrations of thousands of kilometers, returning with remarkable precision to the same several square kilometers of territory on both their breeding and their wintering grounds.

Much of our current knowledge about the mechanisms of animal orientation comes from experimental work with bees (*67*), fish (*68*), and birds (*69*). The routes, destinations, and timing of migration are, however, better known for birds than for any other animal. Investigators can follow navigating birds for long distances by airplanes and radar. Birds also can be made to carry various devices (for example, radio transmitters, miniature cameras, bar magnets, and vision-impeding spectacles). Migrants can be tested in specially designed cages for their orientation response to natural and artificial stimuli (*70*).

We now know that migratory birds can make use of several different cue systems for determining the compass directions of their migratory flights. The position of the sun by day, polarized light patterns at sunset, and the configurations of the stars at night all serve as celestrial cues (*71*). The earth's magnetic field may also provide directional information (*72*). Migrants have proven to be good meteorologists as well. They time their departures to catch favorable tail winds aloft, and they often adjust their flight headings to (at least partially) compensate for the drift that occurs when they encounter crosswinds (*73, 74*). One key finding to emerge from these studies is that birds have considerable redundancy in their orientation systems. If one cue is temporarily unavailable or provides unreliable information, another cue can be used in its place. Although the existence of such redundant "back ups" makes good adaptive sense, it has been difficult to design experiments that can sort out the roles of different cues. Redundancy has also raised stimulating new questions: Does the simultaneous use of multiple cues result in increased accuracy of orientation? Does a hierarchy exist in the relative importance of different types of information? How does each cue system become integrated with or calibrated against the others (*74, 75*)?

Navigation researchers distinguish between a compass "sense," by which the animal orients in a particular direction and maintains that course, and a map "sense," by which the animal determines its position relative to some reference location (usually home). Many species of migratory birds may require only a compass ability; other migrants and homing pigeons use a map sense as well. Whereas several compass cues have been studied in detail, the map sense remains poorly understood. Current information indicates that celestial cues do not provide map information (*76*). Models based on olfactory (*77*) and magnetic (*75, 78*) cues have been advanced, but need more testing. Solving the map riddle remains the major challenge for orientation researchers.

The Interfaces of Physiology, Ecology, and Behavior

Physiological adaptations. Birds as experimental subjects have been important in linking pure regulatory physiology to ecology and behavior. There are many good examples of physiological and behavioral adaptations to the environment. Schmidt-Nielsen (*79*) and Scheid (*80*) showed that the avian lung is unique in being a rigid structure through which air flows in a one-way system regulated by a complex network of air sacs. The result is a remarkably efficient system for oxygen uptake and loss of carbon dioxide, especially for flight, as indicated by the study of birds flying in wind tunnels to simulate the behavior under natural conditions (*81*). Studies have been carried out on such topics as the ability to control body temperature in extremely hot and cold climates for life in deserts and high altitudes and latitudes (*82*), nasal salt glands for secretion of excess sodium chloride in marine and arid environments (*83*), and kidneys that minimize water loss by excreting uric acid (*84*). These studies paved the way for a generation of avian physiological ecologists who, together with investigators of other vertebrate groups, have advanced our knowledge of how organisms maintain homeostasis in environments as diverse as the poles, deserts, oceans, and rain forests.

Adaptation to the environment requires compromise among competing demands for time and energy allocated to different functions (*85*). Studies of avian reproduction, particularly during the periods of incubating the eggs and caring for young, have elucidated the role of environmental and physiological constraints on nestling development and parental care (*86–88*). In particular, growth rate strikes a balance between rapid growth for avoidance of predation and slower growth requiring less energy and nutrients (*20*). Some birds hatch as self-sufficient chicks and others as helpless nestlings, and there are intermediates between these types. These differences show how the rate of growth of the chick is adjusted to allocation of the parent's time between foraging and direct care (*86, 89*). Behavioral studies also revealed the role of food solicitation in regulating parental feeding (*90*), adding sibling competition and parent-offspring conflict to the list of factors that interact to mold avian reproductive patterns (*91*).

Hormones and social behavior. The concept of hormonal regulation of behavior was developed in part by many studies on birds during the period between the 1930s and the 1960s (*92, 93*). A combination of sociobiological and physiological approaches is now adding a new dimension to the study of social behavior. The classical investigations of Lehrman (*94*) and Hinde (*95*) showed that behavior of one individual affects hormone secretion in another with both intra- and intersexual effects. Early investigators assessed hormone secretion indirectly by noting hormone-controlled behavior, or by measuring the state of the gonads and secondary sex characters. With the advent of increasingly sensitive radioimmunoassay techniques for measuring hormone levels in the blood, marked birds can be caught repeatedly with minimum disturbance to collect a small amount of blood for hormone titration.

During aggressive interactions over territory or mates, plasma levels of testosterone increase, especially in monogamous species (Fig. 2) (*96*). Testosterone elevates and enhances the expression of territorial aggression, but at the same time inhibits male parental behavior (*97, 98*). Thus, monogamous species in which males feed young have high testosterone for a very short period in early spring, whereas polygynous species in which males tend not to feed young remain very aggressive and have high testosterone for prolonged periods during the breeding season (*98*). If a "polygynous-like" seasonal profile of testosterone levels is induced in monogamous males, these males also tend to become polygynous (*98*). Such studies offer physiological explanations for social behavior and, by the use of hormone agonists and antagonists to manipulate behavior, provide new tools for field investigations in sociobiology and behavioral ecology.

Hormones and seasonal behavior. Over 60 years ago, Rowan (*99*) discovered that the annual cycle of daylength exerted a regulatory effect on gonadal development and migratory behavior in birds. When he exposed captive, male, dark-eyed juncos (*Junco hyemalis*) to long days in midwinter, they began to show migratory and reproductive behavior, and also testicular development. These discoveries led to a growing field of research on environmental control of reproduction and associated events in all vertebrate groups. For example, King and Farner (*100*) published a series of seminal papers on the regulation of lipid deposition during bird migration. Many others demonstrated a role for changing daylength in the regulation of breeding seasons in over 60 avian

species (*101*). Fewer investigations, however, have tackled the neuroendocrine mechanisms by which photoperiod is perceived and transduced into hormone secretion that in turn regulates reproductive development and associated behaviors.

In a now classic experiment, Follett, Mattocks, and Farner (*102*) showed that exposure of male white-crowned sparrows (*Zonotrichia leucophrys*), to an 8-hour pulse of light given at intervals after the last dawn resulted in a pronounced increase in secretion of luteinizing hormone (one of the gonadotropins associated with reproductive development). This increase occurred, however, only when the light pulse fell within the photosensitive phase of a circadian rhythm, which was between about 12 and 18 hours after dawn. Their experiment remains one of the most convincing demonstrations of an endogenous rhythm of photosensitivity, first postulated by Bünning (*103*). Curiously the photoreceptors by which birds measure daylength are not retinal (*104*), but lie within the brain (*105*). The use of fiber optics to illuminate specific areas of the brain has demonstrated that the photoreceptors are situated in the basal hypothalamus (*106*). In the past 30 years, the role of light as a timing cue for endogenous circannual rhythms has also gained acceptance through research on birds (*107*). Additionally, a state of refractoriness was found in which organisms exposed to a stimulatory photoperiod eventually become insensitive to that stimulus. This process is crucial for the termination of reproductive function and prevents further breeding at times when environmental conditions for survival of young become unfavorable. The mechanisms

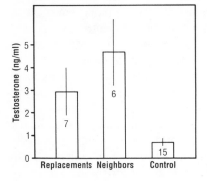

Fig. 2. Plasma levels of testosterone in free-living male song sparrows (*Melospiza melodia*) after removal of a territorial male. Replacements are those males that took over the vacant territory, and neighbors are those males with territories adjacent to the replacements. Controls were captured in a separate area in which territory boundaries were known to be stable. Histograms represent the means and vertical bars are the standard errors. Numbers within the histograms are the sample size. Replacements, $P < 0.005$; neighbors, $P = 0.001$. [Reprinted from (*96*) with permission, © 1985 Academic Press]

of refractoriness to light stimuli are still not clear and remain one of the major unsolved problems of reproductive biology (*108*).

Neurobiology

Hormones and brain. Bird studies have had a leading role in the development of neuroendocrinology. For example, Benoit and Assenmacher (*109*) showed that the hypothalamus of the mallard duck (*Anas platyhynchos*) exerted a profound regulatory role over secretions of the anterior pituitary gland. This discovery came at a time when the question of neuroendocrine secretion and regulation of pituitary function was a major controversy. The work of Oksche, and associates (*110*) on birds was significant in elucidating the morphology of the hypothalamo-pituitary unit. Developments in the isolation and sequencing of hypothalamic regulatory peptides in mammals were supplemented by similar work in birds in which several forms of single type of peptide were identified, suggesting that neuroendocrine regulation of pituitary function is more complex than was realized previously (*111*).

Steroid hormones act on target cells by regulating gene expression and protein synthesis (*112*). The avian oviduct secretes a special protein, ovalbumin, which makes up a major component of the egg white. O'Malley and his associates (*113*) were among the first to discover the advantages of the oviduct of chicks for the study of hormone-mediated protein synthesis. They showed that progesterone binds to intracellular receptors in the oviduct to form a steroid-receptor complex which, in turn, binds to the genome and stimulates transcription of the mRNA for ovalbumin. However, there are alternate pathways for the action of steroid hormones once they entered a cell. Hormones can be metabolized to an active form that binds to receptors or, as in birds, they may be converted to an inactive form like 5β-dihydrotestosterone, derived from testosterone (*114*). Whether this shunt mechanism is widespread in vertebrates remains to be seen.

The discovery of the song control system in the brain by Nottebohm and his associates (*115*) has provided a very useful preparation in which a direct link from hormones to brain and to behavior can be established. Songbirds have special brain areas for the control of song, whereas other birds, with the exception of parrots, lack such areas (*116*). Few vertebrate neural systems controlling particular behaviors form as discrete and readily identified groups of nuclei as the song system. Song is also a seasonal behavior, which is controlled by sex steroids. A subcutaneous implant of testosterone induces song in females, young birds, and males in nonbreeding conditions. This behavioral effect of testosterone is accompanied by rapid physiological and morphological changes in many parts of the song control system. For example, testosterone induces or increases protein synthesis in a song nucleus of female white-crowned sparrows within a few days after hormone implantation (*117*). Castration of male zebra finches (*Poephila guttata*) causes a decrease in the number of acetylcholine receptors in the muscles of their vocal organ and subsequent administration of testosterone restores both the number of the receptors and the activity of the enzymes involved in the synthesis and degradation of acetylcholine (*118*). In female canaries (*Serinus canaria*), both the length of dendrites of a certain class of neurons and the number of synapses in a song nucleus almost doubles in response to administration of testosterone (*119*).

Although these studies artificially manipulated hormone levels, similar changes occur naturally with the season. Plasma levels of testosterone and other sex steroids change seasonally (*120*). Nottebohm (*121*) discovered that the volume of one of the song nuclei of the male canary changes with the season. It is now known that similar changes occur in other song nuclei and other birds (*122*). These volume changes are partly due to changes in the ability of neurons to stain with thionin, which stains Nissl substance found in dendrites and perikarya (*123*). The seasonal volume fluctuations are, therefore, partly due to seasonal variations in the length of dendrites. Changes in dendrites and synapses indicate that the adult brain circuitry is not fixed, although such changes may not involve qualitative alterations in the existing neural circuits (124). The idea that the seasonal appearance and disappearance of song is accompanied by making and unmaking of neural connections may revolutionize our understanding of behavioral plasticity and learning (125).

In many species, the male sings and the female does not. This sexual dimorphism in behavior is correlated with gender differences in brain anatomy. The male song nuclei contain more neurons of larger size than the female counterparts (126). Also, connections between some of the nuclei are missing or rudimentary in the female (127). Such large brain gender differences were not known in any other animal. Subsequent searches have since revealed several cases of large gender differences in the mammalian central nervous system including humankind (128).

Brain development and neural plasticity. The song control system provides ideal materials for the study of the role of hormones in brain development. It was thought that gender differences in the brain arise because the sexually dimorphic areas of the female brain do not grow, whereas the homologous areas in the male grow under the influence of early gonadal steroids (129). In the zebra finch, the male and the female song nuclei are similar in early life. Neurons of the female forebrain song nuclei do not just fail to grow but undergo atrophy and death, whereas the male counterparts grow (127). The involvement of cell death in brain sexual differentiation is now known or suspected in most of the well-documented mammalian examples (130).

The causes of cell death and the conditions for cell survival are the focus of much current research in neuroscience. Most known cases of programmed neuronal death in which cell bodies as well as axons die occur in the spinal cord, ganglia, and the brain stem. The avian song system offers opportunities for investigators to study neuronal death in the forebrain. In the zebra finch, an implant of estrogen in a newly hatched female masculinizes her song control system both by preventing neuronal death in her forebrain song nuclei and by promoting the development of connections between some of the nuclei (131). Although the mechanisms of estrogen action are not known in this case, the possibility to prevent programmed cell death by such a simple agent opens a new avenue of research in neuroscience.

The avian brain displays important similarities with the brains of fishes, amphibians, and reptiles. Growth and regeneration of organs in adult animals are well known among these animals. In birds, for example, new neurons are born in the adult forebrain, and some of them become incorporated in functional circuits including the song system. These new neurons migrate from their birthplace to their destinations closely attached to radial glial cells (132). Radial glia are present in the adult brain among birds and lower vertebrates, whereas these cells disappear in mammals after they guide neuronal migration during the development of the brain. Studies show that new inner ear hair cells grow to replace injured ones in young birds (133). The identification of the conditions necessary for regeneration and adult retention of embryonic attributes will help to understand why mammals lack these properties.

Brain structure and function. Because the forebrains of nonmammalian vertebrates lack the layered structure typical of the mammalian neocortex, it was long assumed that nonmammalian vertebrates do not have a neocortex. The large tissue mass on the surface of the forebrains of birds and some reptiles was thought to be homologous to the striatum of mammals, which is a subcortical structure. Karten (134) reexamined this view by studying the projections of the auditory and visual systems to the forebrain in the pigeon and found that the striatal tissue mass contains large thalamic receiving areas as in the mammalian neocortex. This work led him to a revolutionary theory that areas homologous to the neocortex are present in the brain of nonmammalian vertebrates, but that they form aggregates instead of layers. This theory has provided a possible solution to one of the persistent puzzles in comparative neuroanatomy, that is, the evolution of the neocortex.

The brains of different animals exploit

similar design principles in differing degrees. Animals adapted for extreme conditions or special niches are likely to push certain design principles more than other species. Such "specialists" abound among birds. For example, barn owls (*Tyto alba*) can localize small rodents in total darkness by listening to the rustling sounds they make (*135*). The brain mechanisms of sound localization are better understood in this species than in any other vertebrate (*136*). The nucleus laminaris is a third-order auditory station in the avian brain and thought to be homologous to the medial superior olivary nucleus of mammals. In unspecialized species such as chickens and pigeons, the nucleus laminaris consists of a monolayer of neurons, which form circuits for the measurement of interaural time differences (*137*). In the barn owl, this nucleus is hypertrophied and the neural circuits are more elaborate, although they use the same principles of operation as the monolayer version (Fig. 3). The neural circuits for time measurement contain binaural coincidence detector neurons and axonal delay lines (*138*). The coincidence detectors fire maximally when nerve impulses arrive simultaneously from the right and left sides. When the axonal paths to these neurons from the ears are equal, nerve impulses starting simultaneously from the two ears reach the coincidence detectors at the same time. These coincidence detectors fire maximally when sound reaches the two ears simultaneously. Other coincidence detectors fire maximally only when sound reaches one ear later than the other ear by a particular delay, because the axonal paths to them from the two sides are unequal. Physiological evidence suggests similar circuits in the medial superior olivary nucleus, but they have yet to be identified (*139*).

Mapping of the external sensory world in the brain is another design principle that is used

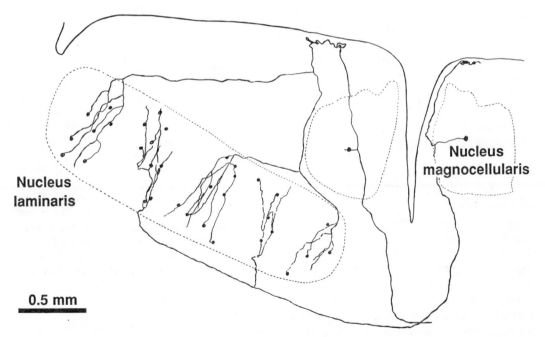

0.5 mm

Fig. 3. Axonal delay lines for measurement of interaural time differences in the barn owl's brainstem. Shown is the innervation of the nucleus laminaris by two typical axons of the magnocellularis cochlear nucleus, an ipsilateral one arriving from the dorsal side (top of the figure) and a contralateral one from the ventral side after crossing the midline of the medulla. These fibers branch out and course across the nucleus laminaris to the side opposite to the side of their entry. As the ipsilateral and contralateral fibers interdigitate, they innervate laminaris neurons, shown by small circles. Conduction time of nerve impulses in these fibers varies systematically as a function of distance from the point of their entry to the nucleus laminaris. Thus, for example, laminaris neurons located near the dorsal side receive impulses from the ipsilateral cochlear nucleus earlier than those from the contralateral nucleus. The nucleus laminaris uses this asymmetry in conduction time for measurement of interaural time differences in the microsecond range. [Reprinted from (*138*)]

by widely different animals. The cortical maps of the retina, the inner ear, and the body surface are well known. These maps are due to topographical projections of the sensory epithelia. Since the inner ear, unlike the retina, does not register the location of stimuli, it was thought unlikely that the auditory system uses a map-like representation of space. Such a map was, however, discovered in the barn owl's midbrain in which each neuron responds maximally only to sound coming from a restricted area in space (*140*). The space-specific response of these neurons is a result of processing within the brain. Thus, the auditory space map of the owl led to the concept of centrally synthesized or computational maps (*141*). A subsequent study revealed a crude map of auditory space in the superior colliculus of the cat (*142*). The auditory space map of the owl projects topographically to the tectum, in which it forms a joint auditory-visual map of space (*143*). This and other findings indicate that computational and projectional maps are functionally equivalent. Therefore, it follows that all maps are made for information processing and encoding and are not the simple consequence of embryological and anatomical design requirements.

Concluding Remarks

Biologists can take advantage of the diversity of life by using special animals, organs, behavior, and social systems to gain insights into general principles. Birds have ranked high in offering particularly useful models. Avian research has been pivotal to the development of subdisciplines of organismal biology such as ecology, sociobiology, and ethology. Neurobiology, including developmental neurobiology, neuroendocrinology, neuroethology, and comparative neuroanatomy, has also derived benefit from bird studies. Research on birds should continue to be important in the future development of biology. For example, bird studies with hormone radioimmunoassay will have a profound impact on our methods of studying social behavior. Long-term studies of avian societies, combined with molecular techniques for as-

signing parentage, will help clarify the evolutionary significance of different mating strategies and patterns of parental care. Premigratory fattening in songbirds and prefledging loss of fat in petrels could provide useful models for studying control of appetite and problems of obesity. Gerontologists might profit from investigating the general delay of the onset of physiological aging in birds. The study of avian brains seems also destined to produce more discoveries of general importance, in particular concerning the proliferation of brain cells throughout the individual's life. Avian contributions to biology extend far beyond the selected topics covered in this article and will continue to be important in many areas of biological research.

References and Notes

1. T. H. Clutton-Brock, Ed., *Reproductive Success* (Univ. of Chicago Press, Chicago, IL, 1988).
2. E. Mayr, *Systematics and the Origin of Species* (Columbia Univ. Press, New York, 1942).
3. F. Cooke and P. A. Buckley, Eds., *Avian Genetics* (Academic Press, Orlando, FL, 1987).
4. P. T. Boag and P. R. Grant, *Science* **214**, 82 (1981); T. D. Price, P. R. Grant, H. L. Gibbs, P. T. Boag, *Nature* **309**, 787 (1984).
5. C. G. Sibley, J. E. Ahlquist, B. L. Munroe, Jr., *Auk* **105**, 409 (1988).
6. S. E. Kingsland, *Modeling Nature: Episodes in the History of Population Ecology* (Univ. of Chicago Press, Chicago, IL, 1985); J. Sheail, *Seventy-Five Years in Ecology, The British Ecological Society* (Blackwell, Oxford, 1987).
7. A. Lotka, *Elements of Physical Biology* (William & Wilkins, Baltimore, MD, 1925); G. F. Gause, *The Struggle for Existence* (William & Wilkins, Baltimore, MD, 1934).
8. D. Lack, *Darwin's Finches* (Cambridge Univ. Press, New York 1947).
9. _____, *Ecological Isolation in Birds* (Harvard Univ. Press, Cambridge, MA, 1971).
10. _____, *The Natural Regulation of Animal Numbers* (Oxford Univ. Press, London, 1954).
11. D. Lack, *Ecological Adaptations for Breeding in Birds* (Methuen, London, 1968).
12. G. E. Hutchinson, *Cold Spring Harbor Symp. Quant. Biol.* **22**, 415 (1957).

13. R. H. MacArthur and R. Levine, *Am. Nat.* 101, 377 (1967).
14. T. W. Schoener, *Science* 185, 27 (1974).
15. D. Simberloff, *Synthese* 43 (no. 3), 79 (1980); D. Simberloff, in *Perspectives in Ornithology*, A. H. Brush and G. A. Clark, Jr., Eds. (Cambridge Univ. Press, New York, 1983), pp. 411–455; J. Wiens, *ibid.*, pp. 355–403; D. R. Strong, Jr., L. A. Szyska, D. S. Simberloff, *Evolution* 33, 897 (1979).
16. R. H. MacArthur and E. O. Wilson, *The Theory of Island Biogeography* (Princeton Univ. Press, Princeton, NJ, 1967); *Evolution* 17, 373 (1963).
17. M. E. Gilpin and J. M. Diamond, *Oecologia* 52, 75 (1982); D. Simberloff, *Annu. Rev. Ecol. Syst.* 5, 161 (1974).
18. J. M. Diamond, *Biol. Conserv.* 7, 129 (1975); *Science* 193, 1027 (1976); J. Terborgh, *ibid.*, p. 1029.
19. J. M. Diamond and M. E. Gilpin, *Oecologia* 64, 75 (1982); E. F. Conner and D. Simberloff, *Ecology* 60, 1132 (1978).
20. G. Hardin, *Science* 131, 1292 (1960).
21. V. C. Wynne-Edwards, *Animal Dispersion in Relation to Social Behaviour* (Oliver and Boyd, Edinburgh, Scotland, 1962).
22. G. C. Williams, *Adaptation and Natural Selection* (Princeton Univ. Press, Princeton, NJ, 1966).
23. D. Lack, *Population Studies of Birds* (Clarendon Press, Oxford, 1966); C. M. Perrins, *J. Anim. Ecol.* 34, 601 (1965); T. E. Martin, *Annu. Rev. Ecol. Syst.* 18, 453 (1987).
24. W. D. Hamilton, *J. Theor. Biol.* 7, 1 (1964); S. C. Stearns, *Q. Rev. Biol.* 51, 3 (1976); W. M. Schaffer, *Ecology* 55, 291 (1974).
25. J. R. Krebs and R. H. McCleery, in *Behavioural Ecology*, J. R. Krebs and N. D. Davies, Eds. (Sinauer Associates, Sunderland, MA, 1984), pp. 91–121; D. W. Stevens and J. R. Krebs, *Foraging Theory* (Princeton Univ. Press, Princeton, NJ, 1986).
26. N. Nur, *Evolution* 43, 351 (1988); D. Reznick, *Oikos* 44, 257 (1985).
27. J. H. Crook, *Behaviour (Suppl.)* 10, 1 (1964).
28. E. S. Reese, *Z. Tierpsychol.* 37, 37 (1975); P. F. Sale, in *Contrasts in Behavior: Adaptation in the Aquatic and Terrestrial Environments*, E. S. Reese and F. J. Lighter Eds. (Wiley, New York, 1987), pp. 313–346.
29. D. Wells, *Anim. Behav.* 25, 666 (1977).
30. J. W. Bradbury and S. L. Vehrencamp, *Behav. Ecol. Sociobiol.* 1, 383 (1977); *ibid.*, 2, 1 (1977).
31. P. J. Jarman, *Behaviour* 48, 215 (1974).

32. J. H. Crook and J. S. Gartlan, *Nature* 210, 1200 (1966); T. H. Clutton-Brock and P. H. Harvey, *J. Zool. (Lond.)* 183, 1 (1977).
33. J. L. Brown, *Wilson Bull.* 76, 160 (1964).
34. F. L. Carpenter and F. E. MacMillen, *Science* 194, 639 (1976); F. B. Gill and L. L. Wolf, *Ecology* 56, 333 (1975); A. Kondric-Brown and J. Brown, *ibid.* 59, 285 (1978); P. W. Ewald and G. H. Orians, *Behav. Ecol. Sociobiol.* 12, 95 (1983).
35. R. E. Kenward, *J. Anim. Ecol.* 47, 449 (1978).
36. P. Ward and A. Zahavi, *Ibis* 115, 517 (1973); C. R. Brown, *Science* 234, 83 (1986); C. H. Brown, *Ecology* 69, 602 (1988); E. Greene, *Nature* 329, 239 (1982).
37. G. H. Orians, *Am. Nat.* 103, 589 (1969).
38. _____, *Ecol. Monogr.* 31, 285 (1961); J. Verner, *Evolution* 18, 252 (1964); J. Verner and M. F. Willson, *Ecology* 47, 143 (1966); M. F. Willson, *Ecol. Monogr.* 36, 51 (1966).
39. S. T. Emlen and L. W. Oring, *Science* 197, 215 (1977).
40. R. Thornhill and J. Alcock, *The Evolution of Insect Mating Systems* (Harvard Univ. Press, Cambridge, MA, 1983).
41. D. I. Rubenstein and R. W. Wrangham, *Ecological Aspects of Social Evolution* (Princeton Univ. Press, Princeton, NJ, 1986).
42. R. L. Trivers, in *Sexual Selection and the Descent of Man*, B. Campbell, Ed. (Aldine, Chicago, IL, 1972).
43. F. McKinney, K. M. Cheng, D. J. Bruggers, in *Sperm Competition and the Evolution of Animal Mating Systems*, R. L. Smith, Ed. (Academic Press, New York, 1984); D. F. Westneat, *Anim. Behav.* 35, 877 (1987); P. W. Sherman and M. L. Morton, *Behav. Ecol. Sociobiol.*; T. Burke, N. B. Davies, M. W. Bruford, B. J. Hatchwell, *Nature* 338, 249 (1989); D. F. Westneat, P. W. Sherman, M. L. Morton, in *Current Ornithology*, D. Power, Ed. (Plenum, New York, in press), vol. 7.
44. S. T. Emlen, in *Behavioural Ecology, An Evolutionary Approach*, J. R. Krebs and N. B. Davies, Eds. (Blackwell Scientific, Oxford, 1984), pp. 305–339; J. L. Brown, *Helping and Communal Breeding in Birds* (Princeton Univ. Press, Princeton, NJ, 1987); P. B. Stacy and W. D. Koenig, *Cooperative Breeding in Birds: Long Term Studies of Avian Demography and Behavior* (Cambridge Univ. Press, Cambridge, MA, 1989).
45. U.-H. Reyer, *Anim. Behav.* 32, 1163 (1984); W. D. Koenig and R. L. Mumme, *Population Ecology of the Cooperatively Breeding Acorn Woodpecker* (Princeton Univ. Press, Princeton,

46. S. T. Emlen and P. H. Wrege, *Behav. Ecol. Sociobiol.* **23**, 305 (1988); *ibid.*, in press. (1989).

47. G. E. Woolfenden and J. W. Fitzpatrick, *The Florida Scrub Jay: Demography of a Cooperative-Breeding Bird* (Princeton Univ. Press, Princeton, NJ, 1984).

48. M. J. West-Eberhard, *Q. Rev. Biol.* **50**, 1 (1975).

49. K. Lorenz, *J. Ornithol.* **83**, 137, 289 (1935).

50. N. Tinbergen, *The Study of Instinct* (Oxford Univ. Press, Oxford, 1951).

51. O. Heinroth, *Verh. Int. Ornithol. Kongr. (Berlin)* **5**, 589 (1911).

52. C. O. Whitman, *Publ. Carnegie Inst.* **257**, 1 (1919).

53. K. Lorenz, *J. Ornithol.* **89**, 194 (1941).

54. A. Kortlandt, *Arch. Zool.* **4**, 401 (1940).

55. N. Tinbergen, *Behaviour* **15**, 1 (1959).

56. P. Marler, *Nature* **176**, 6 (1955).

57. _____, *Behaviour* **11**, 13 (1957).

58. S. T. Emlen, *ibid.* **41**, 130 (1972); P. H. Becker, in *Acoustic Communication in Birds*, D. E. Kroodsma and E. H. Miller, Eds. (Academic Press, New York, 1982), vol. 1, pp. 214–252.

59. J. S. Weeden and J. B. Falls, *Auk* **76**, 343 (1959); P. C. Mundinger, *Science* **168**, 480 (1970); G. Gottlieb, *ibid.* **147**, 1596 (1965); B. Tschanz, *Z. Tierpsychol. Beih.* **4**, 1 (1968).

60. P. Marler, in *Monkeys and Apes: Field Studies of Ecology and Behavior*, I. DeVore, Ed. (Holt, Rinehart and Winston, New York, 1965), pp. 544–584.

61. D. E. Kroodsma and E. H. Miller, *Acoustic Communication in Birds* (Academic Press, New York, 1982), vol. 2; P. Marler, *Trends Neurosci.* **4**, 88 (1981); M. Konishi, *Annu. Rev. Neurosci.* **8**, 125 (1985).

62. W. H. Thorpe, *Bird-Song* (Cambridge Univ. Press, New York, 1961).

63. M. Konishi, *Z. Tierpsychol.* **22**, 770 (1965).

64. Dolphins can imitate simple sound patterns; see D. G. Richards, J. P. Wolz, L. M. Herman, *J. Comp. Psychol.* **98**, 10 (1984).

65. P. Marler, *Am. Sci.* **58**, 669 (1970).

66. P. K. Kuhl, *Human Evol.* **3**, 19 (1988).

67. K. von Frisch, *The Dance Language and Orientation of Bees* (Harvard Univ. Press, Cambridge, MA, 1967).

68. A. Hasler, *Underwater Guideposts* (Univ. of Wisconsin Press, Madison, WI, 1966).

69. D. R. Griffin, *Bird Migration* (Doubleday, New York, 1964); G. V. T. Matthews, *Bird Navigation* (Cambridge Univ. Press, Cambridge, MA, ed. 2, 1968).

70. G. Kramer, *Proc. Int. Ornithol. Congr.* **10**, 269 (1951).

71. E. G. F. Sauer, *Z. Tierpsychol.* **14**, 29 (1957); S. T. Emlen, *Auk* **84**, 309 (1967); *Science* **170**, 1198 (1970); F. R. Moore, *Nature* **275**, 154 (1978); K. P. Able, *ibid.* **299**, 550 (1982).

72. W. Wiltschko, *Z. Tierpsychol.* **25**, 537 (1968); _____, and R. Wiltschko, *Science* **176**, 62 (1972).

73. S. A. Gauthreaux, Jr., and K. P. Able, *Nature* **228**, 476 (1970); K. P. Able, *Anim. Behav.* **25**, 924 (1977); W. J. Richardson, *Oikos* **30**, 224 (1978).

74. S. T. Emlen, in *Avian Biology*, D. S. Farner and J. R. King, Eds. (Academic Press, New York, 19752, vol. 5, pp. 129–219; K. P. Able, in *Animal Migration, Orientation, and Navigation*, S. A. Gauthreaux, Ed. (Academic Press, New York, 1981), pp. 284–373.

75. W. Wiltschko and R. Wiltschko, in *Current Ornithology*, R. F. Johnston, Ed. (Plenum, New York, 1988), vol. 5, pp. 67–121.

76. W. T. Keeton, *Rec. Adv. Behav.* **5**, 47 (1974).

77. F. Papi, P. Ioale, V. Fiaschi, S. Benvenuti, N. E. Baldaccini, *Proc. Int. Ornithol. Congr.* **17**, 569 (1980).

78. C. H. Walcott and A. J. Ledner, in *Perspectives in Ornithology*, A. H. Brush and G. A. Clark, Jr., Eds. (Cambridge Univ. Press, New York, 1983), pp. 513–541.

79. K. Schmidt-Nielsen, *Sci. Am.* **225**, 72 (December 1971).

80. P. Scheid, in *Avian Biology*, D. S. Farner and J. R. King, Eds. (Academic Press, New York, 1982), vol. 6, pp. 405–453.

81. V. A. Tucker, *J. Exp. Biol.* **48**, 67 (1968).

82. W. R. Dawson and J. W. Hudson, in *Comparative Physiology of Thermoregulation*, G. Causey Whittow, Ed. (Academic Press, New York, 1970), pp. 224–310; G. A. Bartholomew, in *Animal Physiology: Principles and Adaptations*, M. S. Gordon, Ed. (Macmillan, New York, 1972), pp. 298–365.

83. K. Schmidt-Nielsen, *Circulation* **21**, 955 (1960).

84. G. A. Bartholomew, *Proc. Int. Ornithol. Congr.* **15**, 237 (1972).

85. R. E. Ricklefs, *Publ. Nuttall Ornithol. Club* **15**, 152 (1974); T. J. Case, *Q. Rev. Biol.* **55**, 243 (1978); R. H. Drent and S. Daan, *Ardea* **68**, 225 (1980).

86. R. E. Ricklefs, *Biol. Rev.* **54**, 269 (1979).

87. _____, *Ecology* **65**, 1602 (1984).

88. R. J. O'Conner, *The Growth and Development of Birds* (Wiley-Interscience, New York,

1984).

89. R. E. Ricklefs, *Avian Biol.* **7**, 1 (1983).
90. A. B. Harper, *Am. Nat.* **128**, 99 (1986); D. J. T. Hussell, *ibid.* **131**, 175 (1988); J. Stamps, A. Clark, P. Arrowood, B. Kus, *Behaviour* **94**, 1 (1985).
91. D. B. Werschkul and J. A. Jackson, *Ibis* **121**, 97 (1979); D. W. Mock, *Science* **225**, 731 (1984); G. R. Bortolotti, *Ecology* **67**, 182 (1986).
92. J. Balthazart, in *Avian Biology*, D. S. Farner, J. R. King, K. C. Parkes, Eds. (Academic Press, New York, 1983), vol. 7, pp. 221–365.
93. C. H. Harding, *Am. Zool.* **21**, 223 (1981).
94. D. S. Lehrman, in *Sex and Behavior*, F. A. Beach, Ed. (Wiley, New York, 1965), pp. 355–580.
95. R. A. Hinde, *ibid.*, pp. 381–415.
96. J. C. Wingfield, *Horm. Behav.* **19**, 174 (1985).
97. B. Silverin, *Anim. Behav.* **28**, 906 (1980); *J. Exp. Zool.* **232**, 581 (1984); M. C. Moore, *Horm. Behav.* **16**, 323 (1982); J. C. Wingfield, G. F. Ball, A. F. Dufty, R. E. Hegner, M. Ramenofsky, *Am. Sci.* **75**, 602 (1987).
98. J. C. Wingfield, *Auk* **101**, 665 (1984); R. E. Hegner and J. C. Wingfield, *ibid.* **104**, 462 (1987).
99. W. Rowan, *Nature* **115**, 494 (1925).
100. J. R. King and D. S. Farner, *Condor* **65**, 200 (1963); D. S. Farner, J. R. King, M. H. Stenton, *Prog. Endocrinol. Excerp. Med. Int. Congr. Ser.* **184**, 152 (1968).
101. D. S. Farner and B. K. Follett, in *Hormones and Evolution*, E. J. W. Barrington, Ed. (Academic Press, New York, 1979), pp. 829–872.
102. B. K. Follett, P. W. Mattocks, D. S. Farner, *Proc. Natl. Acad. Sci. U.S.A.* **71**, 1666 (1974).
103. E. Bünning, *Cold Spring Harbor Symp. Quant. Biol.* **25**, 249 (1960).
104. J. Benoit, *C. R. Soc. Biol.* **120**, 136 (1935).
105. M. Menaker, R. Roberts, J. Elliott, H. Underwood, *Proc. Natl. Acad. Sci. U.S.A.* **67**, 320 (1970).
106. K. Yokoyama, A. Oksche, T. R. Darden, D. S. Farner, *Cell Tiss. Res.* **189**, 441 (1978).
107. E. Gwinner, *Circannual Rhythms* (Springer-Verlag, Berlin, 1986).
108. T. J. Nichols, A. R. Goldsmith, A. Dawson, *Physiol. Rev.* **68**, 133 (1988).
109. J. Benoit and I. Assenmacher, *Arch. Anat. Microsc. Morphol. Exp.* **42**, 334 (1953).
110. A. Oksche and D. S. Farner, *Ergeb. Anat. Entwicklungsgesch.* **48**, 4 (1974).
111. R. P. Millar and J. A. King, *J. Exp. Zool.* **232**, 425 (1984).

112. R. D. Palmiter, E. R. Mulvihill, G. S. McKnight, A. W. Senear, *Cold Spring Harbor Symp. Quant. Biol.* **42**, 639 (1978).
113. B. W. O'Malley, M. R. Sherman, D. O. Toft, *Proc. Natl. Acad. Sci. U.S.A.* **67**, 501 (1970).
114. J. B. Hutchison and T. Steimer, *Science* **213**, 244 (1981).
115. F. Nottebohm, T. M. Stokes, C. M. Leonard, *J. Comp. Neurol.* **165**, 457 (1976).
116. F. Nottebohm, *Prog. Psychobiol. Physiol. Psychol.* **9**, 85 (1980).
117. M. Konishi and E. Akutagawa, *Brain Res.* **222**, 442 (1981).
118. V. Luine, F. Nottebohm, C. Harding, B. McEwen, *ibid.* **192**, 89 (1980); W. V. Bleisch, V. N. Luine, F. Nottebohm, *J. Neurosci.* **4**, 786 (1984).
119. T. J. DeVoogd and F. Nottebohm, *Science* **214**, 202 (1981); F. Nottebohm, *Brain Res.* **289**, 492 (1980).
120. F. Nottebohm, M. E. Nottebohm, L. A. Crane, J. C. Wingfield, *Behav. NeuralBiol.* **47**, 197 (1987).
121. F. Nottebohm, *Science* **214**, 1368 (1981).
122. J. R. Kirn, R. P. Clower, D. E. Kroodsma, T. J. DeVoogd, *J. Neurobiol.* **20**, 139 (1989).
123. M. Gahr, *J. Comp. Neurol.*, in press.
124. R. A. Canady, G. D. Burd, T. D. DeVoogd, F. Nottebohm, *J. Neurosci.* **8**, 3770 (1988).
125. F. Nottebohm, in *Neural Control of Reproductive Function*, J. M. Lakoski, J. R. Perez-Plo, D. K. Rassin, Eds. (Liss, New York, 1989), pp. 583–601.
126. F. Nottebohm and A. P. Arnold, *Science* **194**, 211 (1976).
127. M. Konishi and E. Akutagawa, *Nature* **315**, 145 (1985).
128. S. M. Breedlove and A. P. Arnold, *Science* **210**, 564 (1980); R. A. Gorski, R. E. Harlan, C. D. Jacobson, J. E. Shryne, A. M. Southam, *J. Comp. Neurol.* **193**, 529 (1980); D. F. Swaab and E. Fliers, *Science* **228**, 1112 (1985).
129. R. W. Goy and B. S. McEwen, *Sexual Differentiation of the Brain* (MIT Press, Cambridge, MA, 1980).
130. S. M. Breedlove, *J. Neurobiol.* **17**, 157 (1986); D. Commins and P. Yahr, *J. Comp. Neurol.* **224**, 132 (1984).
131. M. E. Gurney, *J. Neurosci.* **1**, 658 (1981); M. Konishi and E. Akutagawa, *Ciba Found. Symp.* **126**, 173 (1987).
132. S. A. Goldman and F. Nottebohm, *Proc. Natl. Acad. Sci. U.S.A.* **80**, 239 (1983); J. A. Paton and F. Nottebohm, *Science* **225**, 1046 (1984); A. Alvarez-Buylla, N. M. Theelen, F. Nottebohm, *Proc. Natl. Acad. Sci. U.S.A.* **85**,

8722 (1988); A. Alvarez-Buylla and F. Nottebohm, *Nature* **335**, 353 (1988).

133. B. M. Ryals and E. W. Rubel, *Science* **240**, 1774 (1988).

134. H. J. Karten, *Ann. N.Y. Acad. Sci.* **167**, 164 (1969).

135. R. S. Payne, *J. Exp. Biol.* **54**, 535 (1971).

136. M. Konishi, T. T. Takahashi, H. Wagner, W. E. Sullivan, C. E. Carr, in *Auditory Function*, G. M. Edelman, W. E. Gall, W. M. Cowan, Eds. (Wiley, New York, 1988), pp. 721–745.

137. S. R. Young and E. W. Rubel, *J. Neurosci.* **3**, 1273 (1983).

138. C. E. Carr and M. Konishi, *Proc. Natl. Acad. Sci. U.S.A.* **85**, 8311 (1988).

139. J. M. Goldberg and P. B. Brown, *J. Neurophysiol.* **32**, 613 (1969).

140. E. I. Knudsen and M. Konishi, *Science* **200**, 795 (1978).

141. M. Konishi, *Trends. Neurosci.* **9**, 163 (1986); E. I. Knudsen, S. du Lac, S. D. Esterly, *Annu. Rev. Neurosci.* **10**, 41 (1987).

142. J. C. Middlebrooks and E. I. Knudsen, *J. Neurosci.* **4**, 2621 (1984).

143. E. I. Knudsen, *ibid.* **2**, 1177 (1982); E. I. Knudsen and P. H. Knudsen, *J.Comp. Neurol.* **218**, 187 (1983).

144. We thank P. Marler, G. H. Orians, and S. F. Volman for critically reading an early draft of the paper. This work was supported in part by NSF grants BNS 85-17725 to S.T.E. and DCB 8616189 to J.C.W.

14

The Rat as an Experimental Animal

Thomas J. Gill III, Garry J. Smith,
Robert W. Wissler, Heinz W. Kunz

The rat is a major experimental animal in transplantation, immunology, genetics, cancer research, pharmacology, physiology, neurosciences, and aging. The strains and randomly bred stocks that have been used almost exclusively are derived from the Norway rat (*Rattus norvegicus*), which is thought to have originated in the area between the Caspian Sea and Tobolsk, extending as far east as Lake Baikal in Siberia. It spread to Europe and the United States with the development of commerce in the 18th century, and by the middle of the 19th century it was being used extensively for studies in anatomy, physiology, and nutrition. The first inbred lines were developed at the beginning of the 20th century by H. H. Donaldson, W. E. Castle, and their colleagues for studies in basic genetics and in cancer research (*1*). Further development and genetic characterization of inbred, congenic, and recombinant strains occurred in the United States, Japan, and Czechoslovakia (*2*), and several reviews have documented these developments in detail (*3–5*). In addition to its experimental uses, the rat has a worldwide economic and medical impact, since it destroys one-fifth of the world's crops each year, carries many diseases that are pathogenic for humans, and kills many children by direct attack (*6*).

This review will focus on current work utilizing the rat in immunogenetics, transplantation, cancer-risk assessment, cardiovascular diseases, and behavior. In these areas of research, the rat has the advantage of being a well-characterized, intermediate-sized rodent without the disadvantages, both scientific and economic, of larger animals and without many of the technical disadvantages of smaller rodents.

Immunogenetics

Considerable effort has been expended in recent years to develop and characterize inbred, congenic, and recombinant strains of rats, and a wide variety of these genetic resources is now available (*3, 4, 7–9*). Several compilations of basic data have been assembled (*5*), and current developments are regularly updated in the *Workshops on Alloantigenic Systems of the Rat* (*10*) and in the *Rat Newsletter* (*11*). This work

T. J. Gill III is the Maud L. Menten Professor of Experimental Pathology and professor of human genetics at the University of Pittsburgh School of Medicine. H. W. Kunz is associate professor of pathology at the University of Pittsburgh School of Medicine, Pittsburgh, PA 15261. G. J. Smith is associate professor and director of the Carcinogenesis Research Unit at the School of Pathology of the University of New South Wales, Kensington, New South Wales, 2033 Australia. R. W. Wissler is the Donald N. Pritzker Distinguished Service Professor of Pathology, active emeritus, at the University of Chicago School of Medicine, Chicago, IL, 60637. This chapter is reprinted from *Science* 245, 269 (1989).

has also provided insight into the comparative genetics of the major histocompatibility complex (MHC) and of MHC-linked genes affecting growth and development. The level of polymorphism of MHC antigens in the rat is very low compared to that of other species; the class I antigens have been most extensively studied. Nonetheless, the resistance to disease, reproductive capacity, and ecological stability of the rat do not differ from those of other species. Hence, the biological significance of MHC polymorphism remains a mystery.

The structure of the MHC in the rat (*RT1*) based on data from serological, molecular, and functional studies is shown in Fig. 1 (*3, 12, 13*). The general organization of the class I and class II loci is the same as in the mouse but different from that in all other species studied: the class II loci are interspersed between class I loci rather than following them sequentially (*14*). This observation indicates that (i) the rat and the mouse formed separate genera after the divergence of the prototypic Muridae, (ii) the evolutionary conservation of the MHC persists despite internal rearrangements, and (iii) the

function of these loci does not depend, at least to a first approximation, on their specific order or on their polymorphism.

The *RT1.A* and *RT1.E* loci encode classical class I transplantation antigens and appear to be the homologs of the mouse *H-2K* and *H-2D* loci. There are several other class I loci in the vicinity of *RT1.A*, and the best defined are the diallelic *RT1.F* and *Pa* (pregnancy-associated) loci (*3, 13, 16*). The antigen encoded by the *Pa* locus was first identified on the surface of the basal trophoblast in the allogeneic WF(*u*) × DA(*a*) mating by alloantisera and by monoclonal antibodies made by the WF mother (*17*). This antigen carries an epitope that is broadly shared among other class I antigens, but does not have the allele-specific epitope of a classical class I transplantation antigen. Immunohistochemical and electron microscopic studies (*18*) showed that both the Pa and A^a antigens are also on most somatic tissues and that they are carried by separate molecules. The mapping of the *A*, *F*, and *Pa* loci is based on the use of various combinations of inbred, congenic, and recombinant strains; a

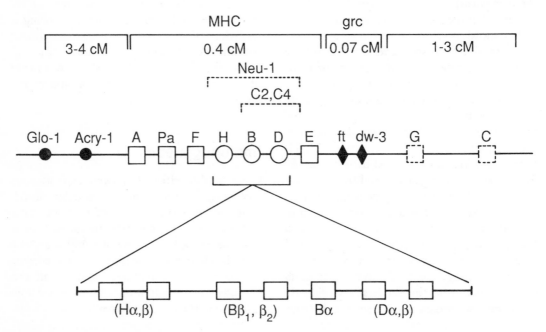

Fig. 1. The major histocompatibility complex of the rat. ☐ Class I major (classical) transplantation antigens; the dashed squares, the class I medial transplantation antigens; O, class II antigens; ●, loci controlling polymorphic proteins (Glo-1, glyoxylase I; Acry-1, α-crystallin-1); and ◆, the loci of the grc (ft, fertility; dw-3, dwarf-3). The loci indicated by brackets have been mapped to the regions indicated (Neu-1, neuraminidase-1; C, complement components). The evidence for this mapping is presented in (*3, 12, 13*). A cytogenetic study (*15*) has placed the MHC on chromosome 14 of the rat.

number of monoclonal antibodies; and specifically designed alloantisera. No recombinants among these loci have yet been found, but immunoprecipitation and peptide mapping studies have demonstrated that they are separate molecules: hence, the order of these loci in Fig. 1 must be considered tentative. The *RT1.G* and *RT1.C* loci encode class I antigens that appear to be homologous to the mouse Qa/TL antigens, but these loci have not yet been well characterized (*19*).

The class II loci *RT1.B* and *RT1.D* were detected serologically and by molecular analysis (*3*), whereas *RT1.H* has been detected only by molecular analysis (*12*). The *B* and *D* loci appear to be homologous to the mouse *A* and *E* loci, and the *H* locus appears to be homologous, in part, to the mouse $\psi A\beta 3$ pseudogene and the human *HLA-DP* locus.

The growth and reproduction complex (*grc*) is closely linked to the MHC (*20*). In the homozygous state, it is semilethal in males and females, causes small body weight in both males and females (*dw-3*), and causes male sterility and reduced female fertility (*ft*). These defects are similar to some of those associated with the *t* haplotypes in the mouse, but the *grc* is not homologous to the *t* genes since it does not cause segregation distortion or suppression of recombination (*3, 20*). The fertility defect occurs at the same stage of gametogenesis in both males and females: there is complete arrest of spermatogenesis at the primary spermatocyte stage, and a partial defect in the maturation of the primary ovarian follicle. The *grc* acts at an early stage of meiotic prophase I; it is not associated with any known chromosomal or hormonal abnormality; and it increases susceptibility to chemical carcinogens in both males and females (*21*). Its effects are probably due to the deletion of a segment of the chromosome close to the MHC (*22*). If so, then the increased susceptibility to cancer may be due to the loss of cancer suppressor genes, or anti-oncogenes, as in retinoblastoma and Wilms' tumor in humans (*23*). Hence, these animals may provide a unique system in which to study the genetics of susceptibility to cancer.

The homozygous *grc* genotype (20 to 25% in utero mortality) can interact with the heterozygous *Tal/* + gene, which is a recessive lethal gene on a different chromosome. The *Tal* gene is not lethal in the heterozygous state but, when homozygous, causes the death of all embryos at 10 to 14 days of gestational age (*24*). This demonstration in mammals of a lethal epistatic interaction, which is the interaction between genes on different chromosomes, provides a useful system in which to study gene interaction during development.

Molecular analysis has delineated the major regions of the rat MHC on the basis of restriction fragment length polymorphisms (RFLPs) (*13, 22, 25*). There are approximately the same number of class I–hybridizing fragments of DNA as in the mouse (*26*), despite the much lower level of serological polymorphism in the rat (*3*). The class II loci have not been examined in any detail yet, but there is a "hotspot" of recombination in the *RT1.H* region.

The biochemical comparisons among the rat, mouse, and human MHC class I and class II antigens are summarized in Table 1. The amino acid sequences of the rat class I and class II antigens are more homologous to those of the mouse than to those of the human, although both levels of homology are fairly high. The homology among antigens encoded by the same class I locus is the same in the rat and the mouse, and both are lower than in the human. The homology between antigens encoded by different class I loci of the same haplotype is much higher in the rat than in the mouse or the human, whereas the interlocus homology for the class II antigens is approximately the same for all three genuses. When one compares the rat with the mouse and the human the most striking difference is in the number of serologically defined class I and class II antigens. This difference has been documented most extensively for the class I antigens in both inbred (*3*) and wild (*27*) populations; it has been less extensively studied for the class II antigens. The class I and class II antigens present in both the inbred and wild populations are serologically and functionally indistinguishable, and there is a high degree of linkage disequilibrium among the loci in the MHC of the rat (*27*).

The difference between the rat and the

Table 1. Amino acid homologies between MHC class I and class II antigens of the rat and those of the mouse and the human (3, 14, 101).

| | | Percentage homologies | | | | | Approximate number of serologically defined alleles* | | |
| | | Rat compared to | | Allelic and interlocus homologies* | | | | | |
		Mouse	Human	Rat	Mouse	Human	Rat	Mouse	Human
Class I	Signal peptide	85	50	68–73 (A)	32–69 (K)	85–95 (A)	12 (A)	92 (K)	24 (A)
	α1 domain	71–73	68		34–57 (D)	93 (B)	2 (E)	63 (D)	52 (B)
	α2 domain	71–78	67	97–98 (A:E)	36–69 (K:D)	79–85 (A:B)	4 (C)	2 (L)	11 (C)
	α3 domain	87	72						
	Transmembrane-cytoplasmic domain	36–46	40						
Class II		80–91	73–81	56–59 (B:D)	52–60 (A:E)	64–66 (DR:DQ)	10 (B,D)	74 (A) 72 (E)	20 (DR) 9 (DQ) 6 (DP)

*Locus or loci compared given in parentheses.

mouse and human in the serological polymorphism of their class I antigens stands in contrast to the similarity of their RFLP patterns (20 to 36 class I–hybridizing fragments) (*3, 22, 25*). This observation might reflect a similarity in the total number of class I genes in all three genuses but a difference in the number of functional genes. The situation with the class II loci in the rat appears to be the same: their serological polymorphism is very low but their RFLP is high (*3, 12*). Thus, the rat is an extremely useful animal in which to study the control of the functional activity of MHC loci and the biological consequences thereof.

The limited MHC antigen polymorphism in the rat raises the question of what the biological significance of MHC polymorphism is (*28*). Neither the host defense mechanisms nor the reproductive capacity of the rat appear to differ from those of the mouse and the human, and the rat has certainly prospered in an otherwise hostile environment (*6*). Current thinking assigns a central role to class I antigens in the presentation of foreign antigens to the host immune system and to class II antigens in the recognition of foreign antigens. If these are, indeed, the primary functions of the MHC antigens, then either the specificities of their antigen-recognizing structures are much broader than those of the antibody combining sites or the extent of their antigen-recognition repertoire is not reflected in their serological polymorphism. There is also the relevant, and intriguing, observation that the MHC polymorphism in the protochordate *Botryllus* is the same as that in the mouse and the human (*29*). Why? Only more extensive structural studies of MHC antigens at both the protein and DNA levels will provide the crucial insights into the biological significance of MHC antigen polymorphism.

Transplantation

The rat is the animal most often used in organ transplantation studies: its size makes surgical procedures feasible, provides large amounts of cells and serum, and allows serial biopsies of the transplanted organ to assess the rejection process. The advances in rat immunogenetics over the past two decades have enhanced considerably its usefulness in transplantation research. The rejection times of various organs in different strain combinations have been documented (*5*), and the roles of the different MHC and non-MHC antigens in this process (*30*) have been examined by the use of different combinations of inbred, congenic, and recombinant strains. Such transplantation studies have been done with skin (*7, 30*), kidney (*31*), heart (*32*), bone marrow (*33*), liver (*34*), small bowel (*35, 36*), pancreas (*37*), and brain (*38, 39*). There are four major areas of current interest in experimental transplantation research, and the rat is the crucial animal in each of them: allotransplantation of the small bowel, heart, and liver; neural transplantation; xenografting; and reproduction.

Allografting

In systemic allotransplantation, grafting of the small bowel is the most pressing area of study (*35, 36*). Loss of function in this organ occurs in a variety of situations and at all stages of life: for example, congenital abnormalities, necrotizing enterocolitis, mesenteric artery thrombosis, and trauma. The problems encountered include the proper preservation and restoration of the physiological function of this delicate organ. The immunological problems are those of the host-versus-graft reaction by the recipient's immune system and the graft-versus-host reaction by the lymphoid tissue in the Peyer's patches of the graft. In this sense, small bowel grafting presents the same type of tissue matching problems as bone marrow grafting, but the offending T cells cannot be removed from the bowel graft as easily as they can from the bone marrow graft.

Two other important areas of research in allografting are heart grafting and liver grafting. The most critical issue in the long-term survival of cardiac transplant patients is the development of atherosclerosis in the coronary arteries of the transplant (*40*). In humans, this process can lead to the loss of the transplant in 5 to 7 years, so an understanding of its pathogenesis will provide a cogent insight into its therapy. In human liver transplantation, the role of histocompatibility (HLA) matching in the survival of the transplant has not been clarified, and there is the suggestion that under certain circumstances matching can reduce the survival of the graft (*41*). The liver transplantation model has been well developed in the rat (*34*), and it should provide the appropriate system in which to explore these questions.

Neural transplantation

The rat has been an important animal in the study of allogeneic and xenogeneic neural transplantation. Embryonic neural tissue can be transplanted into neonatal and adult brains where it can mature and integrate into the host brain. Both allografts and xenografts can survive for prolonged periods, but they are always susceptible to immune rejection either spontaneously or after challenge by related antigens or by mechanical trauma to the central nervous system (*38*). In the rejection process, however it is precipitated, the host astrocytes are induced to express MHC class I and class II antigens, and the control of such expression may be central to the acceptance of the neural transplant. Cyclosporine A can effectively prolong neural grafts (*42*). Recent studies in humans (*43*) suggest that grafts of neuroectodermal origin can be performed, but such grafts have not yet proven to be clinically useful for any significant period of time. The critical factors that affect the success of a neural transplant are the technique and site of the transplant, the amount of disruption of the blood-brain barrier, the size and source of the donor tissue, the vascularization of the transplant, the age of the host and of the donor at the time of transplantation, and the immunogenetic difference between host and donor.

Studies in rats have shown that such transplants can reduce cognitive defects due to frontal cortex lesions (*44*), improve impairment of motor function in aged animals (*45*), and make functional connections in an allogeneic or xenogeneic setting (*46*). These studies are also providing insight into the immunological status of the brain and the immune reactivity in this organ and into the pathogenesis of focal neurodegenerative diseases (*38*).

The potential value of neural grafts in clinical medicine lies in replacement of damaged neural circuits and in the replacement of cells making chemicals that modulate neural function. Neural circuit replacement might be used to treat trauma in adults and congenital neurological defects in children, and it is in the latter that long-term possibilities for the therapeutic use of neural grafting lie. The use of transplanted cells as a substitute for chemical replacement therapy is complicated by the fact that many of the diseases causing such deficits may have an autoimmune basis, so the transplanted cells themselves may fall victim to the underlying disease process. Much basic work must be done to clarify the immunological and neurophysiological aspects of neural transplantation, the development of specific immunosuppressive regimens for neural transplants, and the pathogenesis of the neurodegenerative diseases for which it might be used as therapy. The effort is worthwhile, since transplantation of tissue into the brain is one of the most promising approaches to have come from experimental neurobiology as potential therapy for a variety of disorders involving damage to the central nervous system. Finally, the use of neural xenotransplants in humans is a distinct possibility (*38*), and the ethical dilemmas raised by this procedure must be examined.

Xenografts

The use of grafts from animals of different families and genuses, xenografting, has been explored sporadically (*47*) and has recently had a resurgence because of the interesting basic immunological questions that it raises and because of the possibility of the use of such grafts as neural transplants (*38*) and as temporary expedients ("bridging grafts") in humans.

Each xenograft system has its own peculiarities (*47*): thus, it is not possible, at the present time, to generalize about the nature of the immune response to xenografts. In order to explore systematically the immunobiology and immunogenetics of xenografting, three areas of resarch should be developed. First, xenoantigens should be identified and characterized. The relative immunogenicity of various xenografts should be studied in one donor-recipient combination in order to develop a coherent body of knowledge that can serve as a paradigm for other systems. The rat-mouse combination will be the most useful one to study initially, because both species are immunologically and genetically well defined. This research should explore (i) the possible existence of unique xenoantigenic systems, (ii) the role of donor MHC antigens in eliciting an immune response to the xenograft, (iii) the cumulative effect that weak antigenic systems have in xenograft rejection, and (iv) the genesis and nature of "natural" or "preformed" antibodies. As an extension of this line of work, the role that the evolutionary distance between donor and recipient plays in the magnitude of the immune response to the xenograft should be examined. Second, the immune response to the xenograft should be analyzed systematically and in detail, including an investigation of the origin and specificities of preformed antibodies. The latter study may provide some insight into methods for controlling their formation. Third, the mechanism of xenograft rejection should be compared to that of allograft rejection to determine whether the major differences between them are qualitative or quantitative.

Reproductive immunology and genetics

This area has as its central theme the mechanism by which the fetal allograft survives (48). The rat is an important experimental animal for examining the nature of the trophoblast antigens and the genetic control of their expression. The allele-specific, class I transplantation antigens are not expressed on the trophoblast surface in allogeneic pregnancies, but they are on the surface in syngeneic pregnancies; in both types of pregnancies, they are present in the cytoplasm (18). The Pa antigen is expressed on the trophoblast surface

and in the trophoblast cytoplasm in both allogeneic and syngeneic placentas; class II antigens are not expressed in either type of placenta (18). This differential antigen expression may be an important factor in the maternal acceptance of the allogeneic placenta. Recent work shows that all of the class I antigens expressed in the placenta are of paternal origin, and this is the first example at the antigen level of genomic imprinting, which is a critical process in reproductive genetics (49). The very low level of MHC antigen polymorphism in the rat is crucial to the discrimination needed for these types of studies.

Recessive lethal genes are important causes of fetal death in experimental animals, and they may play an important role in recurrent spontaneous abortion in humans (48, 50). The *grc* in the rat, as discussed above, provides a unique model system in which to study these effects. This area of research is an important bridge between the aspects of reproduction of primary interest in the field of transplantation and the broader field of developmental genetics.

Risk Assessment for Potential Carcinogens

The rat has been used frequently for prediction of the effects of chemicals on humans (51). For studies of teratogenesis, the advantages of the rat include the ease of counting corpora lutea when assessing the effects of chemicals on ovulation and implantation (52), a large litter size, a short gestation period, and a well-studied embryology. However, the susceptibility and sensitivity of rats to particular teratogenic agents may be low when compared with the mouse and the rabbit (52), and there are significant differences from man in the effects of chemicals on the fetus (53). In mutagenesis studies, the rat appears to offer little inherent advantage over several other species (54). It is in the field of carcinogenic risk assessment that the rat has played a prominent role and will continue to do so.

Prediction of carcinogenicity for a given chemical is a major concern for government, the chemical industry, and the public. The development of cancer usually involves, at some stage, an agent or agents foreign to the cell—including xenobiotics, radiation, and oncogenic viruses. Carcinogenesis is a multistep process frequently involving a genotoxic (DNA-altering) step resulting in the alteration of cell division, growth, and differentiation (55). Different chemicals, including some with similar structures, may work by different mechanisms, and the cellular differences among tissues further complicate the process. Often one, or sometimes more, specific activated metabolite of a chemical may be the ultimate carcinogen (56); hence, different tissues and species of animals may respond differently to any given chemical based on their inherent metabolic patterns. The many unknown aspects of the induction of cancer, the long latency period between exposure and overt disease, and the potential for carcinogenesis at low doses of chemicals have made risk assessment an extremely difficult exercise.

Ultimately, it is epidemiologic studies of humans that will confirm the ability of an agent to cause human cancer (57), but such studies are usually performed only after exposure of large populations. This situation has led to the development of carcinogenic risk assessment methodologies that utilize nonhuman test systems (53). Assessment of carcinogenicity involves long-term dietary, parenteral, or topical application of the chemical to various mammalian species (58). The rat features prominently in such studies because of a favorable combination of small body size, ease of breeding, and relatively low spontaneous tumor rates. The choice of the strain of rats that is used is important in view of the variation in spontaneous tumor rates and different responses to chemicals among inbred strains (58). More recently, it has become apparent that such long-term bioassays may occasionally produce conflicting results, as occurred initially with vinylidene chloride (59, 60), or may be used with agents such as arsenic that exhibit sufficient evidence of carcinogenicity in humans but limited evidence in animal tests (60). Furthermore, because the mechanisms of chemical carcinogenesis have become better understood and the potential for simultaneous exposure to

several chemicals has become apparent, chemicals may in the future be assessed for their activity at different stages of the multistep carcinogenic process (61).

The long-term application of a test chemical to animals will continue to be the fundamental method of carcinogenic risk assessment because short-term, and particularly in vitro, tests cannot mimic all of the aspects of animal metabolism and physiology (62). The long-term bioassays should be done over a large part of the life span of the species, starting in utero, in order to eliminate false negative results due to the prolonged latency of carcinogenic effects. In this respect, the rat is a suitable experimental animal because of its relatively short life span.

In view of the important role played by metabolic enzymes in activating chemicals to reactive carcinogens, the question arises as to whether the rat is metabolically an appropriate substitute for humans. Crouch and Wilson (63), using the National Cancer Institute long-term bioassay data and a mathematical formula for carcinogenic potency, demonstrated that the ratio of potency between humans and rats was, on average, within a fivefold range; however, for a given chemical it varied from 1500:1 to 0.02:1. The range of potencies was less divergent between mice and rats, although Bernstein et al. (64) have argued that this lack of divergence may be a statistical artifact inherent in the long-term bioassays. Purchase (65) analyzed 250 chemicals for carcinogenicity in rats and mice based on data from the National Cancer Institute, International Agency for Research on Cancer, and U.S. Public Health Service, and his analysis indicated that a chemical carcinogenic in one species had a 15% chance of not being carcinogenic in the other. These data emphasized the importance of testing chemicals in more than one species in long-term bioassays (58). The rat is clearly an appropriate choice for one of these species because so much is known about its metabolic and physiological patterns and because various classes of chemicals are carcinogenic for rats (53, 59).

Recent studies on mechanisms of chemical carcinogenesis have demonstrated deficiencies in long-term animal carcinogenesis testing when it is used as the sole assessment criterion, because problems may occur with chemicals that are carcinogenic but that cause only moderate tumor incidence in a given tissue in different species (59). Certain chemicals, notably epigenetic (non–DNA altering) ones, may affect a particular stage of the multistep carcinogenic process initiated by another chemical without being themselves active in a long-term bioassay when tested alone. These facts, together with the increasing costs and slowness of long-term testing, have forced consideration of assays that require less time. Weisburger and Williams (59) outlined a decision-point approach to testing whereby chemicals might be analyzed in four increasingly complex classes of carcinogenicity assessment. These classes are as follows: (i) Analysis of the structure of the chemical. This analysis considers the reactivity of the chemical and its metabolites based on structure (66). (ii) Short-term tests in vitro. A battery of tests is used including prokaryotic and mammalian mutagenesis systems and studies of direct effects on DNA and chromosomes. (iii) Limited bioassays in vivo. The formation of preneoplastic lesions or rapid tumor induction is assessed in selected species. (iv) Long-term bioassays in vivo. A positive result in these studies is increased overt tumor formation or tumor-induced death of the animal.

For limited bioassay procedures, the induction of breast cancer in female Sprague-Dawley rats and the induction of altered foci in the rat liver may be useful. Cellular and subcellular preparations from rat livers are also commonly used for metabolic activation of chemicals in short-term carcinogenesis and mutagenesis tests (67, 68). Coculture of rat hepatocytes with liver epithelial-type cells has been reported to sustain high levels of hepatocyte, carcinogen-metabolizing cytochrome P-450 enzymes (69). Such procedures may extend the utility of in vitro hepatocyte cell lines in toxicity testing. The comprehensive assessment proposal of Weisburger and Williams (59) is not an established procedure (58), but rather illustrates potential future directions for carcinogenic risk assessment. The rat plays an important role in short-term in vitro tests and in limited in vivo bioassays.

The rat has been the most frequently studied species in the in vivo bioassay system of altered liver-focus induction. Research into the cellular events in the course of chemically induced tumor formation has characterized many of the changes that precede malignancy (70, 71). Cell populations affected by the carcinogen generally appear as characteristically altered foci detectable by sensitive immunohistochemical reactions, and they appear much earlier than tumor formation. Induction of such foci is not an unequivocal indicator of ultimate malignancy, and their significance in the development of malignancy is debated (70). Nevertheless, this assay has been proposed as a limited in vivo bioassay system in carcinogenicity assessment (59, 70, 72). Pereira and Stoner (73) have reported that the rat liver focus assay exhibited greater sensitivity and fewer false negatives that the strain A mouse lung adenoma assay [some limitations of which are discussed in (53)] in detecting genotoxic carcinogens. Parodi *et al.* (74) concluded that, at least for a small group of chemicals active predominantly in the liver, assays for liver focus and nodule formation were as accurate, and possibly more accurate, in detecting carcinogenicity than was the Ames test. Preneoplastic lesions have been studied in tissues other than the liver, but a systematic evaluation of their use in bioassays has not been reported (75). In view of the large amount of knowledge concerning liver focus formation in the rat (72), it is clear that this species will feature prominently in potential bioassay applications. Strains of rats carrying the growth and reproduction complex (*grc*), which is linked to the MHC, exhibit enhanced focus formation compared to wild-type rats when exposed to chemical carcinogens (21, 76), and they are candidates for development of highly sensitive liver-focus bioassays.

In the future of carcinogenicity assessment, there is increasing interest in subdividing the carcinogenic process and studying individual stages. As more is learned about the multistep mechanisms, it may be possible to develop assays for the identification of agents that predispose cells to malignancy at specific steps in the process; one such system has already been described for the rat (*61*). With the increasing emphasis on genetic mechanisms in carcinogenesis, the availability of randomly bred, outbred, inbred, and congenic strains of rats (*3–5*) will make this species even more useful in risk assessment as well as in studies on the basic mechanisms of carcinogenesis.

Cardiovascular Diseases

The extensive body of knowledge regarding nutrition, endocrinology, metabolism, and physiology; the detailed studies on anatomy and histology; and the convenient size of the rat make it a particularly useful experimental animal for cardiovascular research. Reproducible, genetically determined abnormalities have been discovered in rat populations that have proven useful in examining the cardiovascular effects of hypertension, obesity, diabetes, and other metabolic diseases (*4, 77*) and a variety of congenital abnormalities of the cardiovascular system (*78*).

Early studies indicated that this species was quite different from humans in its serum lipid and lipoprotein constitution and that it was very difficult to produce sustained hyperlipidemia in the rat (*79*). Until approximately 1950, many attempts to produce atheromatous lesions in the rat had failed in spite of the extensive knowledge about the effects of nutritional manipulation in this species. Then, in the early 1950s simultaneous reports from three laboratories indicated that this resistance could be overcome under the proper experimental conditions (*80–82*). Each study was designed to capitalize on the newly emerging concepts of risk factors for atherosclerosis, and each utilized rats whose resistance to atherogenesis was diminished by unique ways of producing hypercholesterolemia. Hartroft and his colleagues (*80*) and Wissler and his group (*81*) fed rats special diets designed to raise their blood cholesterol levels and then induced hypertension or renal disease or fed the rats chemicals such as propyl thiouracil and sodium cholate.

Malinow and his associates (*82*) utilized particularly potent dietary imbalances plus thyroid-depressing agents to induce atherosclerotic lesions. Some of the major findings emerging from these studies were the greater involvement of the coronary arteries than of the aorta, the location of the aortic lesions in the proximal part of the ascending thoracic aorta, and the additive influence of multiple risk factors (*83*). In subsequent studies this model was used to define the influences of various kinds of food fats (*84*) and of metabolic manipulations (*85*) and to delineate the ultrastructural features of these lesions (*86*). In the latter studies, the lesions resemble the foam cell lesions of the rabbit and of other animals in which the blood cholesterol had very high values and in which there was some degree of endothelial injury (*87*). The availability of a wide variety of genetically defined strains of rats will now allow studies such as these to be designed to explore the genetic basis of the various risk factors involved in atherogenesis.

Two inbred strains of rats are particularly useful for studying the pathogenesis of cardiovascular diseases: the SHR (spontaneously hypertensive) strain (*88*) and the BB strain, which spontaneously develops insulin-dependent diabetes mellitus (*89*). The SHR rats develop hypertension that increases with age; is more severe in males; leads to cerebral, myocardial, vascular, and renal lesions; and is responsive to antihypertensive agents. The hypertension is a genetically transmitted trait that is most likely polygenic, and in well-maintained colonies all of the animals develop hypertension between 5 and 10 weeks of age. The inbred, genetically related WKY strain is often used as the normotensive control for the SHR strain. Stroke-prone (*90*) and obese (*91*) substrains of the SHR strain have been developed, but they are difficult to select and maintain because these phenotypic traits most likely have a polygenic basis. The onset of diabetes in the BB rats is rapid, occurs around 90 days of age, affects both males and females, and is under polygenic control, one component of which is linked to the MHC. The clinical syndrome consists of hyperglycemia, hypoinsulinemia, ketosis, polyuria, glycosuria, and weight loss. Pathologic examination shows selective inflammatory destruction of the beta cells of the islets of Langerhans in the pancreas, and the inflammatory process has a substantial immunological component.

Behavior

The rat has been used for studies in behavior since the turn of the century, and a substantial literature has emerged from these studies (*92, 93*). The investigation of the hereditary and environmental aspects of learning began with the introduction of maze experiments by Small (*94*) and led to the development of "maze-bright" and "maze-dull" lines of rats by selective breeding (*95*). Various emotional characteristics have been developed in rats by selective breeding (*93, 96*), and the role of different areas of the brain in behavior has been investigated by stimulation and by extirpation experiments (*44, 45, 97*). Finally, the effects of aging (*93, 98*) and of various pharmacological agents, including alcohol (*99*) and narcotics (*100*), on behavior have been explored.

These studies have provided insights into behavior and into its anatomic and physiologic basis and have led to the development of the field of experimental psychology. However, the lines of rats used were not developed according to the standard rules of genetic inbreeding, and they generally led, at best, to populations with a restricted genetic composition, relative to a randomly breeding population of rats, in which a certain phenotypic characteristic was prominent. This situation has complicated the more detailed genetic interpretation of much of the experimental literature on behavior, and it is particularly acute when examining the relative roles of heredity and environment in learning. One possible approach to developing appropriate strains of rats for behavioral studies may be to select partially inbred rats for their behavioral characteristics and then to breed them for these traits in the context of a mating scheme that would also continue the inbreeding.

Concluding Remarks

The rat is a major experimental animal in all fields of biomedical research and technology, and studies with it have provided much basic and applied knowledge. Its greatest utility has been in those fields broadly classified as experimental pathology and experimental surgery. The extensive work done on the immunology and genetics of the rat in recent decades has greatly enhanced its utility and has contributed substantially to the body of knowledge in immunogenetics. As the constraints on the use of larger animals grow, the rat should provide an excellent alternative to their use. Such a change would also have the advantage of allowing more sophisticated studies to be designed, since so much is known about the biology of the rat, and this would greatly enhance the value of the experiments done.

References and Notes

1. H. H. Donaldson, *J. Acad. Nat. Sci. Philadelphia* 15, 365 (1912); W. E. Castle, *Proc. Natl. Acad. Sci. U.S.A.* 33, 109 (1947).
2. R. D. Owen, *Ann. N.Y. Acad. Sci.* 97, 37 (1962); J. Palm, *ibid.*, p. 57; O. Stark, V. Kren, B. Frenzl, *Folia Biol. (Prague)* 13, 85 (1967); B. Heslop, *Aust. J. Exp. Biol. Med. Sci.* 46, 479 (1968); H. W. Kunz and T. J. Gill III, *J. Immunogenet.* 1, 413 (1974); T. Natori et al., *Transplant. Proc.* 11, 1568 (1979).
3. T. J. Gill III, H. W. Kunz, D. N. Misra, A. L. Cortese Hassett, *Transplantation* 43, 773 (1987).
4. T. J. Gill III, *Physiologist* 28, 9 (1985).
5. J. B. Calhoun, *The Ecology and Sociology of the Norway Rat* (Department of Health, Education and Welfare, Bethesda, MD, 1962); R. Robinson, *Genetics of the Norway Rat* (Pergamon, New York, 1965); *Inbred and Genetically Defined Strains of Laboratory Animals*, part 1, *Mouse and Rat*, P. L. Altman and D. D. Katz, Eds. (Federation of American Societies for Experimental Biology, Bethesda, MD, 1979); M. F. W. Festing, *Inbred Strains in Biomedical Research* (Oxford Univ. Press, New York, 1979); *Spontaneous Animal Models of Disease*, E. J. Andrews, B. C. Ward, M. H.

Altman, Eds. (Academic Press, New York, 1979), vols. 1 and 2; *The Laboratory Rat*, H. J. Baker, J. R. Lindsey, S. H. Weisbroth, Eds. (Academic Press, New York, 1979), vols. 1 and 2; T. J. Gill III, *Am. J. Pathol.* 101, 521 (1980); M. F. W. Festing, in *Immunologic Defects in Laboratory Animals*, M. E. Gershwin and B. Merchant, Eds. (Plenum, New York, 1981), pp. 267–283.

6. T. Y. Canby and J. L. Stanfield, *Natl. Geogr.* 15, 60 (1977).

7. E. Gunther and O. Stark, in *The Major Histocompatibility System in Man and Animals*, D. Goetze, Ed. (Springer-Verlag, New York, 1977), pp. 207–53.

8. M. Aizawa and T. Natori, *Major Histocompatibility Complex of the Rat* Rattus norvegicus (Kokoku Printing Co. Ltd., Sapporo, 1988).

9. J. C. Howard, *Metabolism* 32, 41 (1983).

10. T. J. Gill III and H. W. Kunz, Eds., *International Workshops on Alloantigenic Systems in the Rat*. I. *Transplant. Proc.* 10, 271 (1978); II. *ibid*. 11, 1549 (1979); III. *ibid*. 13, 1307 (1981); IV. *ibid*. 15, 1533 (1983); V. *ibid*. 17, 1793 (1985); VI. *ibid*. 19, 2983 (1987); VII. *ibid*., 21, 3239 (1989).

11. *Rat Newsletter* is published by the Department of Pathology, University of Pittsburgh School of Medicine, Pittsburgh, PA 15261.

12. J. W. F. Watters, J. Locker, H. W. Kunz, T. J. Gill III, *Immunogenetics* 26, 220 (1987); H. Fujii *et al.*, *ibid*. 29, 217 (1989).

13. A. L. Cortese Hassett *et al.*, *Transplant. Proc.* 21, 3244 (1989).

14. J. Klein, *Natural History of the Major Histocompatibility Complex* (Wiley-Interscience, New York, 1986).

15. T. Oikawa *et al.*, *Jpn. J. Genet.* 58, 327 (1983).

16. D. N. Misra, H. W. Kunz, T. J. Gill III, *Immunogenetics* 26, 204 (1987); D. N. Misra, H. W. Kunz, A. L. Cortese Hassett, T. J. Gill III, *ibid*. 25, 35 (1987); D. N. Misra, H. W. Kunz, T. J. Gill III, *Transplant. Proc.* 21, 3271 (1989).

17. A. M. Ghani, T. J. Gill III, H. W. Kunz, D. N. Misra, *Transplantation* 37, 187 (1984); A. M. Ghani, H. W. Kunz, T. J. Gill III, *ibid*., p. 503; T. A. Macpherson, H. N. Ho, H. W. Kunz, T. J. Gill III, *ibid*. 41, 392 (1986); H. N. Ho, T. A. Macpherson, H. W. Kunz, T. J. Gill III, *Am. J. Reprod. Immunol. Microbiol.* 13, 51 (1987).

18. A. Kanbour *et al.*, *J. Exp. Med.* 166, 1861 (1987); A. Kanbour-Shakir *et al.*, *Proc. Natl. Acad. Sci. U.S.A.*, in press.

19. H. W. Kunz, A. L. Cortese Hassett, T. In-omata, D. N. Misra, T. J. Gill III, *Immunogenetics* 30, 156 (1989); K. Wonigeit, H. J. Hedrich, E. Gunther, *Transplant. Proc.* 11, 1584 (1979); W. Stock and E. Gunther, *J. Immunol.* 128, 1923 (1982).

20. H. W. Kunz, T. J. Gill III, B. D. Dixon, F. H. Taylor, D. L. Greiner, *J. Exp. Med.* 152, 1506 (1980); T. J. Gill III, S. Siew, H. W. Kunz, *J. Exp. Zool.* 228, 325 (1983); T. J. Gill III *et al.*, in *Immunoregulation and Fetal Survival*, T. J. Gill III and T. G. Wegmann, Eds. (Oxford Univ. Press, New York, 1987), pp. 137–155.

21. K. N. Rao, H. Shinozuka, H. W. Kunz, T. J. Gill III, *Int. J. Cancer* 34, 113 (1984); M. Melhem, K. N. Rao, M. Kazanecki, T. J. Gill III, H. W. Kunz, *Cancer Res.*, in press.

22. A. L. Cortese Hassett, K. S. Stranick, J. Locker, H. W. Kunz, T. J. Gill III, *J. Immunol.* 137, 373 (1986); A. L. Cortese Hassett, J. Locker, G. Rupp, H. W. Kunz, T. J. Gill III, *ibid*. 142, 2089 (1989).

23. A. G. Knudson, Jr., *Cancer Res.* 45, 1437 (1985); M. F. Hansen and W. K. Cavenee, *ibid*. 47, 5518 (1987).

24. D. J. Schaid, H. W. Kunz, T. J. Gill III, *Genetics* 100, 615 (1982).

25. M. Palmer, P. J. Wettstein, J. A. Frelinger, *Proc. Natl. Acad. Sci. U.S.A.* 80, 7616 (1983); E. Gunther, W. Wurst, K. Wonigeit, J. T. Epplen, *J. Immunol.* 134, 1257 (1985).

26. L. Hood, M. Steinmetz, B. Malissen, *Annu. Rev. Immunol.* 1, 529 (1983); M. C. Carroll *et al.*, *Proc. Natl. Acad. Sci. U.S.A.* 84, 8535 (1987).

27. D. V. Cramer, B. K. Davis, J. W. Shonnard, O. Stark, T. J. Gill III, *J. Immunol.* 120, 179 (1978); E. P. Blankenhorn and D. V. Cramer, *Immunogenetics* 21, 135 (1985); D. K. Wagener, D. V. Cramer, J. W. Shonnard, B. K. Davis, *ibid*. 9, 157 (1979).

28. T. J. Gill III, D. V. Cramer, H. W. Kunz, D. N. Misra, *J. Immunogenet.* 10, 261 (1983).

29. V. L. Scofield, J. M. Schlumpberger, L. A. West, I. Weissman, *Nature* 295, 499 (1982).

30. S. M. Katz, D. V. Cramer, H. W. Kunz, T. J. Gill III, *Transplantation* 36, 463 (1983); S. M. Katz *et al.*, *ibid*., p. 96.

31. A. Paris and E. Gunther, *Immunogenetics* 10, 205 (1980); G. D. Majoor and P. J. C. van Breda Vreisman, *Transplantation* 41, 92 (1986).

32. D. V. Cramer, *CRC Crit. Rev. Immunol.* 7, 1 (1987).

33. M. Pinto, T. J. Gill III, H. W. Kunz, B. D. Dixon-McCarthy, *Transplantation* 35, 607 (1983); M. K. Oaks and D. V. Cramer, *ibid*.

39, 69 (1985); *ibid.,* p. 504; D. Leszczynski, R. Renkonen, P. Hayry, *Am. J. Pathol.* **120,** 316 (1985).

34. M. Kamada, H. F. F. Davis, D. Wight, L. Culank, B. Roser, *Transplantation* **35,** 304 (1983); N. Kamada, *Immunol. Today* **6,** 336 (1985); S. J. Knechtle, J. A. Wolfe, J. Burchette, F. Sanfilippo, R. R. Bollinger, *Transplantation* **43,** 169 (1987).

35. R. L. Kirkman, *Transplantation* **37,** 429 (1984).

36. M. D. Lee, H. W. Kunz, T. J. Gill III, D. A. Lloyd, M. R. Rowe, *ibid.* **42,** 235 (1986); D. Shaffer *et al., ibid.* **45,** 262 (1988).

37. B. Steiniger, J. Klempnauer, K. Wonigeit, *ibid.* **40,** 234 (1985); M. J. Orloff, A. Macedo, G. E. Greenleaf, B. Girard, *ibid.* **45,** 307 (1988).

38. R. D. Lund, K. Rao, H. W. Kunz, T. J. Gill III, *ibid.* **46,** 216 (1988); in *Neuroimmune Networks: Physiology and Disease,* E. J. Goetzl and N. Spector, Eds. (Liss, New York, in press); *Transplant. Proc.* **21,** 3159 (1989); T. J. Gill III and R. D. Lund, *J. Am. Med. Assoc.* **261,** 2674 (1989).

39. S. J. Geyer, T. J. Gill III, H. W. Kunz, E. Moody, *Transplantation* **39,** 244 (1985); M. K. Nicholas *et al., J. Immunol.* **139,** 2275 (1987).

40. B. F. Uretsky *et al., Circulation* **76,** 827 (1987); E. A. Pascoe *et al., Transplantation* **44,** 838 (1987).

41. B. H. Marcus *et al., Transplantation* **46,** 372 (1988).

42. H. Inoue *et al., Neurosci. Lett.* **54,** 85 (1985).

43. I. Madrazo *et al., N. Engl. J. Med.* **316,** 831 (1987).

44. R. Labbe, A. Firl, Jr., E. J. Mufson, D. G. Stein, *Science* **221,** 470 (1983).

45. F. H. Gage, S. B. Dunnett, U. Stenevi, A. Björklund, *ibid.,* p. 966.

46. H. K. Klassen and R. D. Lund, *Exp. Neurol.* **102,** 102 (1988).

47. H. Auchincloss, Jr., *Transplantation* **46,** 1 (1988).

48. T. J. Gill III and C. F. Repetti, *Am. J. Pathol.* **95,** 465 (1979); T. J. Gill III, in *The Physiology of Reproduction,* E. Knobil and J. Neill, Eds. (Raven, New York, 1988), pp. 2023–2042; T. G. Wegmann and T. J. Gill III, Eds., *Immunology of Reproduction* (Oxford Univ. Press, New York, 1983); D. A. Clark and B. A. Croy, Eds., *Reproductive Immunology 1986* (Elsevier, Amsterdam, 1986); T. J. Gill III and T. G. Wegmann, Eds., *Immunoregulation and Fetal Survival* (Oxford Univ. Press, New York, 1987).

49. M. A. H. Surani, S. C. Barton, M. L. Norris, *Biol. Reprod.* **36,** 1 (1987).

50. T. J. Gill III, *Am. J. Reprod. Immunol. Microbiol.* **15,** 133 (1987).

51. C. Maltoni and I. J. Selikoff, Eds., *Ann. N.Y. Acad. Sci.* **543,** 1 (1988); F. Feo, P. Pani, A. Columbano, R. Garcea, Eds., *Chemical Carcinogenesis: Models and Mechanisms* (Plenum, New York, 1988).

52. H. Tuchmann Duplessis, in *Methods in Prenatal Toxicology,* D. Neubert, H. J. Merker, T. E. Kwasigroch, Eds. (Thieme, Stuttgart, West Germany, 1977), pp. 25–34.

53. D. B. Clayson, in *Toxicological Risk Assessment,* D. B. Clayson, D. Krewski, I. Munro, Eds. (CRC Press, Boca Raton, FL, 1985), vol. 1, pp. 105–122; *Mutation Res.* **185,** 243 (1987).

54. S. Venitt and J. M. Parry, Eds., *Mutagenicity Testing: A Practical Approach* (IRL Press, Oxford, 1984).

55. H. C. Pitot, *Cancer Surv.* **2,** 519 (1983).

56. J. Miller and E. Miller, in *Environmental Carcinogenesis,* P. Emmelot and E. Kriek, Eds. (Elsevier, Amsterdam, 1979), pp. 25–50.

57. R. Doll and R. Peto, *J. Natl. Cancer Inst.* **66,** 1191 (1981).

58. J. Sontag, N. Page, U. Saffiotti, *Guidelines for Carcinogen Bioassay in Small Rodents* (DHEW Publ. NIH 76-801, National Cancer Institute, Bethesda, MD, 1976), pp. 1–65; E. L. Anderson, in *Methods for Estimating Risk of Chemical Injury: Human and Non-Human Biota and Ecosystems,* V. B. Vouk, G. C. Butler, D. G. Hoel, D. B. Peakall, Eds. (Wiley, Chichester, United Kingdom, 1985), pp. 405–436; R. Montesano, H. Bartsch, H. Vainio, J. Wilbourn, H. Yamasaki, Eds., *Long-Term and Short-Term Assays for Carcinogens: A Critical Appraisal,* IARC Scientific Ser. no. 83 [International Agency for Research on Cancer (IARC), Lyon, 1986]; U. Mohr and H.-B. Richter-Reichelm, in *Animals in Toxicology Research,* I. Bartosek, A. Guaitani, E. Pacei, Eds. (Raven, New York, 1982), pp. 65–70.

59. J. H. Weisburger and G. M. Williams, *Science* **214,** 401 (1981); in *Casarett and Doull's Toxicology,* C. Klaason, M. Amdur, J. Doull, Eds. (Macmillan, New York, ed. 3, 1986), pp. 99–173.

60. International Agency for Research on Cancer, *IARC Monogr. Suppl.* **7** (IARC, Lyon, 1987), pp. 100–106.

61. T. L. Goldswothy and H. C. Pitot, *J. Toxicol. Environ. Health* **16,** 389 (1985).

62. H. Greim, U. Andrae, W. Goggelmann, L. Schwarz, K. H. Summer, in *Cancer Risks:*

Strategies for Elimination, P. Bannasch, Ed. (Springer-Verlag, Berlin, 1987), pp. 33–46.

63. E. Crouch and R. Wilson, *J. Toxicol. Environ. Health* 5, 1095 (1979).

64. L. Bernstein, L. S. Gold, B. N. Ames, M. C. Pike, D. G. Hoel, *Fund. Appl. Toxicol.* 5, 79 (1985).

65. I. F. H. Purchase, *Br. J. Cancer* 41, 454 (1980).

66. J. Ashby and R. W. Tennant, *Mutat. Res.* 204, 17 (1988).

67. B. Ames, *Science* 204, 587 (1979); _____, R. Magaw, L. S. Gold, *ibid.* 236, 271 (1987).

68. G. Williams, in *Short-Term Tests for Chemical Carcinogens*, R. San and H. Stich, Eds. (Springer-Verlag, New York, 1980), pp. 581–609.

69. J. M. Begue, C. Guguen-Guillouzo, N. Pasdeloup, A. Guillouzo, *Hepatology* 4, 839 (1984).

70. P. Bannasch, H. Enzmann, H. Zerban, in *Cancer Risks: Strategies for Elimination*, P. Bannasch, Ed. (Springer-Verlag, Berlin, 1987), pp. 47–64.

71. E. Farber, *Biochim. Biophys. Acta* 605, 149 (1980).

72. M. A. Moore and T. Kitagawa, *Int. Rev. Cytol.* 101, 125 (1986).

73. M. A. Pereira and G. D. Stoner, *Fund. Appl. Toxicol.* 5, 688 (1985).

74. S. Parodi, M. Taningher, L. Santi, *Anticancer Res.* 3, 393 (1983).

75. P. Bannasch, *Carcinogenesis* 7, 849 (1986).

76. M. Melhem, K. N. Rao, H. W. Kunz, T. J. Gill III, in *Chemical Carcinogenesis: Models and Mechanisms*, F. Feo, P. Pani, A. Columbano, R. Garcea, Eds. (Plenum, New York, 1988), pp. 485–493.

77. L. M. Zucker, *Ann. N.Y. Acad. Sci.* 131, 447 (1965); K. Okamoto, *Int. Rev. Exp. Pathol.* 7, 727 (1969); H. Wolinsky, *Circ. Res.* 26, 507 (1970); *ibid.* 28, 622 (1971); S. Koletsky, *Am. J. Pathol.* 80, 129 (1975); A. V. Chobanian *et al.*, *Diabetes* 31 (*Suppl. 1*), 54 (1982); E. B. Marliss, A. A. F. Sima, A. F. Machooda, in *The Etiology and Pathogenesis of Insulin-Dependent Diabetes Mellitus*, J. M. Martin, R. M. Ehrlich, F. I. Holland, Eds. (Raven, New York, 1982), pp. 251–274; J. C. Russell and R. M. Amy, *Can. J. Physiol. Pharmacol.* 64, 1272 (1986).

78. J. G. Wilson and J. Warkany, *Pediatrics* 5, 708 (1950); J. G. Wilson and J. W. Karr, *Am. J. Anat.* 88, 1 (1951); J. G. Wilson, C. B. Roth, J. Warkany, *ibid.* 92, 189 (1953); R. E. Hudson, *Cardiovascular Pathology* (Williams & Wilkins, Baltimore, MD, 1965), vol. 2, pp. 1647–1653.

79. N. Anitschkow, in *Arteriosclerosis: A Survey of the Problem*, E. V. Cowdry, Ed. (Macmillan, New York, 1933), pp. 271–322; W. C. Hueper, *Arch. Pathol.* 39, 187 (1945); R. Altschul, *Selected Studies on Arteriosclerosis* (Thomas, Springfield, IL, 1950), pp. 66–74; L. N. Katz and J. Stamler, *Experimental Atherosclerosis* (Thomas, Springfield, IL, 1952), pp. 258–261.

80. W. S. Hartroft, J. H. Ridout, E. A. Sellers, C. H. Best, *Proc. Soc. Exp. Biol. Med.* 81, 384 (1952).

81. R. W. Wissler, *Proc. Inst. Med. Chicago* 19, 79 (1952); _____, M. L. Eilert, M. A. Schroeder, L. Cohen, *A.M.A. (Am. Med. Assoc.) Arch. Pathol.* 57, 333 (1954).

82. M. R. Malinow, D. Hojman, A. Pellegrino, *Acta Cardiol.* 9, 480 (1954).

83. L. C. Fillios, S. B. Andrus, G. V. Mann, F. J. Stare, *J. Exp. Med.* 104, 539 (1956); G. F. Wilgram, *ibid.* 109, 293 (1959); S. Naimi, R. Goldstein, M. M. Nothman, G. F. Wilgram, S. Prager, *J. Clin. Invest.* 41, 1708 (1962); W. J. S. Still and R. M. O'Neal, *Am. J. Pathol.* 40, 21 (1962).

84. C. R. Seskind, M. T. Schroeder, R. A. Rasmussen, R. W. Wissler, *Proc. Soc. Exp. Biol. Med.* 100, 631 (1959); C. R. Seskind, V. R. Wheatley, R. A. Rasmussen, R. W. Wissler, *ibid.* 102, 90 (1959).

85. M. S. Moskowitz, A. A. Moskowitz, W. L. Bradford, R. W. Wissler, *Arch. Pathol.* 61, 245 (1956); R. W. Priest, M. T. Schroeder, R. Rasmussen, R. W. Wissler, *Proc. Soc. Exp. Biol. Med.* 96, 298 (1957).

86. I. Joris, T. Zand, J. J. Nunnari, F. J. Krolikowski, G. Majno, *Am. J. Pathol.* 113, 341 (1983).

87. W. J. S. Still, *Arch. Pathol.* 89, 392 (1970).

88. K. Okamoto, *Int. Rev. Exp. Pathol.* 7, 227 (1969); "Spontaneously hypertensive (SHR) rats: Guidelines for breeding, care and use," *ILAR News* 19, G1 (1976).

89. A. A. Like, E. Kislauskis, R. M. Williams, A. A. Rossini, *Science* 216, 644 (1982); R. D. Guttmann, E. Colle, F. Michel, T. Seemeyer, *J. Immunol.* 130, 1732 (1983); M. Angelillo *et al.*, *ibid.* 141, 4146 (1988).

90. K. Okamoto, *Circ. Res. Suppl. I* (1972), p. 143.

91. S. Koletsky, *Exp. Mol. Pathol.* 19, 53 (1973).

92. G. M. Harrington, *Behav. Genet.* 11, 445 (1981).

93. R. E. Wimer and C. C. Wimer, *Annu. Rev. Psychol.* 36, 171 (1985); G. E. McClearn and

T. T. Foch, in *Stevens Handbook of Experimental Psychology*, R. C. Atkinson, R. J. Herrnstein, G. Lindzey, R. D. Luce, Eds. (Wiley, New York, 1988), pp. 677–764.
94. W. S. Small, *Am. J. Psychol.* 11, 80 (1900).
95. R. C. Tyson, *39th Yearb. Natl. Soc. Study Educ.* 1, 111 (1940).
96. C. S. Hall, in *Handbook of Experimental Psychology*, S. S. Stevens, Ed. (Wiley, New York, 1951), pp. 304–329; C. Guenaire, G. Feghali, B. Senault, J. Delacour, *Physiol. Behav.* 37, 423 (1986); R. L. Commissaris, G. M. Harrington, A. M. Ortiz, H. J. Altman, *ibid.* 38, 291 (1986).
97. J. Olds and P. Milner, *J. Comp. Physiol. Psychol.* 47, 419 (1954); N. E. Miller, *Am. Psychol.* 13, 100 (1958).
98. M. Auroux, *Teratology* 27, 141 (1983).
99. F. R. George, *Pharmacol. Biochem. Behav.* 27, 379 (1987); M. A. Linseman, *Psychopharmacology* 92, 254 (1987).
100. T. Suzuki, Y. Koike, S. Yanaura, F. R. George, R. A. Meisch, *Jpn. J. Pharmacol.* 45, 479 (1987); T. Suzuki, K. Otani, Y. Koike, M. Misawa, *ibid.* 47, 425 (1988).
101. R. Sodoyer *et al.*, *EMBO J.* 3, 879 (1984); E. D. Albert, M. P. Baur, W. R. Mayr, Eds., *Histocompatibility Testing 1984* (Springer-Verlag, New York, 1984), pp. 333–341; "Nomenclature for factors of the HLA system, 1987," *Immunogenetics* 28, 391 (1988); A. Radojcic *et al.*, *ibid.* 29, 134 (1989).
102. The work in the authors' laboratories was supported by grants from the National Institutes of Health [CA 18659, HD 09880, HD 08662 (T.J.G. and H.W.K.) and HL 33740, HL 07237, LM 0009 (R.W.W.)]; the Tim Caracio Memorial Cancer Fund, the Beaver County Cancer Society, and the Pathology Education and Research Foundation (T.J.G. and H.W.K.); the New South Wales State Cancer Council and a Yamigawa-Yoshida Memorial International Cancer Study Grant from the International Union Against Cancer (G.J.S.); and the Nutrition and Heart Disease Study (R.W.W.).

15

Transgenic Animals

Rudolf Jaenisch

The introduction of genes into the germ line of mammals is one of the major recent technological advances in biology. Transgenic animals have been instrumental in providing new insights into mechanisms of development and developmental gene regulation, into the action of oncogenes, and into the intricate cell interactions within the immune system. Furthermore, the transgenic technology offers exciting possibilities for generating precise animal models for human genetic diseases and for producing large quantities of economically important proteins by means of genetically engineered farm animals.

The first animals carrying experimentally introduced foreign genes were derived by microinjection of simian virus 40 (SV40) DNA into the blastocyst cavity (*1*). The presence of the injected DNA in a number of somatic tissues derived from mice that had been injected as embryos was demonstrated by DNA reassociation kinetics. However, integration of the viral DNA into the germ line was not demonstrated in these early experiments. A later study suggested that some of the SV40 DNA remained episomal in somatic tissues (*2*). Germ line transmission of foreign DNA was detected in subsequent studies when mouse embryos were exposed to infectious Moloney marine leukemia retrovirus (M-MuLV), which resulted in the generation of the first transgenic mouse strain (*3*).

Infection of mouse embryos with retroviruses constitutes one method of genetically manipulating mouse embryos. Another more commonly used technique for generating transgenic animals is the direct microinjection of recombinant DNA into a pronucleus of the fertilized egg (*4*). Finally, a recently developed technique involves the introduction of DNA by viral transduction or transfection into embryonic stem cells (ES cells), which are able to contribute to the germ line when injected into host blastocysts (*5*). In this article I will not attempt to give a comprehensive review of the field but rather emphasize principles and recent developments. Several detailed review articles have been published over the last 2 years (*6*) that cover the earlier work on transgenic animals.

Methods for Introducing Genes into Animals

Microinjection of DNA into pronucleus

Microinjection of cloned DNA directly into a pronucleus of a fertilized mouse egg has been the most widely and successfully used method for generating transgenic mice. Typically, multiple DNA molecules arranged in a head-to-tail array integrate stably into the host genome. It is thought that the injected DNA molecules

R. Jaenisch is an investigator at the Whitehead Institute for Biomedical Research and professor of biology at Massachusetts Institute of Technology, Cambridge, MA 02141. This chapter is reprinted from *Science* 240, 1468 (1988).

associate by homologous recombination before integration and in most cases insert subsequently at a single chromosomal site. It has been proposed that random chromosome breaks, possibly caused by repair enzymes that are induced by the free ends of the injected DNA molecules, may serve as integration sites of the foreign DNA (7). Frequently, rearrangements, deletions, duplications (8), or translocations (9) of the host sequences occur at the insertion sites. However, the injected DNA does not always integrate into the host genome. Bovine papilloma virus (BPV), for example, either integrates stably into the genome of transgenic mice or is maintained as an episome depending on the structure of the injected DNA (10). Episomal replication and transmission to the offspring have also been reported for a rearranged plasmid coding for the polyoma virus large T antigen, although the mechanism responsible for the episomal state has not been resolved (11).

The principal advantage of direct microinjection of recombinant DNA into the pronucleus is the efficiency of generating transgenic lines that express most genes in a predictable manner. However, one disadvantage of this method is that it cannot be used to introduce genes into cells at later developmental stages. Moreover, the cloning of the chromosomal insertion site may be difficult because of the multiple copy inserts and the host sequence rearrangements.

Retrovirus infection

In contrast to microinjected DNA, retroviruses integrate by a precisely defined mechanism into the genome of the infected cell. Only a single proviral copy is inserted at a given chromosomal site and no rearrangements of the host genome are induced apart from a short duplication of host sequences at the site of integration (12). Preimplantation stage mouse embryos can be exposed to concentrated virus stocks (3) or cocultivated on monolayers of virus-producing cells (13). Methods also have been devised to introduce virus into postimplantation embryos between days 8 and 12

of gestation (14). While this allows infection of cells from many somatic tissues, germ cells are infected with a low frequency (15). Similarly, when chicken embryos were exposed at the blastodisk stage to avian leukemia virus, infection of germ cells was inefficient (16). Because the chick pronucleus cannot be microinjected with DNA, retrovirus infection is probably the only feasible method for generating transgenic chickens.

The main advantage of the use of retroviruses or retroviral vectors for gene transfer into animals is the technical ease of introducing virus into the embryos at various developmental stages. Furthermore, it has proved much easier to isolate the flanking host sequences of a proviral insert than those flanking a DNA insert derived from pronuclear injection. This is of considerable advantage when attempting to identify the host gene disrupted by insertion of the proviral DNA. The main disadvantages of the use of retroviruses for gene transfer are the size limitation for transduced DNA and the unresolved problems of reproducibly expressing the transduced gene in the animal.

Embryonic stem cells

Embryonic stem cells are established in vitro from explanted blastocysts and retain their normal karyotype in culture (17). When injected into host blastocysts, ES cells can colonize the embryo and contribute to the germ line of the resulting chimeric animal (5). Genes can be efficiently introduced into ES cells by DNA transfection or by retrovirus-mediated transduction, and the cell clones selected for the presence of foreign DNA retain their pluripotent character. By means of this approach, mice have been generated from cell clones that were selected in vitro for a specific phenotype (18). This opens exciting possibilities for deriving mouse strains carrying specific mutations. Although only a few laboratories have reported successful germ line contribution of ES cells to date, it is likely that this approach will receive increasing attention for the genetic manipulation of mice.

Expression of Genes in Transgenic Animals

Pronuclear injection of cloned genes

The crucial problem has been the predictable and tissue-specific expression of the injected genes. In early experiments only low or extremely variable expression, dependent on the chromosomal position of the inserted gene, was seen. It was soon realized that the presence of prokaryotic vector sequences is highly inhibitory to the appropriate expression of certain genes including β-globin, α-actin, and α-fetoprotein (19–21). Therefore, most investigators now remove the prokaryotic vector sequences before injection of the gene into embryos to avoid any possible perturbing effects these sequences might have on gene expression. In contrast, genes such as those encoding immunoglobulin (Ig), elastase, and collagen appear to be less sensitive to the presence of vector sequences and are often expressed independently of the chromosomal position (22–25). Table 1 summarizes examples of the successful expression of genes in a wide variety of specific tissues.

Much effort has been directed toward understanding the basis for developmental activation of genes. Transgenic mice have been instrumental in localizing cis-acting sequence elements responsible for tissue-specific gene regulation. Such elements have been found in the 5′ flanking sequences, either proximal or distal to the promoter, within the gene itself, or in the 3′ flanking sequences. In the case of some promoters the tissue-specific enhancer elements have been identified by systematically introducing DNA constructs into embryos with different lengths of flanking sequences. The elastase (23), the γ-crystallin (26), and the protamine (27) genes all have a compact promoter where all information necessary for tissue-specific gene expression is contained within a few hundred base pairs upstream of the start site. For other genes, tissue-specific enhancer elements are spread over considerable distance. For example, an element located 10 kb upstream of the albumin promoter is indispensable for liver-specific expression (28), and an element localized 5 kb 3′ of the T cell receptor gene controls T cell–specific expression (29). Three distinct enhancer elements responsible for tissue-specific expression of the α-

Table 1. Expression of transgenes in specific tissues.

Tissue	Gene or promoter	Reference
Brain	MBP, *Thy*-1, NFP, GRH, VP	(40, 42, 69, 102)
Lens	Crystallin	(26, 63)
Mammary epithelial cells	β-Lactoglobulin, WAP	(45, 48)
Spermatids	Protamine	(27)
Pancreas	Insulin, elastase	(23, 67, 71, 103)
Kidney	Ren-2	(104)
Liver	Alb, AGP-A, CRP, α2u-G, AAT, HBV	(28, 39, 105)
Yolk sac	α-Fetoprotein	(21)
Hemopoietic tissues		
Erythroid cells	β-Globin	(19, 30, 31)
B cells	κ Ig, μ Ig	(22)
T cells	T cell receptor	(29)
Macrophages	M-MuLV LTR	(53)
Connective tissue	MSV LTR, collagen, vimentin	(9, 24, 106)
Muscle	α-Actin, myosin light chain	(20, 107)
Many tissues	H-2 (HLA), β2-m; CuZn SOD	(75, 108)

AAT, α1-antitrypsin; AGP-A, α1-acid glycoprotein; Alb, albumin; α2u-G, α2u globulin; β2-m, β2-microglobulin chain; CRP, C-reactive protein; CuZn SOD, Cu/Zn-superoxide dismutase; GRH, gonadotropin-releasing hormone; HBV, hepatitis B virus; HLA, histocompatibility antigen class; MBP, myelin basic protein; MSV, murine sarcoma virus; NFP, neurofilament protein; Ren-2, renin-2; VP, vasopressin; WAP, whey acidic protein.

fetoprotein gene are contained within 7 kb of upstream sequences (21).

Transgenic animals have been particularly useful in the analysis of developmental activation of the β-globin gene family. During development, distinct "embryonic," "fetal," and "adult" globin genes are sequentially expressed in erythroid cells. Cloned fetal and adult globin genes introduced into the mouse germ line were expressed correctly (19), and the stage-specific activation was dependent on enhancer elements at the 3' as well as the 5' end of the gene (30). Nevertheless, serious problems with β-globin expression remained. Expression of globin genes was always found to be dependent on the site of chromosomal integration and independent of the copy number, and the transgenes were never expressed at a level comparable to the expression of the endogenous gene. When sequences located at distances 20 to 50 kb from either side of the β-globin gene were included, the injected gene was expressed at a level comparable to the endogenous globin gene and directly related to the copy number (31). It is therefore possible that these sequences control the accessibility of the β-globin locus to tissue-specific trans-acting factors or that they contain nuclear matrix–binding sites or represent enhancer elements that may exert their effects over very long distances. Genes other than those encoding β-globin also frequently show expression that is position-dependent and not related to copy number. Consequently, it is possible that sequence elements acting at a distance in controlling gene expression are not unique to globin genes and may have to be identified before high expression can be achieved reproducibly.

The tissue in which a particular gene is expressed is frequently determined by the combination of a tissue-specific enhancer with a particular promoter. For example, the Ig enhancer directs Ig or *myc* expression to B cells (22) but enhances expression of the SV40 T antigen in many tissues (32). That sequences localized in introns may also be significant in determining the level of transcription in animals, but not in cultured cells, was demonstrated in a recent study when pairs of genes, either with or without introns (as complementary DNA), were microinjected into embryos (33).

For many experimental purposes, it would be highly desirable to be able to modulate expression of a transgene with some external stimulus. Promoters of genes subject to modulation by hormonal or other environmental stimuli have been shown in several instances to properly control expression of transgenes. The metallothionein promoter, for example, has been used to direct expression of many different reporter genes, and in some cases expression could be stimulated by feeding the animals with heavy metals (34). Hormone-inducible promoters that function in transgenic mice include the mouse mammary tumor virus (MMTV) long terminal repeat (LTR) (35, 36), the transferrin gene (37), the H-2 E_α gene (38), and two liver-specific genes (39). These results are promising and indicate that transgene expression can be modulated in vivo by external signals. However, at present, the stimuli used to activate inducible genes show toxic side effects that limit their experimental utility.

The ability to introduce and express genes in the animal has opened the door to efforts designed to correct genetic defects. So far "gene therapy," that is, the repair of a mutated gene, has not been accomplished in the animal. Genes introduced into animals invariably integrate at a site distant from the resident defective gene. Therefore, mutant and introduced normal genes will segregate independently in the next generation. Nevertheless, successful corrections of hereditary disorders at the phenotypic level have been accomplished and include hormone deficiencies (40), β-thalassemia (41), and a myelination defect (42). These types of experiments will help to elucidate the molecular deficiency causing the respective hereditary disorder.

Genes have also been microinjected into rabbit, sheep, and pig embryos (43). The success rate of generating transgenic domestic farm animals is, however, much lower than that obtained with mice, in large part because of technical difficulties in visualizing the pronucleus in the embryos of these species. A human growth hormone gene successfully introduced into the pig in spite of these difficulties was nevertheless unable to increase growth, perhaps because the human hormone was not biologically active in pigs (*43*). Although the importance of genetic engineering for improving livestock has been questioned (*44*), it is likely that transgenic farm animals may become a source of economically valuable proteins. For example, medically relevant proteins, whose expression has been targeted to the mammary epithelial cells, may be harvested from the milk of transgenic cows as has been shown to be possible for transgenic mice (*45*).

Retroviruses

Early experiments have shown that preimplantation mouse embryos (*46*) or embryonal carcinoma (EC) cells (*47*) are not able to support the expression and replication of retroviruses. In contrast, virus is expressed efficiently in later stage embryos (*48*) or in differentiated EC cells. Virus replication in early embryonic cells is restricted because the viral LTR, which contains the viral promoter and transcriptional control elements, does not function at that stage. Transcriptional inactivity has been correlated with de novo methylation of the provirus (*47, 48*), with nonfunctioning of the viral enhancer or downstream sequences (*49*), with the presence of repressor activity, and with lack of activating factors (*50*) in the nonpermissive embryonal cells. In most cases, the block, once established in the embryo, is maintained by a cis-acting mechanism at later stages of development (*48*). This likely involves DNA methylation, because injection of postnatal animals with the drug azacytidine (aza C) leads to expression of the previously inactive provirus (*51*). However, some proviruses carried in transgenic mice are expressed in specific tissues (*52*) or at specific stages of development, and it has been shown that virus activation in those cases is influenced by the chromosomal position of the provirus (*13*). A puzzling observation concerns the expression of genes under LTR control in certain tissues when introduced by pronuclear injection (*9, 53, 54*). This suggests that the viral control elements are able to respond to trans-acting transcription signals in the developing organism when they are introduced by a mechanism other than the normal, viral enzyme–mediated integration process.

The inactivity of the viral LTR in embryonic cells has reduced the utility of retroviral vectors for gene transfer into the germ line (*55*). An alternative to expressing virus-transduced genes from the viral LTR is to place the gene of interest under the control of an internal promoter. Transgenic mice carrying viral vectors with the human β-globin gene under control of its own promoter (*52*) or the *neo* gene under that of the thymidine kinase promoter (*56*) expressed the transduced gene in hemopoietic cells or in many tissues, respectively, as would be expected from the two types of promoters. These results are promising and suggest that genes transduced into embryos by viral infection can be expressed when controlled by an appropriate internal promoter rather than the viral LTR.

Applications of Transgenic Technology

Models for oncogenesis and diseases

The potential for using specific promoters or enhancers to direct expression of heterologous genes to a specific cell type has stimulated numerous attempts to change the physiology of an animal experimentally. The transgenic technology has been particularly valuable for studying the consequences of oncogene expression in the animal (*57*). With the use of transgenic mice, problems can be addressed that cannot be approached satisfactorily in cell culture: for example, the spectrum of tissues that are susceptible to the transforming activity of an oncogene, the relation between multistep oncogenesis and cooperativity of oncogenes, and the effect of oncogenes on growth and differentiation.

When different promoters or enhancers were used to direct *myc* or *ras* oncogene expression to different tissues, tumor formation resulted in most cases. Oncogenes were expressed frequently in many tissues and this usually preceded tumor formation by many months. However, many tissues in which a given oncogene was expressed never developed tumors. For example, long latencies and variable penetrance were observed when *myc* or *ras* expression was controlled by the MMTV LTR (*36*), the Ig enhancer (*22*), a fusion between the Ig enhancer and the SV40 promoter (*32*), or the whey acidic protein (WAP) promoter (*58*). This is consistent with the concept of oncogenesis as a multistage process (*59*) with activation of an oncogene likely to be the first of several steps in tumorigenesis. However, even though coexpression of *myc* and *ras* resulted in accelerated tumor formation (*60*), additional somatic events appeared necessary for the realization of the malignant phenotype. Whereas pancreatic acinar cells seemed highly susceptible to the transformation by *ras*, no tumors developed when *myc* was expressed from the same promoter (*61*). Conversely, neoplastic transformation of mammary gland cells was efficiently induced by *myc* expression but rarely by expression of *ras* (*58*). This strengthens the view that the consequences of oncogene expression also depend on the particular cell type and may differ dramatically between one tissue and another.

When a cellular gene is introduced into the germ line under the control of a heterologous promoter, it is assumed that the phenotype arising in the transgenic animals will reveal not only the pathological consequences of unregulated or ectopic expression of the transgene; it will also help the analysis of its normal function in development and differentiation. Experiments of this kind have been done with proto-oncogenes, growth factor genes, and genes encoding cell surface antigens. For example, deregulated expression of the *fos* proto-oncogene in many tissues failed to induce tumors but rather interfered with normal bone development (*62*), whereas expression of the *mos* gene in the lens resulted in disturbance of lens fiber formation (*63*). Similarly, unregulated expression of a hemopoietic growth factor in macrophages (*53*) or of the *Thy-1* cell surface antigen in many tissues (*64*) caused fatal proliferative abnormalities. Also, expression of

a mutant dihydrofolate reductase gene resulted in general growth abnormalities (*65*). Almost any cell type appears to be susceptible to transformation by SV40 T antigen, including cells of the exocrine and endocrine pancreas (*66*), the choroid plexus (*67*), the lens (*68*), the thymus, and the pituitary gland (*32, 69*). T antigen-induced proliferation has also been used as a convenient marker for the study of pancreatic differentiation and the development of immune tolerance (*70*).

Phenotypes induced in transgenic mice by expression of various viral gene products are providing model systems for pathological conditions, some of which resemble human diseases. The small DNA tumor viruses, BPV and polyoma virus, appear to be less promiscuous in their transforming potential than SV40. Introduction of BPV into the germ line resulted in skin tumors (*9*), whereas polyoma virus caused hemangiomas (*71*). A demyelinating disease that resembles progressive multifocal leukoencephalopathy appeared as a consequence of JC virus expression, and a neurofibromatosis resembling von Recklinghausen's disease was observed in mice carrying the human T cell leukemia/lymphoma virus (HTLV-I) *tat* gene (*54*). The latter two viruses are thought to be neurotropic in humans, and the relevant transgenic mice are likely to provide valuable model systems for studying the pathology of virus-induced diseases of the central nervous system.

Immune system

Transgenic mice have also been important to the study of Ig gene expression. Several groups showed functionally rearranged Ig genes introduced into the germ line to be correctly activated and to alter the expression of the endogenous immunoglobulin repertoire (*22*). These and similar types of studies indicate that light and heavy chains, when expressed at a sufficient level, may interfere by some feedback mechanism with further Ig gene rearrangement. Recent evidence also obtained with transgenic mice suggests that expression of functional Ig genes can cause complex abnormalities in the immune system, that allelic exclusion may be mediated by expression of the membrane-bound form of the human μ chain, and that somatic mutations and gene rearrangement are not concomitant processes (72). Chicken or rabbit Ig genes in germ line configuration become rearranged in transgenic mice and form functional hybrid Ig molecules (73), suggesting that the production of interspecies monoclonal antibodies may be possible in genetically engineered mice. A detailed review discussing the implications of these results for our understanding of the development of the immune system has recently been published (74).

Transgenic mice have also been used to study the function of both class I and class II genes. Introduction of porcine or murine class I major histocompatibility antigens into mice showed that the foreign protein was able to associate in both cases with the endogenous murine β2-microglobulin chain and could form a functional transplantation antigen, whereas the human counterpart was unable to do this (75). Similarly, murine class II genes were correctly expressed and functional in transgenic mice (38). Furthermore, transfer of a functionally rearranged T cell receptor β-chain gene into transgenic mice showed that expression of the transgene inhibited rearrangement of the endogeneous β genes in analogy to results obtained with Ig genes (29).

Lineage marker

A central issue in contemporary biology is the construction of fate maps to assess cell ancestry, cell location, and cell commitment in the developing embryo. Visual observation and injected lineage tracers have been used to study the early stages of mammalian development because the preimplantation embryo is easily amenable to experimental manipulation (76). However, the inaccessibility of the embryo once it has implanted in the uterus impedes the study of cell lineage at later stages and has prevented the use of direct lineage tracers.

Therefore, the study of cell lineage in the postimplantation embryo necessitated the development of stable markers of individual progenitor cells that leave the embryo undisturbed.

The introduction of exogenous DNA into embryos after the one-cell stage generates genetic mosaicism that may be used to analyze such lineage relations. In the first study of this type, preimplantation mouse embryos were infected with retroviruses that served as genetic markers for the progeny of an infected blastomere (15). Quantitation of proviral copies in somatic tissues and the germ line of the resulting mosaic animals indicated approximately eight cells to be allocated to the formation of the embryo with each of them contributing equally to all somatic tissues. Germ cells appear to be set aside before allocation of the somatic lineages. Similar lineage studies were performed with genetically mosaic mice generated by microinjection of plasmid DNA into a pronucleus (77).

Another approach often used in lower vertebrates or in invertebrates in cell lineage studies has been the removal of cells by microdissection or laser ablation. Recently, specific cell lineages were genetically ablated in transgenic mice by expressing the A chain of diphtheria toxin (DTA) under the control of lineage-specific enhancers. Mice carrying an elastase promoter/toxin construct lacked a normal pancreas as a result of expression of the toxin in pancreatic acinar cells, whereas expression of the DTA gene under the control of the γ-2 crystallin promoter resulted in mice with lens defects (78). This strategy should allow the elimination of any cell type for which a specific promoter or enhancer can be used to express the toxin, and live animals should be obtained as long as nonessential lineages such as lens or pancreatic cells are ablated. Drug-inducible ablation of specific lymphoid cells was recently reported in mice expressing the herpes thymidine kinase gene under Ig gene promoter control (79). Because dose and time of drug delivery can be experimentally controlled, this approach should permit the ablation of any lineage, including those that are essential for survival of the animal.

Markers for chromosomal regions

Inserted foreign DNA sequences may serve as convenient molecular markers for the flanking host loci for which no probes would be available otherwise. For example, a proviral genome integrated into the pseudoautosomal region of the mouse sex chromosomes proved to be a unique molecular marker for the analysis of this region, which is composed of highly repetitive sequences (*80*). The genetic analysis revealed a high frequency of unequal crossing-over as well as double cross-over events in the pairing region of the sex chromosomes. Transgenes carried on the X chromosome were found either to escape the normal X-inactivation process or to behave like an X-linked gene (*81*). The cloning and characterization of the host sequences flanking these inserts may contribute to an understanding of the molecular control mechanisms of chromosome pairing and mammalian X inactivation.

Chromosomal markers provided by transgenic mice have also given clues as to the molecular nature of genomic imprinting, that is, the hypothesis of differential expression of maternal and paternal genomes and the requirement for both in mammalian embryogenesis (*82*). When methylation and transcription of foreign DNA sequences in transgenic lines were analyzed, differences in the extent of DNA modification and of gene expression that depended on the parental origin of the respective sequences were found (*83*). These results provide the first evidence for functional and heritable molecular differences between maternally and paternally derived alleles on mouse chromosomes. Cloning of the host sequences flanking the exogenous DNA should reveal whether these differences are due to differential imprinting of the host locus or are a consequence of the insertion of the foreign DNA.

Mutations in transgenic mice

The insertion of foreign DNA sequences into the cellular genome can cause mutational changes by disrupting the function of an endogenous gene. Most insertional mutations in transgenic mice are recessive and have been induced by infection of embryos or ES cells with retroviruses or by microinjection of recombinant DNA into the pronucleus (Table 2). The majority of the mutant strains have an embryonic lethal phenotype (*8, 22, 84, 85*). Other phenotypes include defects in limb formation (*9, 86*), transmission distortion (*87*), or disturbance of kidney function (*88*). Insertional mutations can also occur spontaneously by germ line insertions of endogeneous viruses that are activated in certain strain combinations (*89*). We have identified four retrovirus-induced mutations after inbreeding of 70 transgenic mouse strains (*84, 85, 88*), and similar frequencies have been observed in other laboratories (*90*). Thus, the available data suggest that retroviruses induce mutations at an

Table 2. Mutations in transgenic mice.

Type of mutation and procedure	Phenotype (number of mutants)	Identified mutant genes	Reference
Recessive (insertional mutagenesis)			
Retrovirus infection of			
Embryos	Embryonic lethals (3)	α1(I) collagen	(*84, 85*)
	Kidney failture (1)		(*88*)
EC cells	Enzyme defect (1)	HGPRT	(*17*)
Microinjection of DNA	Embryonic lethals (5)		(*8, 21*)
	Limb disturbance (3)		(*9, 86*)
	Transmission distortion (1)		(*87*)
Dominant (expression of variant subunit in multimeric protein)			
Microinjection of mutant gene	Perinatal lethal (1)	α1(I) collagen	(*24*)

overall frequency of 5 to 6%; pronuclear injection of plasmid DNA may be slightly more mutagenic (*91*). This figure is likely to be an underestimate, because some mutations may have only subtle phenotypes that are not easily detected.

The generation of mutants by insertional mutagenesis is attractive because the introduced DNA can serve as a probe for isolating the integration site and the flanking host sequences. Flanking sequences have in fact been cloned for a number of mutants but the mutated gene has been identified in only one case, the Mov13 strain, which carries an M-MuLV proviral genome in the first intron of the α1(I) collagen gene (*84*). The proviral insertion induces changes in the methylation pattern and chromatin conformation of the collagen gene and was shown to interfere with transcriptional initiation, causing a complete block in type I collagen synthesis in homozygous embryos (*92*). This results in death at midgestation after the rupture of major blood vessels (*93*). The Mov13 strain has been useful for studying the role of collagen in development and for the molecular analysis of structural mutations in the α1(I) collagen gene (*24, 94*). In another mutant strain, Mov34, the host sequences flanking the proviral insertion have been isolated and were shown to correspond to an abundantly expressed gene that has not yet been further characterized (*85*). Like Mov13, the provirus insertion interfered with transcription of the gene, presumably causing the lethal phenotype.

Recent evidence indicates that retrovirus integration is not entirely random but occurs preferentially into regions close to deoxyribonuclease I–hypersensitive sites (*95*) that are characteristic for active genes (*96*). It is possible therefore that the chromatin conformation of a given gene can influence its chance of being mutated. Expressed genes with an opened chromatin conformation may represent a more likely target for integration and insertional mutagenesis than inactive heterochromatic genes.

Many integration sites in mutants induced by DNA microinjection have been cloned.

However, in contrast to retrovirus-induced mutations, no transcripts of host sequences corresponding to the mutated gene have been reported to date. The analysis of junction fragments has revealed that deletions, duplications, rearrangements, and translocations have frequently occurred at the site of integration (*8, 9*). This contrasts with retrovirus integration, which leads to a short direct duplication of host sequences at the site of the single proviral insert but does not result in other rearrangements in the host genome (*12*). The sequence rearrangements seen in insertional mutants induced by microinjection of DNA are likely to complicate the analysis of the primary molecular defect that caused the mutant phenotype. This is particularly serious if the rearrangements involve genes distant from the integration site of the exogenous DNA.

Many mutant mouse strains express phenotypes that resemble genetic diseases in humans, but the molecular basis of the mutation is understood in only a few cases. Insertional mutagenesis by introduction of exogenous DNA into the germ line provides a means for inducing new mutations whose molecular defect is more easily analyzed. Such an approach, however, has serious limitations for those interested in developing a systematic mutational dissection of the mammalian genome, because the gene to be mutated cannot be specified. Recently, strategies have been developed that may substantially improve our ability to generate mutants with a predetermined defect. In initial experiments designed to mimic a human disease, ES cells were mutagenized and clones selected that had lost the ability to produce hypoxanthine-guanine phosphoribosyltransferase (HPRT). However, the HPRT-deficient mice generated from these cells were phenotypically normal, in contrast to patients with Lesch-Nyhan syndrome, a severe neurological condition caused by HPRT deficiency in humans (*18*). This disappointing result most likely reflects differences in the purine metabolism between humans and mice. The result is nevertheless highly encouraging since it demonstrates that mutant mouse strains can indeed be developed from cell clones selected in vitro. Gene targeting by homologous recombination should allow mutation of any chosen gene and has been used in ES cells either to correct a mutated HPRT gene or to disrupt the wild-type HPRT gene (97). In each case, the desired cell clones were obtained by using selection for or against HPRT activity. Attempts at mutating genes in the absence of selection schemes require sensitive screening procedures to identify cell clones carrying the exogenous DNA in the target gene. One possible strategy may be the analysis of DNA from pooled cells by the polymerase chain reaction (98).

A new strategy for generating mutants with a precisely predetermined phenotype is to alter a cloned gene by site-directed mutagenesis so that it encodes a mutant product capable of inhibiting the function of the wild-type gene. Such mutations have been termed "antimorphs" or, more recently, "dominant negative mutations" (99). In the case of multimeric proteins, such mutations may cause the formation of nonfunctional multimers (100). The main advantage of this strategy is that it requires only expression of the mutant gene product and not the inactivation of the endogenous wild-type gene in order to realize the mutant phenotype in a cell. To test the feasibility of this approach in the animal, a point mutation analogous to mutations seen in patients with osteogenesis imperfecta II was introduced into the murine $pro\alpha1(I)$ collagen gene in vitro. Substitution of a single glycine residue in the $pro\alpha1(I)$ collagen gene was shown recently to be associated with this dominant perinatal lethal disease in humans (101). When introduced into transgenic mice, expression of as little as 10% mutant RNA of the total $pro\alpha1(I)$ collagen RNA caused a dominant perinatal lethal phenotype that resembled the human condition (24). This kind of approach is likely to be useful for the genetic analysis of many other proteins that form multimeric structures, such as proteins of the cytoskeleton, and could provide defined animal models for human diseases in the absence of mutations in the endogenous gene of interest.

Conclusions and Prospects

The last few years have witnessed an extraordinary increase in the use of transgenic animals. Methods of manipulating embryos and transferring genes have been refined and now constitute standard procedures used for a variety of purposes. Each of the three methods for generating transgenic animals has distinct advantages for some and disadvantages for other applications. Pronuclear injection of recombinant DNA is the method of choice for obtaining expression of a foreign gene in almost any specific tissue (Table 1). Retroviruses or retroviral vectors are superior when genetic tagging of chromosomal loci, for example, for insertional mutagenesis, or of cells for lineage studies are desired. Finally, the most recently developed method of generating transgenic animals from ES cells allows in principle the derivation of mice with any genetic or phenotypic characteristics for which in vitro screening or selection methods are available.

It is likely that rapid advances will occur in the following areas. (i) It will be important to isolate and characterize chromosomal regulatory elements controlling developmental gene activation over large distances (31). Inclusion of such elements in gene constructs should guarantee predictable and efficient expression independent of the chromosomal integration site. This will be particularly important for genetic engineering of large farm animals where cost constraints limit the number of transgenic lines that can be generated and evaluted. (ii) The various possibilities of marking early embryonic cells or ablating specific lineages give experimental access to stages of mammalian development as yet not amenable to easy experimental manipulation. This undoubtably will accelerate our understanding of the complex cell interactions in mammalian development. (iii) The prospect for generating recessive or dominant mutations in preselected genes not only will permit the derivation of precise animal models for human hereditary diseases but also will mark the beginning of a systematic genetic dissection of developmental processes that will radically change the future of experimental mammalian genetics.

References and Notes

1. R. Jaenisch and B. Mintz, *Proc. Natl. Acad. Sci. U.S.A.* 71, 1250 (1974); R. Jaenisch, *Cold Spring Harbor Symp. Quant. Biol.* 39, 375 (1975).
2. K. Willison, C. Babinet, C. Boccara, F. Kelly, in *Cold Spring Harbor Conf. Cell Proliferation* 10, 307 (1983).
3. R. Jaenisch, *Proc. Natl. Acad. Sci. U.S.A.* 73, 1260 (1976); *Cell* 12, 691 (1977).
4. J. W. Gordon, G. A. Scangos, D. J. Plotkin, J. A. Barbosa, F. H. Ruddle, *Proc. Natl. Acad. Sci. U.S.A.* 77, 7380 (1980); R. L. Brinster *et al.*, *Cell* 27, 223 (1981); F. Costantini and E. Lacy, *Nature* 294, 92 (1981); E. F. Wagner, T. A. Stewart, B. Mintz, *Proc. Natl. Acad. Sci. U.S.A.* 78, 5016 (1981); K. Harbers, D. Jähner, R. Jaenisch, *Nature* 293, 540 (1981).
5. A. Bradley, M. Evans, M. H. Kaufman, E. Robertson, *Nature* 309, 255 (1984); A. Gossler, T. Doetschman, R. Korn, E. Serfling, R. Kemler, *Proc. Natl. Acad. Sci. U.S.A.* 83, 9065 (1986).
6. J. Gordon and F. Ruddle, *Gene* 33, 121 (1985); R. Palmiter and R. Brinster, *Annu. Rev. Genet.* 20, 465 (1986); E. Wagner and C. Stewart, in *Experimental Approaches to Mammalian Embryonic Development*, J. Rossant and R. Pederson, Eds. (Cambridge Univ. Press, New York, 1986), pp. 509–549.
7. R. L. Brinster, H. Y. Chen, M. E. Trumbauer, M. K. Yagle, R. D. Palmiter, *Proc. Natl. Acad. Sci. U.S.A.* 82, 4438 (1985).
8. W. Mark, K. Signorelli, E. Lacy, *Cold Spring Harbor Symp. Quant. Biol.* 50, 453 (1985); L. Covarrubias, Y. Nishida, B. Mintz, *Proc. Natl. Acad. Sci. U.S.A.* 83, 6020 (1986).
9. K. A. Mahon, P. A. Overbeek, H. Westphal, *Proc. Natl. Acad. Sci. U.S.A.* 85, 1165 (1988).
10. M. Lacey, S. Alpert, D. Hanahan, *Nature* 322, 609 (1986); A. Elbrecht, F. DeMayo, M. Tsai, B. O'Malley, *Mol. Cell. Biol.* 7, 1276 (1987).
11. M. Rassoulzadegan, P. Léopold, J. Vailly, F. Cuzin, *Cell* 46, 513 (1986); S. Potter and J. Lloyd, *Nucleic Acids Res.* 15, 5482 (1987); P. Léopold, J. Vailly, F. Cuzin, M. Rassoulzadegan, *Cell* 51, 885 (1987).
12. H. Varmus, in *RNA Tumor Viruses*, R. Weiss, N. Teich, H. Varmus, J. Coffin, Eds. (Cold Spring Harbor Laboratory, Cold Spring Harbor, NY, 1982), pp. 369–512.
13. D. Jähner and R. Jaenisch, *Nature* 287, 456 (1980); R. Jaenisch *et al.*, *Cell* 24, 519 (1981).
14. R. Jaenisch, *Cell* 19, 181 (1980).
15. P. Soriano and R. Jaenisch, *ibid.* 46, 19 (1986).
16. D. Salter, E. Smith, S. Hughes, S. Wright, L. Crittenden, *Virology* 157, 236 (1987).
17. T. Doetschman, H. Eistetter, M. Katz, W. Schmidt, R. Kemler, *J. Embryol. Exp. Morphol.* 87, 27 (1985).
18. M. R. Kuehn, A. Bradley, E. J. Robertson, M. J. Evans, *Nature* 326, 295 (1987); M. Hooper, K. Hardy, A. Handyside, S. Hunter, M. Monk, *ibid.*, p. 292.
19. K. Chada *et al.*, *Nature* 314, 377 (1985); T. Townes, J. Lindgrel, H. Chen, R. Brinster, R. Palmiter, *EMBO J.* 4, 1715 (1985); K. Chada, J. Magram, F. Costantini, *Nature* 319, 685 (1986); G. Kollias, N. Wrighton, J. Hurst, F. Grosveld, *Cell* 46, 89 (1986).
20. M. Shani, *Mol. Cell. Biol.* 6, 2624 (1986).
21. R. E. Hammer, R. Krumlauf, S. A. Camper, R. L. Brinster, S. M. Tilghman, *Science* 235, 53 (1987).
22. K. A. Ritchie, R. L. Brinster, U. Storb, *Nature* 312, 517 (1984); U. Storb, R. L. O'Brien, M. D. McMullen, K. A. Gollahon, R. L. Brinster, *ibid.* 310, 238 (1984); R. Grosschedl, D. Weaver, D. Baltimore, F. Costantini, *Cell* 38, 647 (1984); D. Weaver, F. Costantini, T. Imanishi-Kari, D. Baltimore, *ibid.* 42, 117 (1985); S. Rusconi and G. Köhler, *Nature* 314, 330 (1985); U. Storb *et al.*, *J. Exp. Med.* 164, 627 (1986); K. Yamamura *et al.*, *Proc. Natl. Acad. Sci. U.S.A.* 83, 2152 (1986); J. M. Adams *et al.*, *Nature* 318, 533 (1985); W. Y. Langdon, A. W. Harris, S. Cory, J. M. Adams, *Cell* 47, 11 (1986); W. Alexander, J. Schrader, J. Adams, *Mol. Cell. Biol.* 7, 1436 (1987); A. Iglesias, M. Lamers, G. Köhler, *Nature* 330, 482 (1987).
23. G. H. Swift, R. E. Hammer, R. J. MacDonald, R. L. Brinster, *Cell* 38, 639 (1984); D. M. Ornitz *et al.*, *Nature* 313, 600 (1985); B. Davis and R. J. MacDonald, *Genes Dev.* 2, 13 (1988).
24. A. Stacey *et al.*, *Nature* 332, 131 (1988).
25. J. S. Khillan, A. Schmidt, P. A. Overbeek, B. de Crombrugghe, H. Westphal, *Proc. Natl. Acad. Sci. U.S.A.* 83, 725 (1986).
26. D. R. Goring, J. Rossant, J. Clapoff, M. L. Breitman, L.-C. Tsui, *Science* 235, 456 (1987); H. Kondoh *et al.*, *Dev. Biol.* 120, 177

(1987).

27. J. J. Peschon, R. R. Behringer, R. L. Brinster, R. D. Palmiter, *Proc. Natl. Acad. Sci. U.S.A.* **84**, 5316 (1987).

28. C. Pinkert, D. Ornitz, R. Brinster, R. Palmiter, *Genes Dev.* **1**, 268 (1987).

29. P. Krimpenfort *et al.*, *EMBO J.*, **7**, 745 (1988); Y. Uematsu *et al.*, *Cell* **52**, 831 (1988).

30. M. Trudel and F. Costantini, *Genes Dev.* **1**, 954 (1987); R. R. Behringer, R. E. Hammer, R. L. Brinster, R. D. Palmiter, T. M. Townes, *Proc. Natl. Acad. Sci. U.S.A.* **84**, 7056 (1987); M. Trudel, J. Magram, J. Bruckner, F. Costantini, *Mol. Cell. Biol.* **7**, 4024 (1987).

31. F. Grosveld, G. van Assendelft, D. R. Greaves, G. Kollias, *Cell* **51**, 975 (1987).

32. Y. Suda *et al.*, *EMBO J.* **6**, 4055 (1987).

33. R. L. Brinster, J. M. Allen, R. R. Behringer, R. E. Gelinas, R. D. Palmiter, *Proc. Natl. Acad. Sci. U.S.A.* **85**, 836 (1988).

34. R. D. Palmiter, G. Norstedt, R. E. Gelinas, R. E. Hammer, R. L. Brinster, *Science* **222**, 809 (1983); K. Choo, K. Raphael, W. Adam, M. Peterson, *Nucleic Acids Res.* **15**, 871 (1987); E. B. Crenshaw III, A. F. Russo, L. W. Swanson, M. G. Rosenfeld, *Cell* **49**, 389 (1987).

35. Y. Choi, D. Henrad, I. Lee, S. Ross, *J. Virol.* **61**, 3013 (1987); T. Stewart, P. Hollingshead, S. Pitts, *Mol. Cell. Biol.* **8**, 473 (1988).

36. A. Leder, P. K. Pattengale, A. Kuo, T. A. Stewart, P. Leder, *Cell* **45**, 485 (1986).

37. R. Hammer, R. Idzerda, R. Brinster, S. McKnight, *Mol. Cell. Biol.* **6**, 1010 (1986).

38. M. Le Meur, P. Gerlinger, C. Benoist, D. Mathis, *Nature* **316**, 38 (1985); K. Yamamura *et al.*, *ibid.*, p. 67; C. Pinkert *et al.*, *EMBO J.* **4**, 2225 (1985).

39. G. Ciliberto, R. Arcone, E. Wagner, U. Rüther, *EMBO J.* **6**, 4017 (1987); L. Dente, U. Rüther, M. Tripodi, E. Wagner, R. Cortese, *Genes Dev.*, **2**, 259 (1988).

40. R. E. Hammer, R. D. Palmiter, R. L. Brinster, *Nature* **311**, 65 (1984); A. J. Mason *et al.*, *Science* **234**, 1372 (1986).

41. F. Costantini, K. Chada, J. Magram, *Science* **233**, 1192 (1986).

42. C. Readhead *et al.*, *Cell* **48**, 703 (1987); B. Popko *et al.*, *ibid.*, p. 713.

43. R. E. Hammer *et al.*, *Nature* **315**, 680 (1985); R. Hammer *et al.*, *J. Anim. Sci.* **63**, 269 (1986); G. Brem *et al.*, *Zuchthygiene (Berlin)* **20**, 251 (1985).

44. R. Land and I. Wilmut, *Theriogenology* **27**, 169 (1987).

45. K. Gordon *et al.*, *Biotechnology* **5**, 1183 (1987); J. P. Simons, M. McClenaghan, A. J. Clark, *Nature* **328**, 530 (1987).

46. R. Jaenisch, H. Fan, B. Croker, *Proc. Natl. Acad. Sci. U.S.A.* **72**, 4008 (1975).

47. C. L. Stewart, H. Stuhlmann, D. Jähner, R. Jaenisch, *ibid.* **79**, 4098 (1982); J. W. Gautsch and M. C. Wilson, *Nature* **301**, 32 (1983).

48. D. Jähner *et al.*, *Nature* **298**, 623 (1982); D. Jähner and R. Jaenisch, in *DNA Methylation*, A. Razin, H. Cedar, A. Riggs, Eds. (Springer-Verlag, New York, 1984), pp. 189–219.

49. B. Davis, E. Linney, H. Fan, *Nature* **314**, 550 (1985); E. Barklis, R. C. Mulligan, R. Jaenisch, *Cell* **47**, 391 (1986); H. Weiher, E. Barklis, R. Jaenisch, *J. Virol.* **61**, 2742 (1987); T. Loh, L. Sievert, R. Scott, *Mol. Cell. Biol.* **7**, 3775 (1987).

50. C. M. Gorman, P. W. J. Rigby, D. P. Lane, *Cell* **42**, 519 (1985); F. Flamant, C. Gurin, J. Sorge, *Mol. Cell. Biol.* **7**, 3548 (1987).

51. R. Jaenisch, A. Schnieke, K. Harbers, *Proc. Natl. Acad. Sci. U.S.A.* **82**, 1451 (1985).

52. P. Soriano, R. D. Cone, R. C. Mulligan, R. Jaenisch, *Science* **234**, 1409 (1986).

53. R. A. Lang *et al.*, *Cell* **51**, 675 (1987).

54. J. A. Small, G. A. Scangos, L. Cork, G. Jay, G. Khoury, *ibid.* **46**, 13 (1986); J. A. Small, G. Khoury, G. Jay, P. M. Howley, G. A. Scangos, *Proc. Natl. Acad. Sci. U.S.A.* **83**, 8288 (1986); M. Nerenberg, S. H. Hinrichs, R. K. Reynolds, G. Khoury, G. Jay, *Science* **237**, 1324 (1987); S. H. Hinrichs, M. Nerenberg, R. K. Reynolds, G. Khoury, G. Jay, *ibid.*, p. 1340.

55. D. Jähner, K. Haase, R. Mulligan, R. Jaenisch, *Proc. Natl. Acad. Sci. U.S.A.* **82**, 6927 (1985); H. van der Putten *et al.*, *ibid.*, p. 6148; D. Huszar *et al.*, *ibid.*, p. 8587.

56. C. Stewart, S. Schuetze, M. Vanek, E. Wagner, *EMBO J.* **6**, 383 (1987).

57. D. Hanahan, in *Oncogenes and Growth Control*, P. Kahn and T. Graf, Eds. (Springer-Verlag, New York, 1986), pp. 349–361; B. Groner, C. Schönenberger, A. Andres, *Trends Genet.* **3**, 306 (1987).

58. C. Schönenberger *et al.*, *EMBO J.*, **7**, 169 (1988); A. C. Andres *et al.*, *Proc. Natl. Acad. Sci. U.S.A.* **84**, 1299 (1987).

59. R. A. Weinberg, *Science* **230**, 770 (1985); J. M. Bishop, *ibid.* **235**, 305 (1987).

60. E. Sinn *et al.*, *Cell* **49**, 465 (1987).

61. C. J. Quaife *et al.*, *ibid.* **48**, 1023 (1987).

62. U. Rüther, C. Garber, D. Komitowski, R. Müller, E. F. Wagner, *Nature* **325**, 412 (1987).

63. J. Khillan *et al.*, *Genes Dev.* 1, 1327 (1987).
64. J. W. Gordon *et al.*, *Cell* 50, 445 (1987); G. Kollias *et al.*, *ibid.* 51, 21 (1987); S. Chen, F. Botteri, H. van der Putten, C. P. Landel, G. A. Evans, *ibid.*, p. 7.
65. J. Gordon, *Mol. Cell. Biol.* 6, 2158 (1986).
66. D. Hanahan, *Nature* 315, 115 (1985); D. M. Ornitz, R. E. Hammer, A. Messing, R. D. Palmiter, R. L. Brinster, *Science* 238, 188 (1987).
67. R. D. Palmiter, H. Y. Chen, A. Messing, R. L. Brinster, *Nature* 316, 457 (1985); T. Dyke *et al.*, *J. Virol.* 61, 2029 (1987); C. Pinkert, R. Brinster, R. Palmiter, C. Wong, J. Butel, *Virology* 160, 169 (1987); J. Small, D. Blair, S. Showalter, G. Scangos, *Mol. Cell Biol.* 5, 642 (1985).
68. K. A. Mahon *et al.*, *Science* 235, 1622 (1987).
69. F. Botteri, H. van der Putten, D. Wong, C. Sauvage, R. Evans, *Mol. Cell. Biol.* 7, 3178 (1987); D. Murphy *et al.*, *Am. J. Pathol.* 129, 552 (1987).
70. G. Teitelman, S. Alpert, D. Hanahan, *Cell* 52, 97 (1987); S. Faas, S. Pan, C. Pinkert, R. Brinster, B. Knowles, *J. Exp. Med.* 165, 417 (1987); T. E. Adams, S. Alpert, D. Hanahan, *Nature* 325, 223 (1987).
71. V. L. Bautch, S. Toda, J. A. Hassel, D. Hanahan, *Cell* 51, 529 (1987); R. L. Williams, S. A. Courtneidge, E. F. Wagner, *ibid.* 52, 121 (1988).
72. L. A. Herzenberg *et al.*, *Nature* 329, 71 (1987); M. C. Nussenzweig *et al.*, *Science* 236, 816 (1987); R. L. O'Brien, R. L. Brinster, U. Storb, *Nature* 326, 405 (1987).
73. D. Bucchini *et al.*, *Nature* 326, 409 (1987); M. Goodhardt *et al.*, *Proc. Natl. Acad. Sci. U.S.A.* 84, 4229 (1987).
74. U. Storb, *Annu. Rev. Immunol.* 5, 151 (1987).
75. W. I. Frels, J. A. Bluestone, R. J. Hodes, M. R. Capecchi, D. S. Singer, *Science* 228, 577 (1985); C. Bieberich, T. Yoshioka, K. Tanaka, G. Jay, G. Scangos, *Mol. Cell. Biol.* 7, 4003 (1987); P. Krimpenfort *et al.*, *EMBO J.* 6, 1673 (1987); F. Kievits, P. Ivanyi, P. Krimpenfort, A. Berns, H. L. Ploegh, *Nature* 329, 447 (1987).
76. R. Pederson, in *Experimental Approaches to Mammalian Embryonic Development*, J. Rossant and R. Pederson, Eds. (Cambridge Univ. Press, New York, 1986), pp. 3–33.
77. T. Wilkie, R. Brinster, R. Palmiter, *Dev. Biol.* 118, 9 (1986).
78. R. D. Palmiter *et al.*, *Cell* 50, 435 (1987); M. L. Breitman *et al.*, *Science* 238, 1563 (1987).
79. E. Borrelli, R. Heyman, M. Hsi, R. Evans, *Proc. Natl. Acad. Sci. U.S.A.* 85, 757 (1988).
80. K. Harbers, P. Soriano, U. Müller, R. Jaenisch, *Nature* 324, 682 (1986); P. Soriano *et al.*, *Proc. Natl. Acad. Sci. U.S.A.* 84, 7218 (1987).
81. M. A. Goldman *et al.*, *Science* 236, 593 (1987); R. Krumlauf, V. M. Chapman, R. E. Hammer, R. Brinster, S. M. Tilghman, *Nature* 319, 224 (1986).
82. A. Surani, in *Experimental Approaches to Mammalian Embryonic Development*, J. Rossant and R. Pederson, Eds. (Cambridge Univ. Press, New York, 1986), pp. 402–435.
83. W. Reik, A. Collick, M. L. Norris, S. C. Barton, M. A. Surani, *Nature* 328, 248 (1987); C. Sapienza, A. C. Peterson, J. Rossant, R. Balling, *ibid.*, p. 251; M. Hadchouel, H. Farza, D. Simon, P. Tiollais, C. Pourcel, *ibid.* 329, 454 (1987); J. L. Swain, T. A. Stewart, P. Leder, *Cell* 50, 719 (1987).
84. R. Jaenisch *et al.*, *Cell* 32, 209 (1983); A. Schnieke, K. Harbers, R. Jaenisch, *Nature* 304, 315 (1983); K. Harbers, M. Kuehn, H. Delius, R. Jaenisch, *Proc. Natl. Acad. Sci. U.S.A.* 81, 1504 (1984).
85. P. Soriano, T. Gridley, R. Jaenisch, *Genes Dev.* 1, 366 (1987).
86. R. P. Woychik, T. A. Stewart, L. G. Davis, P. D'Eustachio, P. Leder, *Nature* 318, 36 (1985); P. A. Overbeek, S.-P. Lai, K. R. Van Quill, H. Westphal, *Science* 231, 1574 (1986); J. McNeish, W. Scott, S. Potter, personal communication.
87. T. Wilkie and R. Palmiter, *Mol. Cell. Biol.* 7, 1645 (1987).
88. H. Weiher and R. Jaenisch, unpublished observations.
89. N. A. Jenkins and N. G. Copeland, *Cell* 43, 811 (1985); J.-J. Panthier, H. Condamine, F. Jacob, *Proc. Natl. Acad. Sci. U.S.A.* 85, 1156 (1988).
90. S. E. Spence, D. J. Gilbert, D. A. Swing, N. G. Copeland, N. A. Jenkins, personal communication.
91. T. Gridley, P. Soriano, R. Jaenisch, *Trends Genet.* 3, 162 (1987).
92. M. Breindl, K. Harbers, R. Jaenisch, *Cell* 38, 9 (1984); D. Jähner and R. Jaenisch, *Nature* 315, 594 (1985); S. Hartung, R. Jaenisch, M. Breindl, *ibid.* 320, 365 (1986).
93. J. Löhler, R. Timpl, R. Jaenisch, *Cell* 38, 597 (1984).
94. K. Kratochwil, M. Dziadek, J. Löhler, K. Harbers, R. Jaenisch. *Dev. Biol.* 117, 569 (1986); H. Wu and R. Jaenisch, unpublished observations.

95. S. Vijaya, D. Steffen, H. Robinson, *J. Virol.* **60**, 683 (1986); H. Rhodewold, H. Weiher, W. Reik, R. Jaenisch, M. Breindl, *ibid.* **61**, 336 (1987).

96. R. Reeves, *Biochim. Biophys. Acta* **782**, 343 (1984).

97. K. R. Thomas and M. R. Capecchi, *Cell* **51**, 503 (1987); T. Doetschman *et al.*, *Nature* **330**, 576 (1987).

98. S. J. Scharf, G. T. Horn, H. A. Erlich, *Science* **233**, 1076 (1986).

99. I. Herskowitz, *Nature* **329**, 219 (1987).

100. P. Novick and D. Botstein, *Cell* **40**, 405 (1985).

101. D. H. Cohn, P. H. Byers, B. Steinmann, R. E. Gelinas, *Proc. Natl. Acad. Sci. U.S.A.* **83**, 6045 (1986); J. Bateman, D. Chan, I. Walker, J. Rogers, W. Cole, *J. Biol. Chem.* **262**, 7021 (1987).

102. G. Kollias *et al.*, *Proc. Natl. Acad. Sci. U.S.A.* **84**, 1492 (1987); J. Julien, I. Tretjakoff, L. Beaudet, A. Peterson, *Genes Dev.* **1**, 1085 (1987).

103. R. F. Selden, M. J. Skoskiewicz, K. B. Howie, P. S. Russell, H. M. Goodman, *Nature* **321**, 525 (1986); D. Bucchini *et al.*, *Proc. Natl. Acad. Sci. U.S.A.* **83**, 2511 (1986).

104. D. Tronik, M. Dreyfus, C. Babinet, F. Rougeon, *EMBO J.* **6**, 983 (1987).

105. V. Costa Soare, R. Gubits, P. Feigelson, F. Costantini, *Mol. Cell. Biol.* **7**, 3749 (1987); U. Rüther, M. Tripodi, R. Cortese, E. Wagner, *Nucleic Acids Res.* **15**, 7519 (1987); G. Kesey, S. Povey, A. Bygrave, R. Lovell-Badge, *Genes Dev.* **1**, 161 (1987); R. Sifers *et al.*, *Nucleic Acids Res.* **15**, 1459 (1987); F. V. Chisari *et al.*, *Science* **230**, 1157 (1985); H. Farza *et al.*, *Proc. Natl. Acad. Sci. U.S.A.* **84**, 1187 (1987).

106. P. Krimpenfort *et al.*, *EMBO J.* **7**, 941 (1988).

107. P. Einat, Y. Bergman, D. Yaffe, M. Shani, *Genes Dev.* **1**, 1075 (1987); M. Shani, I. Dekel, O. Yaffe, *Mol. Cell. Biol.* **8**, 1006 (1988).

108. C. J. Epstein *et al.*, *Proc. Natl. Acad. Sci. U.S.A.* **84**, 8044 (1987).

109. I thank all my colleagues, particularly D. Baltimore, D. Huszar, A. Sharpe, S. G. Waelsch, and R. Weinberg, for critically reading the manuscript. I also thank many colleagues for communicating their results prior to publication.

16

Cetaceans

Bernd Würsig

The taxonomic order Cetacea consists of about 75 living species of mammals (*1*) that evolved from a carnivorous stock of ungulates over 50 million years ago. They lost external hindlimbs, gained a paddle-like tail for propulsion, and developed a streamlined body without insulating hair but with thick padding of a modified fatty tissue, the blubber (*2*). Blubber allows small cetaceans to survive in the heat-sapping environment of even polar regions and allows large cetaceans to store energy supplies that permit fasts of many months as the animals migrate and breed away from feeding grounds (*3*).

Two suborders comprise the order Cetacea, the mysticetes (baleen whales) and the odontocetes (toothed whales) that are probably descendants of a common ancestral group, the archaeocetes. The baleen whales lost their teeth and grew thin, long keratinous plates of ectodermal origin, like fingernails and hair, that hang from their upper jaws. These baleen plates strain out invertebrates and small fishes from water taken into the mouth (*4*). The larger filtering apparatus allowed more efficient capture and retention of aggregating prey. The baleen whales evolved disproportionately large mouths and large body size. Large size permitted huge fat reserves which made possible seasonal migrations to take advantage of food blooms and to fast for prolonged periods while reproducing in warmer but less productive waters.

The toothed whales evolved sophisticated echolocation, in which rapid sequences of high-frequency clicks are sent into the environment and the whale listens for the modified return of echoes to discriminate distances, sizes, shapes, and textures of objects (*5*). This allows feeding in the absence of light, and probably allows efficient predator detection. Echolocation may be the single most important new adaptation that has helped odontocetes to expand through all ocean basins and into several major rivers.

Anatomy and Physiology

Much of cetacean success in adapting to life in the seas is due to straightforward adaptations of ancestral stock. For example, (i) the kidneys have increased capacity for handling ionic concentrations of solutes due to compartmentalization into many interconnecting renal units, which increases the total surface area over which ions can be transferred; (ii) blood has a high concentration of hemoglobin, and muscle is richly supplied with myoglobin for storage of large amounts of oxygen (*6*); and (iii) soft body and skeletal morphology has been adapted for

B. Würsig is a professor of marine biology at the Moss Landing Marine Laboratories, California State University, Moss Landing, CA 95039-0450, and a professor of marine biology and director of the Marine Mammal Research Program, Texas A&M University at Galveston, Galveston, TX 77553-1675. This chapter is reprinted from *Science* **244**, 1550 (1989).

propulsion in a dense, three-dimensional medium, sustained breath-holding, temperature regulation, and storage of energy.

Diving

Dolphins dive repeatedly to depths of at least 300 m (7) and sperm whales (*Physeter macrocephalus*) to over 1 km (8). The baleen whales are probably not deep divers, although humpback whales (*Megaptera novaeangliae*) dive to at least 148 m and fin whales (*Balaenoptera physalus*) to 355 m (9).

Dolphins and whales dive with full lungs. Considerable attention has centered on how they avoid the bends. Several physiologic adaptations seem responsible. First, alveolar collapse probably occurs below a depth of about 70 m and stops respiratory gas exchange in the lungs. As air volume decreases, blood-absorbing bundles of arterioles and venules fill in the spaces in the head, preventing tissue compression. The lungs can collapse because of a flexible rib cage and an oblique diaphragm that can cave in the chest cavity. Second, an extensive bundle of blood vessels (the thoracospinal "retia mirabilia") may trap gas bubbles that are generated, and release them between dives more slowly than in other swimming vertebrates. Blood chemistry differences may also be involved (7, 10).

Our incomplete knowledge of cetacean diving physiology stems from two major problems. First, it is logistically difficult and expensive to monitor cetaceans. The ability to measure physiologic parameters on a free-swimming animal and then either store the information on a recoverable device or telemeter it to a base station promises to expand our knowledge in the future (11). Second, the Marine Mammal Protection Act (12) makes it difficult for U.S. researchers to obtain permits to study cetaceans by even slightly invasive methods. Thus, studies of cetacean physiology have lagged behind those of other animals for which research is not similarly restrained.

Brain size

Large brain size may be linked to a life strategy of great longevity, slow maturation, single births after a long gestation, and late weaning (13, 14), rather than to metabolic rate per se (15) or to dive duration (16).

What do cetaceans accomplish with those large brains? A practical definition of intelligence (and its relationship to brain size) still eludes us (17), but numerous other behaviorally related reasons for large brain size have been proposed: the complexity of foraging strategies and the possible need for forming and remembering social relationships in complex societies that have to avoid predation and secure food in three-dimensional space may be likely candidates. Interspecific comparisons of EQ [encephalization quotient; a measure of brain size as the ratio of brain to body mass (18)], as made for several terrestrial groups (19), could help correlate brain size of the Delphinidae with home range foraging strategies and social relationships. It has also been suggested that sophisticated click-based echolocation, which requires analysis of multiple echo parameters for fine-scale object discrimination, resulted in large brain size (13, 20). Most species of dolphins are capable of rapidly learning tricks in captivity and often cooperate in securing food in apparent sophisticated fashion (21). Bottlenose dolphins can also learn through observation, extrapolate information from basic symbols, and remember for long periods of time. This was shown in experiments on two captive dolphins in Hawaii over 10 years and may help in correlation of behavior with brain size and complexity (22).

Sound Production

Sound is an important production and sensory modality, although the senses of taste, touch, and sight are also well developed in dolphins and whales (23). The two suborders have different suites of sounds that are produced in a different manner.

Sounds are picked up by an underwater microphone, or hydrophone, and recorded onto audiotape. Analysis of the frequency, duration, and repetition rate with a sound spectrograph device provides a visual representation. Relative intensity can also be measured. To ascribe functions to cetacean sounds, sound-behavior correlations are studied. Sound descriptions and some sound function deductions have been carried out on numerous whales and dolphins in the past 10 years. Despite the general absence of playback studies, which are used to ascertain sound function in terrestrial animals, the behavioral importance of sound to cetaceans is becoming clear.

Mysticetes

The baleen whales generally produce sounds below 5000 Hz that differ greatly among the three families and, although well described, are of incompletely known function. An individual mysticete producing a given sound can now be pinpointed with sophisticated hydrophone arrays (24). The largest of whales, the blue (*Balaenoptera musculus*) and fin whales of the balaenopterid family, produce intense moans, from 1 second to many seconds long, in the 10- to 20-Hz range. These sounds probably serve to communicate to conspecifics over tens of kilometers, although hypothesized long-distance communication covering entire ocean basins is more questionable (25). Fin whale moans may serve a reproductive function (26). Higher frequency sounds, to several hundred hertz, are also produced by the balaenopterid whales.

The humpback whale has an extensive and varied sound repertoire, with both chirp and moan-like social sounds (27), and trains of repeating sounds termed song, reaching to 5000 Hz in frequency (28). Songs are produced primarily by adult males on the breeding grounds, and, like song in birds, crickets, and gibbon monkeys, probably communicate their species, sex, location, readiness to mate, and readiness to engage in agonistic behavior with other males (29), perhaps by

length or complexity of song (30). More analysis of concurrent behavior and song are necessary before true function of song is explained. Especially valuable will be an analysis of song from individuals of known age or size, perhaps by combining photogrammetric size-measuring techniques (31) with recording sessions.

Gray whales (*Eschrichtius robustus*) produce relatively simple series of pulses and moans, with no known behavior correlate. The short duration pulses that are heard most frequently have the appropriate physical structure to serve as a rudimentary echolocation function (32). Gray whales may stay in contact with each other by sound, but appropriate research on sounds and behavior has not been done. Gray whales migrate close to the shores of the Pacific Northwest in large numbers twice yearly, and it is now apparent that fruitful sound studies can be carried out.

The Balaenidae, right (*Eubalaena* sp.) and bowhead (*Balaena mysticetus*) whales, produce complex sets of moans in the several hundred hertz frequency range, and they appear to be vocal at almost all times on the mating grounds, on the feeding grounds, and during migration. Certain sound types correlate with certain social contexts in right and bowhead whales (33). Bowhead whales also have repeating notes of sound, or song, which may serve a similar mating-related function as in humpback whales (34). Baleen whales do not have vocal cords, and it is probable that sounds are made by air passing through the tracheal region.

Odontocetes

Toothed whale sounds are rich and varied and tend to have higher frequencies than sounds of the mysticetes. There are two basic sound types: pulsed clicks and click series, and unpulsed, frequency-modulated whistles. Both types are used for communication, and a subset of pulsed sounds is used for echolocation as well. Whistles have not been reported for certain odontocetes such as sperm whales, river dolphins, and harbor porpoises (35), although sperm whales produce some tonal sounds.

It is likely that both click and whistle sounds are generated near the nasal region, where a muscular plug and several sets of air sacs connected to the narial passages allow for shunting of air back and forth in the odontocete forehead and cause the vibration of tissue (*36*). The entire process probably evolved from the sibilant nasal sounds made by ungulates on land, transposed during evolution to an internal shunting of air due to the necessity of keeping the nares closed underwater. Some delphinids can produce clicks and whistles simultaneously, possibly by bilateral use of the paired air sac system (*37*).

Sperm whales produce intense clicks in series termed "codas." These may have some individuality (serving a signature function) and also are different between different populations, thus helping to identify individuals of a group. Since sperm whale clicks often occur between animals alternately, or in duetting fashion, they probably serve for communication. Whether clicks are also used for echolocation is not known. An untested hypothesis is that, because of the peculiar head of the sperm whale, different sized whales may produce differently spaced click trains that would communicate their size. Data are needed to correlate click trains from known individuals with their overall size—difficult information to obtain in nature (*38*).

Most odontocete sound studies are done on bottlenose dolphins in captivity, and it was in this species that echolocation was first noted (*39*). Echolocation frequencies can be over ten times our upper hearing of 20 kHz; dolphins can discriminate objects differing by a size ratio of about 1.1 (*40*). A single 13.5-cm fish can be detected at a distance of at least 9 m, and a school of fish may be detected at a range of 100 m or more (*35*). The similar skull morphology among the odontocete families suggests that echolocation was an early adaptation of evolving toothed whales and is probably shared by all present groups.

Some high-intensity click sounds (230 dB at 1 μPa at 1 m for bottlenose dolphins) by bottlenose dolphins, beaked whales, and sperm whales may serve to debilitate prey by overloading fish lateral lines, ears, or shattering bony ossicles and other tissue (*41*). No direct evidence exists for prey stunning, but it may explain the good health of whales and dolphins with badly deformed jaws, or those missing a lower jaw, although tests have not been conclusive (*42*).

Non-echolocatory sounds are also complex and varied, and the cacophony of whistling and squawking sounds heard from schools of dolphins gave rise to the notion that they possess a language. They do communicate with each other, and particular sound types correlate with rest, social-sexual behavior, travel, alarm, or feeding (*43*). However, there is no evidence of a language-like syntactical structure to sequences of sounds. A recently developed telemetry device that attaches to a dolphin's head and identifies which dolphin in a group is vocalizing makes possible an analysis of vocal interactions between individuals of a group (*44*). The device, called a "vocalight" because it lights up during sound production, promises to become a useful tool in sound-related studies.

Dolphins have individual "signature whistles" that may identify members of a social group (*45*). Whistles and pulsed calls are mimicked; mimicry may be useful in dialect formation, as in discrete pods of killer whales in Puget Sound (*46*). Dialects may be important to populations where ranges or possible territories overlap; but, as in humans and white-crowned sparrows, dialects may simply be the result of relatively isolated populations of animals that learn sound repertoires from each other (*47*).

Behavior and Societies

Cetaceans are gregarious mammals, almost always traveling as a pod or school, in which cumulative sensory awareness presumably enhances prey and predator detection, and where long-term social affiliations and interactions can be played out (*48*). Group size and structure are guided by an incompletely understood matrix of factors, including foraging type, need for predator detection and avoidance, social and sexual interactions, and the care and maintenance of developing young.

Foraging strategies help to determine the social life of animals. However, only when cetaceans feed at or near the surface can we directly describe prey taken and method of capture. Researchers rely on indirect cues of feeding, such as long-duration dives, scats at the surface, or a correlation of the potential prey distribution to the stomach samples of net-caught or beached animals. Data on depth of dives is particularly important. This can be obtained from gray whales when they feed at the bottom because they surface with mud streaming from their mouths, by fish-finding sonar traces of whales below the surface, or by electronic dive-telemetry packages attached to whales or dolphins (49). The latter technique is used for seals (50), and its use with cetaceans will give us much better information on when and at what depth individuals feed. This information, coupled with prey availability studies and behavioral observations of feeding groups, will enable us to describe foraging habits, capabilities, and constraints and make better comparisons between foraging strategies and social structure (51). Behavioral information can also be gleaned by noninvasive techniques such as photo-identification of individuals, theodolite tracking groups for information on movement patterns relative to prey distribution, and by high-resolution video for later analysis of behavior (52, 53).

Mysticetes

Each family of baleen whales exhibits a morphology and behavioral repertoire suited to its feeding type. The right and bowhead whales are passive sievers of small invertebrate prey such as calenoid copepods and euphausiid crustaceans (or "krill"). They have large mouths with long, finely fringed baleen that allow huge volumes of water to be filtered out of the sides of the mouth while the prey stay inside. While feeding, the balaenids move forward slowly, passing water continuously through the baleen plates. They have large heads and relatively chunky bodies, indicative of a mode of feeding not requiring bursts of speed. Although much feeding takes place individually, whales often

travel staggered in a V-formation, like the echelon of flying geese, apparently so that adjacent bodies can serve as walls to cut off the escape of prey to the sides (54).

The gray whale is also a relatively passive feeder, but with a small mouth and short, coarse baleen. The gray whale sucks from the substrate inbenthic invertebrate fauna such as tube-dwelling ampeliscid amphipods. It then winnows the edible from inedible and pushes a muddy slurry out of the sides of the mouth, often while surfacing after a feeding dive. It leaves behind visible traces of disturbed benthos, and the impact of feeding on habitat can be most completely assessed for this species (55). Gray whales generally feed in loose aggregations; they do not appear to coordinate feeding activities. They graze on productive stands of benthos, and it is possible that individual feeding ranges (56), and even defense of pasture, may have evolved.

The rorqual whales consist of six streamlined species ranging in size from the 9-m minke whale (*Balaenoptera acutorostrata*) to the 25-m blue whale. They, too, are generally gregarious, although minke and Bryde's (*B. edeni*) whales may live alone in certain nearshore areas (57). The rorquals have short baleen in mouths that can open to greater than a 90° angle between the lower and upper jaw. Rorquals lunge into their prey, which may consist of clouds of krill or schools of small fishes. They gulp up to 70 tons of food and water into mouths that distend as accordion-like furrows or pleats in the throat expand and, by contraction of throat muscles, push water out of a partially closed mouth while prey is trapped against the slit of baleen (58). They may also push part of the head above water and thereby allow water to cascade out of the mouth by gravity. Feeding is a highly active, lunging affair requiring bursts of speed to trap especially maneuverable krill and fishes. Cooperative foraging occurs as up to a dozen or more whales lunge simultaneously side by side or in a tight cluster. Humpback whales use a variety of strategies to control and contain prey, such as blowing a bubble net around the prey, circling rapidly, or smacking and apparently stunning prey with the tail (59).

Baleen whale species are gregarious to various degrees; basic ecological rules discovered for terrestrial animals probably apply. There are correlations between increasing body size, greater degree of dietary specialization, larger home ranges, and increasing group size in antelope, deer, and to some degree in primates (*60*). When the smaller fish-feeding minke and Bryde's whales of the Gulf of California, Mexico, are compared to the larger plankton-feeding fin and blue whales, whales with increased body size tend to have larger ranges, feed on fewer types of prey, and occur in larger groups than the smaller whales (*61*). One major difference may be the constraint of obtaining enough energy to support a large body, perhaps done most efficiently by feeding low on the trophic level, and ranging widely in search of large concentrations of food (*62*). Antarctic minke whales, which feed largely on plankton instead of fish, occur in aggregations and travel widely, as do fin and blue whales of the Gulf of California (*63*). However, although the size-food-lifestyle relationship is a useful first approximation to explain general social patterns, it is by itself overly simplistic (*60, 62, 63*). Research on whales killed by the whaling industry provided information on calving seasonality, gestation time, size at first ovulation, and so forth (*64*). We know very little, however, about mating systems and strategies. The most detailed descriptions of sexual activity have been made from circling aircraft, and sex can be determined when whales roll ventrum up at the surface (*65*).

In right, bowhead, and gray whales, several males often jostle to get close to a female. Males may cooperate to allow one male to copulate (*53*), but it is more likely that each is attempting to get to the female first. The female may be practicing mate choice by rolling belly up at the surface, swimming rapidly away, and otherwise making it difficult for a male to insert his penis, for the largest or most aggressive male is often able to copulate with the female as the others fall behind. Females that have copulated may then copulate with other males, suggesting that the system is not purely polygynous, where one male inseminates several females.

Further evidence for nonpolygyny is that, in general, mammalian testes are larger relative to body size than would be predicted when the breeding system is a multimale-multifemale one tending toward promiscuity (*66*). Right, bowhead, and gray whales have extremely large testes, capable of producing large quantities of sperm, and it is suggested that they practice a form of sperm competition which allows males to attempt to displace or outnumber rival sperm in the vagina and uterus (*67*).

Very little is known about the mating behavior of most rorquals except that humpback, right, and bowhead whales have surface-active groups where animals butt, ram, and inflict bloody wounds on each other (*68*). This may represent aggression between males as they battle for access to a female. Direct competition may make sperm competition unnecessary, possibly because only one male gets to inseminate a female, and, indeed, the testes of humpbacks (and those of all other rorquals) are correspondingly small (*67*). Male humpback whales also sing on mating grounds, which may be an alternative mating strategy, either to mediate male-male competition or to attract the attention of females (*69*). There is no evidence that rorquals are monogamous; however, a lack of extreme secondary sexual characteristics and the relatively low levels of differential male-female mortality in utero suggest that these whales are not extremely polygynous (*70*).

Several techniques will allow the rapid advance of our knowledge of mysticete (and odontocete) social and mating structure. One, used extensively in the past 10 years, involves long-term reidentification of whales and dolphins from natural marks (*71*). This technique is used in conjunction with sex and relationship information obtained by small biopsy samples from whales in nature. Samples are collected by shooting (with crossbow or compound bow) a hollow dart connected to a fishing reel. Sex is determined from karyotypes and genetic relationships assessed from protein, chromosomes, and DNA polymorphisms (*72*). Correlation between behavior, sex, and relationship is still in its infancy, but is likely to grow rapidly in the next several years.

Odontocetes. The six families of toothed

whales include the small harbor porpoise (*Phocoena*) to the mighty sperm whale (*Physeter macrocephalus*). Almost all species are gregarious and probably all echolocate. Most species feed on fish and squid; and most feeding, social activity, mating, and rearing of young take place in the group or school.

In general, animals travel in flocks and schools for enhancement of predator detection, confusion, and avoidance (*73*), as well as for foraging efficiency. Polarized schools (that is, with all members pointed in the same direction) may have increased vigilance, where each individual is able to react to danger faster than if alone. But a nonpolarized school (or social aggregation) has more freedom of movement,

perhaps essential to corralling food and to social-sexual activity. We see dolphins vacillate between these two school types as they rest in polarized fashion and as they interact socially (including feeding and sex) in nonpolarized fashion (*74*) (Fig. 1).

There is a relationship between habitat and group size in odontocetes, with nearshore species usually occurring in groups of a dozen or so individuals, while offshore species tend to have larger groups. This reaches its extreme form in the small pelagic delphinids, where groups of several thousand are not uncommon, perhaps like herds of African plains antelopes (*75*).

Nearshore odontocetes generally have an

Fig. 1. Hawaiian spinner dolphins rest near the bottom of a 20-m-deep bay on the island of Lanai. Rest and social activity occur nearshore in daytime; apparently the shallow nearshore habitat helps dolphins avoid deep water shark predation. Feeding occurs offshore at night, because only then are mesopelagic fishes and squid, which are several hundred meters deep during the day but rise towards the surface at night, available to dolphins. Daytime activity occurs over sandy bottom, not rocks or coral, probably so that predators can be more easily detected over an unbroken visual field.

open group structure. That is, dolphins live in groups that number only one or a few dozen animals, but they often exchange members with groups that have adjacent or overlapping ranges. We therefore must consider a larger pool of animals as the population unit (*76, 77*). In bottlenose dolphins and pilot whales (*Globicephala* sp.), mothers and female offspring form long-term bonds; males account for the fluidity of associations. Group sizes can change daily, which allows for the most efficient group size in the face of changing environments. For example, bottlenose dolphins in Argentina had larger group sizes when they fed on anchovy in an apparent coordinated manner and smaller group sizes when they fed on nonschooling, rock- and bottom-dwelling prey in a more individual manner (*78*).

Dolphins in the open ocean live entirely within the large school envelope. The extent of fluidity in associations between members of subgroups within the school is unknown. Although the physical scale is different, fluidity between subgroups is probably similar to that of the nearshore environment. Males cannot be assured of paternity for their young, especially in a nonmonogamous society; we therefore expect that males and their offspring have no special bonds.

Adult male dolphins in captivity are often aggressive toward younger members, to the point of killing youngsters (*79*). Bateson studied the dominance relationships of a mixed group of spinner and spotted (*Stenella* sp.) dolphins at Sea Life Park, Hawaii, and noted a pecking order that consisted of nonlinear relationships in a matrix of associations. One dolphin could bite a second, that could bite a third, and the reverse did not occur. But dolphin number three could have similar access as number one to food or a mate.

A similar imperfect hierarchy exists in baboons (*76, 80*). Dominance-subservience relationships in nature probably result in a structuring of the society: the young is relegated to a particular physical space, possibly for its own safety. However, infanticide is common in a number of species and could occur in dolphin species in nature (*81*).

Toothed whales appear to cooperate with each other during foraging; for example, killer whales (*Orcinus orca*) herding pinnipeds and fish, and bottlenose dolphins herding fish (*82*). Dusky dolphins (*Lagenorhynchus obscurus*) of coastal water in Argentina herd anchovy (*Engraulis* sp.) into tight balls against the surface. When the fish become lethargic, possibly due to oxygen deprivation in the tight cluster, dolphins take turns feeding on the fish. Thus, temporary restraint from feeding appears to be practiced by all for the common goal of debilitating the prey. We do not know whether differential age and sex roles are played during herding, nor whether a feeding hierarchy exists (*83*).

Fish and sea bird predators that cooperate can more effectively take schooling prey than when alone or in very small groups (*84*). This can lead to the alternative strategies of forming groups when taking prey larger than oneself or prey that needs to be herded, and operating singly when foraging for small or nongrouping prey (*85*). Dusky dolphins of Argentina look for prey schools in many spread-out, small groups of dolphins, aggregate when prey is found, cooperatively herd and socialize as an aggregated group, and then split again. They thus display a fission-fusion society directly related to the taking of schooling prey (*83, 86*). However, dusky dolphins in deep water off New Zealand, where there are no appropriate schooling fishes, feed instead on mesopelagic, loosely or nonaggregated fishes and squid. The dolphins seldom split into small groups, but remain in the safety of the school of several hundred animals, feeding singly and apparently noncooperatively on nonschooling prey (*87*). In this case, the large school serves as the important stable social unit and probably has little direct relationship to foraging efficiency.

Dolphins copulate often and engage in much homosexual activity in the wild and in captivity. Copulation may be an important part of social bonding and of reinforcing hierarchical structure (*43*) (Fig. 2). Since we have difficulty discerning when a female dolphin is physiologically ready for conception, we cannot determine which copulations are for procreation and which are simply part of the social repertoire. We are, therefore, unsure of mating systems, but because of the fluidity of associations of males and the observation that females mate with more than one male, most delphinids are probably polygynous tending toward promiscuity. This is supported by long-term observations of a bottlenose dolphin population in Florida, where associations between animals of known sex and age have been monitored for about 18 years (*88*). The promiscuous mating hypothesis is being tested by measuring the relative reproductive success of different males; chromosome banding patterns of calves, mothers, and possible fathers are being correlated with mating. Further indirect evidence for promiscuity comes from the extremely large testes of delphinids which, like

the baleen whales discussed above, may indicate sperm competition (*89*).

Movements and Migrations

Large whales undergo extensive annual migrations that place them in winter on low-latitude mating and calving grounds and in summer on productive high-latitude feeding grounds. The specifics of movements by particular populations are being elucidated by photographic recognition of whales at different times of the year (*90*). Migrations of the large baleen whales and of male sperm whales tend to be on the order of about 3,000 to 5,000 km one way,

Fig. 2. Two dusky dolphins leap during social-sexual activity within a 12-member group in deep water off Kaikoura, New Zealand. Such dual leaps often end in brief copulations after the animals reenter the water. They may represent chases or may merely be a part of sexual play, but at all time they signify a large degree of social interactions in a dolphin group. Other acrobatic leaps such as somersaults and spins are performed by lone animals; they also indicate high excitement but are not necessarily sexual. Other surface activity may consist of tail slaps, flipper slaps, and low breaches with body side slaps. In dusky dolphin societies, these are often associated with various stages of cooperative surface herding of schooling fish prey.

although the gray whale moves up to 10,000 km between Mexico and the Chukchi Sea (*91*). Since humpback males of the same population sing the same song in any one breeding season, one can determine which whales spend at least part of the year together. Humpbacks that breed near the Revillagigedo Islands of Mexico, the Hawaiian Islands, and Japan may feed together along the Aleutian chain and elsewhere in high latitudes (*92*). Alternatively, enough animals travel between these widely separated mating grounds in order to carry the song between areas.

Migrations of the baleen and sperm whales may occur to allow feeding in productive waters and calving in warmer waters. But there seems no a priori reason for calving to necessitate warm waters; bowhead whales breed and calve near the ice edge and a few of the smaller toothed whales, which should be physically more susceptible to cold than the large baleen whales, can live near ice as well.

Migrations may be a holdover from when lower latitude waters near breeding grounds were more productive. Whales moved farther from those grounds as prey abundances shifted but kept their traditional mating grounds since those are important aggregation points (*93*). Although this is somewhat fanciful, without substantial support in all but a few cases, it may have some merit along with the thermodynamic hypothesis. Predation should be considered as another possible force toward migration. Killer whales are less abundant in lower latitudes, and movement toward such areas may be a useful ploy for protecting the newly born young. A similar force for the extensive migrations of wildebeest and other terrestrial mammals has been proposed (*94*).

Not all whales migrate. Some individual gray, fin, Bryde's, minke, and blue whales may stay year-round in particular, usually protected, nearshore areas. However, except for the often sedentary minke whale, it is not known to what degree such pockets of animals constitute reproducing units, or whether they are mainly composed of juveniles, postreproductive, or nonmating resting members of more extensive populations.

Toothed whales smaller than sperm whales often show seasonal movements, although they are not necessarily in a north-south direction. Thus, in the north Pacific, northern right whale dolphins (*Lissodelphis borealis*) move northward and shoreward in spring, southward and offshore in autumn; Pacific white-sided dolphins (*Lagenorhynchus obliquidens*) make onshore movements in autumn and offshore ones in spring; and harbor porpoises appear to be closer to shore in summer than in winter (*95*). Toothed whales follow seasonal shifts in food supplies since they do not have the fasting capability of the baleen whales. Movements have been related directly to prey concentrations (*96*) and to indirect indicators of prey, such as temperature variations (*97*), sea-surface chlorophyll concentrations (*98*), and such stationary features as depth and bottom type (*99*, *100*).

Dolphins nearshore tend to show diurnality in feeding and predator avoidance. Argentine dusky dolphins, Hawaiian spinner dolphins, and Bay of Fundy harbor porpoises show offshore nighttime and onshore daytime movements, albeit for different major prey with different behavior patterns (*101*). In contrast, Florida bottlenose dolphins are more sedentary, probably because of opportunistic feeding on several prey species, and the general presence of mullet [*Mugil cephalus* (*102*)]. The nearshore preference while resting or socializing of dolphins that feed farther from shore probably occurs to reduce predation by deepwater sharks and killer whales (*103*). In New Zealand and Argentina, I have on four separate occasions observed dusky dolphins retreat to very shallow water when killer whales approached, usually successfully hiding in the turbulent surf zone or behind shallow-water rocks. The situation appears generally similar to baboons that feed in open, exposed habitats but retreat to trees or cliffs to rest or when threatened by predators (*104*).

Short-distance movements to follow food supplies and avoid predation can be explained by direct experience and by learning favorable aspects of all parts of the range. As in terrestrial animals, dolphins and whales likely know their immediate milieu well. But longer distance movements and seasonal migrations, especially those away from shore, must pose substantial problems of orientation and navigation (which implies knowledge of a map). Our knowledge of how marine mammals obtain locational cues from their environment is incomplete. How, for example, do humpback whales that feed in southeast Alaska find the relatively small Revillagigedo or Hawaiian Islands in a huge expanse of sea? Sun orientation (105), cueing onto underwater topography by passive listening (99), echolocation of the bottom (106), detection of thermal structure (107), chemoreception by tasting of water masses and current systems (108), and magnetoreception (109) have all been proposed. Animals in general make use of a multiplicity of cues available to them, and it is possible that all cues above can come into play for various species and areas (110). Recent hypotheses on magnetic sensing, however, suggest that this sensory modality may be used by long-distance migrating cetaceans. It may help to explain the well-known incidence of mass live strandings by several migratory pelagic odontocete cetaceans, such as sperm whales, pilot whales, and Atlantic white-sided dolphins (*Lagenorhynchus acutus*) (111). However, magnetic orientation or navigation, even if cetaceans use it, is probably not the final answer for marine mammal strandings, nor is it likely to be the only method used in navigation. Further insight into migratory cues may come from long-distance satellite-tracking studies, used only experimentally on cetaceans to date but with promise of becoming a major tool as electronic telemetry packages are reduced in size (112).

Concluding Remarks

The study of cetaceans is of intrinsic interest as an exploration of the morphological and behavioral adaptations of mammals to a totally aquatic environment. Cetacean adaptations such as diving, communication, and sociality may help us to better understand aspects of mammalian morphology and behavior. Research on cetaceans has reached the "mainstream" of science. Whereas 10 years ago we were continually apologizing for knowing so very little, we now are on a more confident footing. We are pursuing lines of research in sophisticated and innovative manners: radiotelemetry for physiologic and behavioral data, chromosome and protein analyses for relationship information, and new methods of digitizing and analyzing sounds to describe communication and echolocation abilities.

Early information on morphology, life history, and behavior of baleen and sperm whales came from observations of the whaling industry during hunts and after death. For the smaller toothed whales, most of our knowledge of social interactions, use of sound, and physiology came from dolphins in captivity. Few researchers went out into the field and spent months (and years) observing wild animals, as is now common. Much of the older data helped build a solid foundation for our present knowledge; some of it has been discarded in light of modern findings under more natural conditions. This is less true for morphologic and life history information and more true for data on foraging behavior, social-sexual behavior, and society structure. In the latter realm, several primate researchers are presently comparing the social behavior of coastal bottlenose dolphins and the society structure of terrestrial social mammals, and this cross-fertilization between land and sea mammal research cannot fail to provide new insight into both systems (113).

Large whales are no longer hunted commercially, except for small "local" fisheries and an ongoing small hunt for minke and fin whales, thinly disguised as "scientific whaling" (114). For the time, at least, no large whale is in danger of extinction, although right whales are still scarce in the Northern Hemisphere. However, various species of pelagic dolphins have been killed in great numbers in the eastern tropical Pacific in the past 25 years, in association with purse-seining for tuna, and the killing continues unabated. Exact amounts of population declines are not known, although estimates place spinner dolphins, spotted dolphins, and common dolphins at only a fraction of their preexploitation numbers (*115*). Local populations of nearshore dolphins and whales are also affected by human harvesting and by habitat degradation (*116*). The Gulf of California harbor porpoise (*Phocoena sinus*) and the river dolphins, most notably the Chinese river dolphin (*Lipotes vexillifer*), are in imminent danger of extinction. Their plight is due to habitat destruction, including overfishing, and due to incidental kills from fish hook and net operations (*117*).

Organochlorine pollutants such as polychlorinated biphenyls and DDE accumulate in long-lived cetaceans, and their levels in whales and dolphins have been used to trace movement patterns (*118*), indicate reproductive parameters (*119*), and provide an overall index to the chronology of organochlorine toxin input into the ecosystem (*120*). As such, marine mammals may be used as tracers of pollutants, especially if the traced animals have a well-known, perhaps relatively confined, range. These pollutants are harmful when bioaccumulation is great, and there is a direct link between lowered testosterone levels and raised DDE concentrations in Dall's porpoises (*121*). Similar results in other species are likely to follow as appropriate measurements are made.

Many nearshore areas, and some of the world's major river systems, have become so ecologically degraded as to drive several cetacean species to near extinction. We should be warned by these more visible marine mammals, which come to the surface to breathe and thereby may be seen and counted, and realize that due to human action on the seas as well as on shore, other more ecologically encompassing disasters are taking place beneath the surface.

References and Notes

1. L. G. Barnes, D. Domning, C. Ray, *Mar. Mamm. Sci.* 1, 15 (1985).
2. J. H. Lipps and E. D. Mitchell, *Paleobiology* 2, 147 (1976); R. E. Fordyce, *Palaeogeogr. Palaeoclimat. Palaeoecol.* 21, 265 (1977); *ibid.* 31, 319 (1980).
3. P. F. Brodie, *Ecology* 56, 152 (1975).
4. The gill rakers of sharks, the tongue papillae and ridges on the roof of the mouth of some birds that feed on plankton, and the highly modified lobulated teeth of crabeater seals serve the same function.
5. W. N. Kellogg, *J. Acoust. Soc. Am.* 31, 106 (1959).
6. H. T. Andersen, *The Biology of Marine Mammals* (Academic Press, New York, 1969).
7. S. H. Ridgway, B. L. Scronce, J. Kanwisher, *Science* 166, 1651 (1969).
8. B. C. Heezen, *Deep-Sea Res.* 4, 105 (1957).
9. W. F. Dolphin, *Can. J. Zool.* 65, 354 (1986). Humpback whales diving deeper than 60 m may go into anaerobic metabolism and thus decrease the efficiency of energy uptake per unit time for deep dives. See also Scholander, *Hvalradets Skr.* 22, 1 (1940).
10. S. H. Ridgway, in *Diving in Animals and Man*, A. O. Brubakk, J. W. Kanwisher, G. Sundness, Eds. (Royal Norwegian Society of Science and Letters, Trondheim, Norway, 1986), pp. 33–62; _____ and R. Howard, *Science* 206, 1182 (1979); J. Kanwisher and S. H. Ridgway, *Sci. Am.* 248, 110 (June 1983).
11. R. L. Gentry and G. L. Kooyman, *Fur Seals: Maternal Strategies on Land and Sea* (Princeton Univ. Press, Princeton, NJ, 1986); B. J. LeBoeuf, D. P. Costa, A. C. Huntley, S. D. Feldkamp, *Can. J. Zool.* 66, 446 (1987).
12. *Marine Mammal Protection Act of 1972, Annual Report, 1981–1982* (National Marine Fisheries Service, Washington, DC, 1982).
13. G. A. J. Worthy and J. P. Hickie, *Am. Nat.* 128, 445 (1986).
14. M. D. Pagel and P. H. Harvey, *Evolution* 42, 948 (1988); L. Partridge and P. H. Harvey, *Science* 241, 1449 (1988).
15. E. Armstrong, *Neurosci. Lett.* 34, 101 (1982); *Science* 220, 1302 (1983); in *Size and Scaling in Primate Biology*, W. L. Jungers, Ed. (Plenum, New York, 1985), pp. 115–146.
16. E. D. Robin, *Perspect. Biol. Med.* 16, 369 (1973); M. A. Hofman, *Quart. Rev. Biol.* 58, 496 (1983).
17. R. K. Thomas, *Brain Behav. Ecol.* 17, 454 (1980).
18. H. J. Jerison, *Evolution of the Brain and Intelligence* (Academic Press, New York, 1973).
19. P. Pirlot and J. Pottier, *Rev. Can. Biol.* 36, 321 (1977); J. F. Eisenberg and D. E. Wilson, *Evolution* 32, 740 (1978). Fruit-eating bats may need larger brains to deal with complex temporal and spatial distribution of food. T. H. Clutton-Brock and P. H. Harvey [*J. Zool. London* 190, 309 (1980)] found no difference

in EQ between terrestrial and arboreal primates.

20. H. J. Jerison, in *Whales and Whaling,* S. Frost, Ed. (Auto. Govt. Publ. Serv., Canberra, 1978), vol. 2, pp. 159–197; F. G. Wood and W. E. Evans, in *Animal Sonar Systems,* R. G. Busnel and J. F. Fish, Eds. (Plenum, New York, 1980), pp. 381–425; S. H. Ridgway, in *Dolphin Cognition and Behavior: A Comparative Approach,* R. J. Schusterman, J. A. Thomas, F. G. Wood, Eds. (Erlbaum, Hillsdale, NJ, 1986), pp. 31–59.

21. B. Würsig in *Dolphin Cognition and Behavior: A Comparative Approach,* R. J. Schusterman, J. A. Thomas, F. G. Woods, Eds. (Erlbaum, Hillsdale, NJ, 1986), pp. 347–359; L. M. Herman, D. G. Richards, J. P. Wolz, *Cognition* 16, 129 (1984).

22. L. M. Herman and P. H. Forestell, *Neurosci. Biobehav. Rev.* 9, 667 (1985).

23. P. E. Nachtigall, in *Dolphin Cognition and Behavior: A Comparative Approach,* R. J. Schusterman, J. A. Thomas, F. G. Wood, Eds. (Erlbaum, Hillsdale, NJ, 1986), pp. 79–114; A. J. E. Cave, *J. Zool. London* 214, 307 (1988); A. D. G. Dral, *Z. Säugetierkunde* 53, 55 (1988).

24. W. A. Watkins, *Deep-Sea. Res.* 23, 175 (1976); C. W. Clark, *J. Acoust. Soc. Am.* 68, 508 (1980).

25. W. E. Schevill, in *Marine Bio-Acoustics,* W. N. Tavolga, Ed. (Pergamon, Oxford, 1964), pp. 307–316; R. Payne and D. Webb, *Ann. N.Y. Acad. Sci.* 188, 110 (1971); W. C. Cummings, P. O. Thompson, S. J. Ha, *U.S. Fish. Bull.* 84, 359 (1986).

26. W. A. Watkins, P. Tyack, K. E. Moore, J. E. Bird, *J. Acoust. Soc. Am.* 82, 1901 (1987).

27. G. Silber, *Can. J. Zool.* 64, 2075 (1986); C. G. D'Vincent, R. M. Nelson, R. E. Hanna, *Sci. Rep. Whales Res. Inst.* 36, 41 (1985).

28. R. S. Payne and S. McVay, *Science* 173, 587 (1971).

29. P. Tyack, *Behav. Ecol. Sociobiol.* 8, 105 (1981).

30. J. Darling, *Oceans* 17(2), 3 (1984); K. Chu and P. Harcourt, *Behav. Ecol. Sociobiol.* 19, 309 (1986).

31. W. L. Perryman, M. D. Scott, P. S. Hammond, *Rep. Int. Whal. Comm.* (special issue), 6, 482 (1984); J. C. Cubbage and J. Calambokidis, *Mar. Mamm. Sci.* 3, 179 (1987).

32. M. E. Dahlheim, H. Dean Fisher, J. D. Schempp, in *The Gray Whale, Eschrichtius robustus,* M. L. Jones, S. L. Swartz, S. Leatherwood, Eds. (Academic Press, New York, 1984), pp. 511–541; N. L. Crane, thesis, University of California, Santa Cruz (1986).

33. C. W. Clark and J. M. Clark, *Science* 207, 663 (1980); C. W. Clark, *Anim. Behav.* 30, 1060 (1982); _____ and J. H. Johnson, *Can. J. Zool.* 62, 1436 (1984); W. I. Ellison, C. W. Clark, and G. C. Bishop [*Rep. Int. Whal. Comm.* 37, 329 (1987)] suggest that certain bowhead whale sounds are used to navigate through arctic ice.

34. D. K. Ljungblad, P. O. Thompson, S. E. Moore, *J. Acoust. Soc. Am.* 71, 477 (1982); C. W. Clark, *The Living Bird* 7, 10 (1988).

35. A. N. Popper, in *Cetacean Behavior: Mechanisms and Function,* L. M. Herman, Ed. (Wiley, New York, 1980), pp. 1–52; C. W. Turl, R. H. Penner, and W. L. Au, [*J. Acoust. Soc. Am.* 82, 1487 (1987)] show that white whales, *Delphinapterus leucas,* can detect echolocation signals in greater noise environments than bottlenose dolphins.

36. K. S. Norris, in *Evolution and Environment,* E. T. Drake, Ed. (Yale Univ. Press, New Haven, CT, 1968), pp. 297–324. For a dissenting opinion that argues that sounds are produced by the larynx, see P. E. Purves and G. Pilleri, *Echolocation in Whales and Dolphins* (Academic Press, London, 1983).

37. J. C. Lilly and A. M. Miller, *Science* 133, 1689 (1961).

38. Basic descriptions of sperm whale codas by W. A. Watkins and W. E. Schevill, *Acoust. Soc. Am.* 62, 1485 (1977); W. E. Watkins, K. E. Moore, P. Tyack, *Cetology* 49, 1 (1985). Sperm whale sound production mechanism hypothesis by K. S. Norris and G. W. Harvey, in *Animal Orientation and Navigation,* S. R. Galler, K. Schmidt-Koenig, G. J. Jacobs, R. E. Belleville, Eds. (NASA Document SP-262, Government Printing Office, Washington, DC, 1972), pp. 397–417.

39. W. E. Schevill and B. Lawrence, *Breviora* 53, 1 (1956).

40. R. N. Turner and K. S. Norris, *J. Exp. Anal. Behav.* 9, 535 (1966).

41. V. M. Bel'kovich and A. B. Yablokov, *Nauka Zhizn',* 30, 61 (1963); K. S. Norris and B. Møhl, *Am. Nat.* 122, 85 (1983).

42. R. S. Mackay and J. Pegg, *Mar. Mamm. Sci.* 4, 356 (1988); K. Marten, K. S. Norris, P. W. B. Moore, K. A. Englund, in *Animal Sonar: Processes and Performance,* P. Nachtigall, Ed. (Plenum, New York, in press). It is not only intensity but also peak intensity duration that dictate energy transmitted, and the short pulses of dolphins may not reach enough energy to make prey stunning by sound possible.

43. K. S. Norris *et al., The Behavior of the Hawaiian Spinner Dolphin, Stenella longirostris* (NMFS, La Jolla, CA, 1985).

44. P. Tyack, *J. Acoust. Soc. Am.* **78**, 1892 (1985).

45. M. C. Caldwell and D. K. Caldwell, *Nature* **207**, 434 (1965).

46. P. Tyack, *Behav. Ecol. Sociobiol.* **18**, 251 (1986); J. K. B. Ford and H. D. Fisher, in *Communication and Behavior of Whales,* R. Payne, Ed. (Westview, Boulder, CO, 1983), pp. 129–162.

47. P. Marler and M. Tamura, *Science* **146**, 1483 (1964); L. F. Baptista, *Can. J. Zool.* **63**, 1741 (1985).

48. K. S. Norris and C. R. Schilt, *Ethol. Sociobiol.* **9**, 149 (1988); T. J. Thompson, H. E. Winn, P. J. Perkins, in *Behavior of Marine Animals,* vol. 3, *Cetaceans,* H. E. Winn and B. L. Olla, Eds. (Plenum, New York, 1979), pp. 403–432; C. W. Clark and J. M. Clark, *Science* **207**, 663 (1980); P. Tyack, *Behav. Ecol. Sociobiol.* **13**, 49 (1983).

49. M. K. Nerini, in *The Gray Whale, Eschrichtius robustus,* M. L. Jones, S. L. Swartz, S. Leatherwood, Eds. (Academic Press, New York, 1984), pp. 423–450; W. F. Dolphin, *Can. J. Zool.* **65**, 354 (1986); W. E. Evans, *Ann. N.Y. Acad. Sci.* **188**, 142 (1971).

50. R. L. Gentry and G. L. Kooyman, Eds., *Fur Seals: Maternal Strategies on Land and at Sea* (Princeton Univ. Press, Princeton, NJ, 1986); B. J. LeBoeuf, D. P. Costa, A. C. Huntley, S. D. Feldkamp, *Can. J. Zool.* **66**, 446 (1988).

51. J. Guerrero [thesis, San Jose State University, San Jose, CA (1989)] provides a thorough correlation between feeding behavior and prey type for a baleen whale by combining behavioral data with benthic and hyperbenthic prey availability information from a parallel study by J. S. Oliver, P. N. Slattery, M. A. Silberstein, E. F. O'Connor, *Can. J. Zool.* **62**, 41 (1984).

52. B. Würsig and M. Würsig, *Science* **198**, 755 (1977); *U.S. Fish. Bull.* **77**, 399 (1979).

53. R. S. Payne, *Behavior of Southern Right Whales* (Univ. of Chicago Press, Chicago, IL, in press).

54. B. Würsig *et al., U.S. Fish. Bull.* **83**, 357 (1985); W. M. Hamner, G. S. Stone, B. S. Obst, *ibid.* **86**, 143 (1988).

55. G. C. Ray and W. E. Schevill, *Mar. Fish. Rev.* **36**, 31 (1974); K. R. Johnson and C. H. Nelson, *Science* **225**, 1150 (1984); J. S. Oliver and P. N. Slattery, *Ecology* **66**, 1965 (1985).

56. B. Würsig, R. S. Wells, D. A. Croll, *Can. J. Zool.* **64**, 611 (1986).

57. E. M. Dorsey, *ibid.* **61**, 174 (1983).

58. A. Pivorunas, *Am. Sci.* **67**, 432 (1979); R. H. Lambertsen, *J. Mamm.* **64**, 76 (1983); L. S. Orton and P. F. Brodie, *Can. J. Zool.* **65**, 2898 (1987).

59. C. Jurasz and V. Jurasz, *Sci. Rep. Whales Res. Inst.* **31**, 69 (1979); S. H. W. Hain *et al., U.S. Fish. Bull.* **80**, 259 (1982).

60. P. J. Jarman, *Behaviour* **58**, 215 (1974); T. H. Clutton-Brock, S. D. Albon, P. H. Harvey, *Nature* **285**, 565 (1980); T. H. Clutton-Brock and P. H. Harvey, *J. Zool.* **183**, 1 (1977).

61. B. Tershy, thesis, San Jose State University, San Jose, CA, in preparation.

62. P. F. Brodie, *Ecology* **56**, 152 (1975).

63. S. I. Ohsumi, Y. Masaki, A. Kawamura, *Sci. Rep. Whales Res. Inst.* **22**, 75 (1970).

64. W. F. Perrin, R. L. Brownell, D. P. DeMaster, Eds., *Reproduction in Whales, Dolphins, and Porpoises* (Int. Whal. Comm. Spec. Publ. 6, Cambridge, U.K., 1984).

65. R. S. Payne, personal communication; R. Everitt and B. Krogman, *Arctic* **32**, 277 (1979).

66. G. J. Kenagy and S. C. Trombulak, *J. Mamm.* **67**, 1 (1986).

67. R. L. Brownell, Jr., and K. Ralls, *Rep. Int. Whal. Comm.* (special issue), **8**, 97 (1986).

68. P. Tyack and H. Whitehead, *Behaviour* **83**, 132 (1983); C. S. Baker and L. M. Herman, *Can. J. Zool.* **62**, 1922 (1984).

69. P. Tyack, *Behav. Ecol. Sociobiol.* **8**, 132 (1981).

70. J. Seger and R. L. Trivers, *Proceedings of 5th Biennial Conference on the Biology of Marine Mammals* (New England Aquarium, Boston, MA, 1983), vol. 5, p. 95.

71. For example, S. Kraus and S. Katona, *Humpback Whales in the Western North Atlantic: A Catalog of Identified Individuals* (College of the Atlantic Press, Bar Harbor, ME, 1977); B. Würsig and M. Würsig, *Science* **198**, 755 (1977).

72. G. Worthen, *NMFS Rep. LJ-81-02C* (NMFS-SWFC, La Jolla, CA, 1981); D. A. Duffield, S. H. Ridgway, L. H. Cornell, *Can. J. Zool.* **61**, 930 (1983); A. J. Jeffreys, V. Wilson, S. L. Thein, *Nature* **314**, 67 (1985); L. W. Andersen, *Can. J. Zool.* **66**, 1884 (1988); E. A. Mathews *et al., Mar. Mamm. Sci.* **4**, 1 (1988).

73. For example, W. D. Hamilton, *J. Theor. Biol.* **31**, 295 (1971).

74. K. S. Norris and T. D. Dohl, in *Cetacean Behavior: Mechanisms and Processes,* L. M. Herman, Ed. (Wiley, New York, 1980), pp. 211–

261; K. S. Norris and C. R. Schilt, *Ethol. Sociobiol.* **9**, 149 (1988).

75. D. Au and W. Perryman, *U.S. Fish. Bull.* **83**, 623 (1980).

76. C. K. Tayler and C. S. Saayman, *Ann. Cape Provincial Mus. Nat. Hist.* **9**, 11 (1972).

77. G. S. Saayman and C. K. Tayler, in *The Behavior of Marine Animals,* vol. 3, *Cetaceans,* H. E. Winn and B. L. Olla, Eds. (Plenum, New York, 1979), pp. 165–226; B. Würsig and M. Würsig, *U.S. Fish. Bull.* **77**, 399 (1979); K. S. Norris and T. D. Dohl, *ibid.* **77**, 821 (1980).

78. B. Würsig, *Sci. Am.* **240**, 136 (March 1979).

79. M. C. Tavolga and F. S. Essapian, *Zoologica* **42**, 11 (1957); K. S. Norris, in *Aggression and Defense: Neural Mechanisms and Social Patterns,* C. C. Clemente and D. B. Lindsley, Eds. (Univ. of Calif. Press, Berkeley, CA, 1967), pp. 225–241.

80. G. Bateson, in *Mind in the Waters,* J. McIntyre, Ed. (McClelland and Stewart, Toronto, Canada, 1974), pp. 146–165.

81. G. Hausfater and S. Blaffer-Hrdy, Eds., *Infanticide: Comparative and Evolutionary Perspectives* (Aldine, New York, 1984).

82. T. G. Smith, D. B. Siniff, R. Reichle, S. Stone, *Can. J. Zool.* **59**, 1185 (1981); J. C. Lopez and D. Lopez, *J. Mamm.* **66**, 181 (1985); W. F. Dolphin, *Can. Field-Nat.* **101**, 70 (1987); see B. Würsig, in (*21*).

83. B. Würsig and M. Würsig, *U.S. Fish. Bull.* **77**, 871 (1980).

84. P. F. Major, *Animal Behavior* **26**, 760 (1978); F. G. Cotmark, D. W. Winkler, M. Andersson, *Nature* **319**, 589 (1986).

85. R. J. Schmitt and S. W. Strand [*Copeia* **3**, 714 (1982)] describe alternative herding strategies in marine piscivorous yellowtails (*Serriola lalandei*), which herd schooling open-water fishes to shore and reef-hiding fishes into open water.

86. Human hunter-gatherers also display fission-fusion groups depending on food supply: C. Turnbull, *The Forest People* (Simon and Schuster, New York, 1962).

87. F. Cipriano, *Mauri Ora* **12**, 151 (1985).

88. R. S. Wells, M. D. Scott, A. B. Irvine, in *Current Mammalogy,* H. H. Genoways, Ed. (Plenum, New York, 1987), vol. 1, pp. 247–305.

89. S. H. Ridgway and R. F. Green, *Norsk Hvalfangst. Tid.* **1**, 1 (1967).

90. R. Payne *et al.*, in *Communication and Behavior of Whales,* R. S. Payne, Ed. (Westview, Boulder, CO, 1983), pp. 371–446; H. Whitehead and J. Gordon, in *Behaviour of Whales in Relation to Management,* G. P. Donovan, Ed. (Int. Whal. Comm. Spec. Publ. 8, Cambridge, U.K., 1986), pp. 149–165.

91. C. Scammon [*The Marine Mammals of the Northwestern Coast of North America* (J. H. Carmany and Co., San Francisco, CA, 1874)] was the first to describe the extent of the gray whale's movements.

92. J. D. Darling and D. J. McSweeney, *Can. J. Zool.* **63**, 308 (1985); J. D. Darling, personal communication.

93. P. G. H. Evans, *The Natural History of Whales and Dolphins* (Facts on File, New York, 1987), p. 207.

94. J. M. Fryxell, J. Greever, A. R. E. Sinclair, *Am. Nat.* **131**, 6 (1988); W. Booth, *Science* **242**, 868 (1988).

95. S. Leatherwood and W. A. Walker, in *Behavior of Marine Animals,* vol. 3, *Cetaceans,* H. E. Winn and B. S. Olla, Eds. (Plenum, New York, 1979), pp. 85–141; S. Leatherwood *et al., Sci. Rep. Whales Res. Inst.* **35**, 129.

96. D. E. Gaskin and A. P. Watson, *U.S. Fish. Bull.* **83**, 427 (1985); J. R. Heimlich-Boran, *Can. J. Zool.* **66**, 565 (1988).

97. D. E. Gaskin, *N. Z. J. Mar. and Freshwater Res.* **2**, 527 (1968); D. W. K. Au and W. L. Perryman, *U.S. Fish. Bull.* **83**, 623 (1985).

98. R. C. Smith *et al., Mar. Biol.* **91**, 385 (1986).

99. W. E. Evans, *Ann. N.Y. Acad. Sci.* **188**, 142 (1971).

100. G. J. D. Smith and D. E. Gaskin, *Ophelia* **22**, 1 (1983); C. A. Hui, *U.S. Fish. Bull.* **83**, 472 (1985); L. A. Selzer and P. M. Payne, *Mar. Mamm. Sci.* **4**, 141 (1988).

101. B. Würsig, F. Cipriano, M. Würsig, *Dolphin Societies: Methods of Study,* K. Pryor and K. S. Norris, Eds. (Univ. of Calif. Press, Berkeley, CA, in press).

102. A. B. Irvine, M. D. Scott, R. S. Wells, J. H. Kaufmann, *U.S. Fish. Bull.* **79**, 671 (1981).

103. K. S. Norris and T. P. Dohl, in *Cetacean Behavior: Mechanisms and Functions,* L. M. Herman, Ed. (Wiley, New York, 1980), pp. 211–261.

104. R. I. M. Dunbar and E. P. Dunbar, in *Primate Ecology: Problem-Oriented Field Studies,* R. W. Sussman, Ed. (Wiley, New York, 1979).

105. G. Pilleri and J. Knuckey, *Experientia* **24**, 394 (1968).

106. W. E. Evans, in *The Whale Problem: A Status Report,* W. E. Schevill, Ed. (Harvard Univ. Press, Cambridge, MA, 1974), pp. 385–394.

107. A. G. Tomilin, *Akademia Nauk.* **2**, 3 (1960).

108. L. Arvi and G. Pilleri, in *Investigations on Cetacea,* G. Pilleri, Ed. (University of Berne,

Switzerland, 1970).

109. M. Klinowska, *Aquatic Mamm.* 1, 27 (1985);
J. L. Kirschvink, A. E. Dizon, J. A. Westphal,
J. Exp. Biol. 120, 1 (1986).

110. H. E. Adler, in *Development and Evolution of Behavior: Essays in Memory of T. C. Schneirla*,
L. R. Aronson, E. Toback, D. S. Lehrman, J.
S. Rosenblatt, Eds. (Freeman, San Francisco,
CA, 1970), pp. 303–336.

111. M. Klinowska, in *Research on Dolphins*, M. M.
Bryden and R. Harrison, Eds. (Clarendon,
Oxford, U.K., 1986), pp. 401–432; J. R.
Geraci, *Oceanus* 21, 1 (1978).

112. S. Tanaka, *Nippon Suisan Gakkaishi* 53, 1327
(1987).

113. A summary is provided by W. Booth, *Science*
240, 1273 (1988).

114. W. F. Perrin, *Dolphins, Porpoises and Whales:
An Action Plan for the Conservation of Biological
Diversity, 1988–1992* (IUCN, Gland, Switzer-
land, 1988).

115. T. D. Smith, *U.S. Fish. Bull.* 81, 1 (1983); M.
A. Hall and S. D. Boyer, *Meeting Document #
SC/39/SM12* (Int. Whal. Comm., Cambridge,
U.K., 1987); T. Steiner, D. Phillips, M. J.
Palmer, *The Tragedy Continues: Killing of Dol-
phins by the Tuna Industry* (Whale Center,
Oakland, CA, 1988).

116. For example, S. M. Dawson and E. Slooten,
Rep. Int. Whal. Comm. (special issue), 9, 315
(1988); D. E. Gaskin *et al.*, Report to IUCN
(IUCN, Gland, Switzerland, 1987); T.
Kasuya, in *Marine Mammals and Fisheries*, J.
R. Beddington, R. J. H. Beverton, D. M.
Lavigne, Eds. (Allen and Unwin, London,
1985), pp. 253–275.

117. R. L. Brownell, Jr., *Endangered Species Techni-
cal Bull.* 13, 7 (1988); G. K. Silber, *J. Mamm.*
69, 430 (1988); R. Chen and Y. Hua, in
Biology and Conservation of the River Dolphins,
W. F. Perrin, R. L. Brownell, Jr., K. Zhou, J.
Liu, Eds. (IUCN Species Survival Commis-
sion Occasional Paper #5, IUCN, Gland,
Switzerland, 1989).

118. An. Subramanian, S. Tanabe, Y. Fujise, R.
Tatsukawa, *Mem. Nat. Inst. Polar Res.* 44, 167
(1986).

119. A. Aguilar and A. Borrell, *Mar. Env. Res.* 25,
195 (1988); An. Subramanian, S. Tanabe, R.
Tatsukama, *ibid.*, p. 161.

120. A. Aguilar, *Can. J. Fish. Aq. Sci.* 41, 840
(1984); S. Katona and H. Whitehead,
Oceanogr. Mar. Biol. Ann. Rev. 26, 553
(1988).

121. An. Subramanian *et al.*, *Mar. Poll. Bull.* 18,
643 (1987).

122. I thank C. Clark, J. Guerrero, T. Jefferson, W.
Koski, S. Ridgway, B. Tershy, P. Tyack, M.
Würsig, and two anonymous reviewers for
useful comments on this manuscript. Dolphin
research has been supported by the National
Geographic Society, the New York Zoological
Society, the National Institutes of Health, the
National Science Foundation, the Center for
Field Research, and the World Wide Fund for
Nature; whale research by the U.S. Minerals
Management Service. I thank faculty, staff,
and students of the Moss Landing Marine
Laboratories for providing a nurturing and
supportive environment for long-term field
studies.

17

Primates

Frederick A. King, Cathy J. Yarbrough, Daniel C. Anderson,
Thomas P. Gordon, Kenneth G. Gould

Our genetic relationship to apes and monkeys, the consequence of our relatively recent common evolutionary ancestry, underlies the similar structure, organization, development, and functioning of humans and primates. The term primates actually includes humans, although for purposes of simplicity and distinction we will not refer to them as primates in this review. There are almost 200 primate species which include the great apes (chimpanzees, gorillas, and orangutans), lesser apes (gibbons and siamangs), Old World monkeys, New World monkeys, and prosimians. Ninety-eight percent of human DNA can be found in the genes of chimpanzees, the most widely studied of the great apes. The New World monkeys of tropical America and Old World monkeys of Africa and Asia share 85% and 92% of their DNA, respectively, with humans (1). However, while the numerical differences in nucleic acids may be small, differences in genes and gene control mechanisms are the bases for the obvious phenotypic distinctions between humans and primates (2).

These similarities in the biological mechanisms of humans and primates underlie the value of these animals for research in a broad range of disciplines. Monkeys and, to a lesser extent, chimpanzees often serve as the final test system for the safety and efficacy of treatments, preventive agents, and vaccines developed in studies with other laboratory animals. In many basic and applied studies, primates are the only appropriate animal model when other species are not susceptible to the disease under study, or when primates possess the biological or behavioral characteristics needed to investigate the scientific question most effectively.

The most commonly used primates in research are the Old World species from Africa and Asia, which have been studied for many years and adapt to and reproduce well in captivity. The rhesus monkey, followed by the long-tailed macaque and the baboon, are the most frequently used Old World species, while the squirrel monkey is the most popular New World monkey from the American tropics in research (3). Approximately 30 primate species are currently used in biomedical and behavioral research.

Because of their aforementioned similarities to humans and demonstrated value to medical and scientific research, primates often

F. A. King is the director of the Yerkes Regional Primate Research Center, professor of anatomy and cell biology, adjunct professor of psychology, and associate dean of medicine, Emory University, Atlanta, Georgia 30322. C. J. Yarbrough is the administrative associate for special projects, Yerkes Regional Primate Research Center. D. C. Anderson is an associate research professor of pathobiology and immunology, Yerkes Regional Primate Research Center. T. P. Gordon is an assistant research professor of behavioral biology, coordinator of the Yerkes Center Field Station, and adjunct professor of psychology at Emory University. K. G. Gould is a research professor and chief of reproductive biology at Yerkes Regional Primate Research Center and professor of biology at Emory University. This chapter is reprinted from *Science* 240, 1475 (1988).

are an investigator's first choice as an animal model. However, of the approximately 20 million laboratory animals studied annually by U.S. scientists, only about 60,000 or 3.0% are primates, and most of these are used in multiple research programs of a noninvasive nature. Approximately 90% of laboratory animals in the United States are rodents (4). Several factors explain the relatively limited use of primates in research. These include cost, restricted supply, and (perhaps the most important factor) appropriateness. Primates are clearly not appropriate or necessary for all studies. Because of their relatively limited numbers, primates should be used judiciously. While federal regulations require that Institutional Animal Care and Use Committees insure humane treatments of research animals, many research institutions also evaluate proposals for scientific value as well, particularly in the case of primates, because they are an exceptionally valuable resource. Table 1 lists several of the questions considered for each proposal placed before the Yerkes Regional Primate Research Center's Committee in the interest of the humane and parsimonious use of primates as well as science of high quality.

Scientists are developing research procedures to avoid major surgery and retain reproductive capacity of the primates used in research. For example, scientists have developed techniques to reduce levels of sex steroid hormones without having to remove the ovaries or testes. A mini-pump is implanted subcutaneously to provide continuous release of a gonadotropin-releasing hormone agonist which decreases the sex steroids to near immature values. At the conclusion of the study when the pump is removed, sex steroids are secreted normally (5).

The limited supply of some primates can severely restrict or preclude their use in research. In the 1970s, the governments of India and Bangladesh placed an embargo on the export of rhesus monkeys which is still in effect. It can be argued that from a wildlife conservation viewpoint the embargo was a fortunate occurrence, although the cost of these monkeys has greatly increased as a consequence of having to breed them in this country. However, captive-born animals are typically healthier than their wild-born counterparts, which are often infected with parasites and other infectious agents. The embargo also prompted use of other primate species as research models and led to the development of captive breeding programs for more than 30 primate species in this country (6).

Successful breeding programs insure that primates used in terminal experiments, or which die as a result of natural causes, are replaced. It should be understood that most primates in research are not involved in terminal experiments, nor in studies that com-

Table 1. Criteria for evaluating primate research proposals, Yerkes Regional Primate Research Center, Emory University.

1. Are primates necessary for the proposed study, or can the work be as well conducted with another species or an alternative, nonanimal method?
2. Is the particular primate species selected appropriate biologically or behaviorally for the proposed investigation?
3. Is the study likely to contribute significantly to scientific knowledge or to human or animal health?
4. Is the investigator scientifically and technically qualified to conduct the study?
5. Will the study be conducted in a humane fashion, with proper consideration for the welfare of the animal, and in compliance with existing regulations?
6. If invasive procedures or others likely to produce pain or discomfort are proposed, are they essential to the study?
7. In proposals involving potentially painful procedures or surgery, has provision been made for elimination or minimization of pain or discomfort including proper anesthesia, analgesia, and round-the-clock post-operative care and surveillance?
8. If the research is replication of previous or other ongoing studies, is it justified and needed?
9. Is the number of animals to be used and the research design adequate to procure clearly interpretable results, but not excessive?
10. Will the study limit reproductive capacity in a way that will be injurious to breeding in the particular primate colony or to the species itself?

promise their use in subsequent investigations (6) or affect their breeding. Recently the National Institutes of Health (NIH) initiated a program to establish a stable supply of chimpanzees for essential biomedical and behavioral research without depleting the captive population of these animals. The program also will perpetuate the chimpanzee population through the birth and maintenance of physically healthy and behaviorally normal animals for future generations. While chimpanzees are among the laboratory animals of least number (the research population in the U.S. totals 1200), they are essential to (i) the development and final safety and efficacy testing of certain vaccines (7), (ii) the investigation of diseases which only they share with humans, as well as to (iii) behavioral studies such as those on cognition and language development that have already provided benefits for handicapped human children (8). In the following sections, the opportunities and in some cases limitations for primate research are highlighted. This article is not intended to be a comprehensive review of the many kinds of studies in which primates are used, or the accomplishments of research. Our intention is to convey an understanding of the range of biological, behavioral, and medical disciplines in which primates play an important or essential role.

Neuroscience, Neuropsychology, and Neurological Disorders

In the neurosciences and neuropsychology, primates are often the most appropriate subjects for research in the identification of mechanisms underlying human sensory and motor capacities, perception, learning, memory, reasoning, cognition, and cerebral dominance (9). Primates have played a major role in increasing our knowledge of the structure, organization, chemistry, and physiology of the human brain (10). The complexity of the primate brain and its similarity to that of humans makes primates excellent subjects for the study of motivational states such as hunger, thirst, sexual behavior, and emotion (11).

Certain primates are regarded as prime animal models for research on human vision because the morphology and responses of the eyes and central nervous visual pathways closely resemble those of the human (12). For example, the center of the retina, the macula, of both humans and primates has several morphologic features that allow a high degree of visual acuity, color discrimination, and complex central neural processing (13). Primates are the only species other than humans known to have true Schlemn's canals, highly developed trabecular meshworks, and scleral spurs. In both humans and primates there is a sophisticated functional relationship between the ciliary muscles and the eye's outflow mechanisms. These similarities make the primate an ideal model for research on trauma, ocular defense mechanisms, and the relationship between visual accommodation and aging (12).

The understanding and treatment of children's visual disorders benefit from basic and applied research with young primates. Recent behavioral and anatomical studies (14) suggest major similarities of visual development in humans and primates, with rough parity at birth. The rate of visual development occurs about four times faster in Old World monkeys than in humans, which makes them a convenient model. Studies (in monkeys) designed to understand human visual development have been strengthened by new techniques and methods, such as near retinoscopy and photorefraction; by modern variants of behavioral methods, such as preferential looking and operant techniques (15); and by extended wear contact lenses which can be worn by monkeys in vision studies (16).

Despite its many visual similarities to humans, the primate is not the only animal model for eye research. Because there is not a great deal of diversity among mammals in the basic biochemistry or gross morphological organization of tissue, the rabbit has been regarded as possessing the basic structure of the human visual system (12). However, it is virtually impossible to conduct experimental procedures on the rabbit eye without inducing an ocular irritative response that includes blood-aqueous barrier breakdown, pupillary miosis, increased intraocular pressure, and anterior uveal hyperemia (*12*). These major differences between rabbit and human eye are not a question in primates, and hence monkeys are more suitable subjects for research on aqueous humor dynamics and related conditions, such as glaucoma.

Primates provide significant advantages over rodents and other lower laboratory animals in studying other aspects of the nervous system as well. The organization of the primate central nervous system, especially the forebrain, is much more complex than in rodents and other laboratory species. This advanced development underlies the expression of higher order motor behaviors and their fine control, such as distal limb and digit movement, in primates. The large, convoluted cerebral cortex, with great areas devoted to associational activities, is almost certainly responsible for the primate's ability to learn highly complex cognitive tasks (*17*) beyond the capacity of species other than humans.

Because primates' brains share with humans a high degree of plasticity, their cognitive and social behaviors are heavily dependent on learning and the environment, as is the human behavioral repertoire. Hence in studies of the relationship of neural plasticity and the emergence of behaviors dependent upon social learning primates are often the subjects of choice. Primates are suitable models for studying the mechanisms underlying various neurological disorders, such as epilepsy (*18*), and for the development and testing of treatments. They have been used extensively in investigations of the genesis and spread of the epileptic seizure and the nature of its focus. In certain species of baboons there is a significant incidence of naturally occurring epilepsy. In these baboons seizures appear to have a genetic basis, since geographical distribution is a major determinant of incidence.

Much of the new knowledge about Parkinsonism and the resurgence of interest in research on this disorder are due to the use of the MPTP (N-methyl-4-phenyl-1,2,3,6-tetrahydropyridine)–treated primate as a model. The chronic administration of MPTP in monkeys produces the major neuropathologic features of Parkinson's disease. These features include the bilateral lesions of the substantial nigra and loss of striatal dopamine as well as motor abnormalities that are the hallmark of Parkinson's disease in humans (*19*). The MPTP primate model of Parkinsonism provides scientists with a system for studying the behavioral anomalies, the specific biochemical characteristics, and the pathological manifestations that occur in humans with the disorder. The primate Parkinson-like syndrome also provides a model for designing and testing therapeutic strategies including pharmacological agents and neural grafts (*20*).

In 1985, scientists reported that surgical implants of dopamine-producing fetal monkey brain cells survived and established cellular connections with preexisting tissue in the brains of rhesus monkeys with MPTP-induced Parkinson's-like disorder. Other behavioral and biochemical changes were noted. Within a year several other scientific teams announced replication of the results in several different primate species, strengthening the possibility that this treatment approach may prove safe and effective in patients with Parkinson's disease (*19*). Clinical studies soon were undertaken to evaluate adrenal medullary grafts to treat Parkinson's disease. However, without previous primate studies on adrenal tissue as a guide, the clinical studies have struggled to determine efficacy. Adrenal tissue implants are now being studied in primates (*21*).

The MPTP primate model also is an excellent example of the value of conducting research with primates. The rodent, the initial model for MPTP research, appears to be refractory to the neurotoxic effects of MPTP. If scientists had not proceeded to the monkey, research on Parkinson's disease would have been severely hampered. In addition, at the same time that the toxic effects of MPTP were being recognized in primates, MPTP was being studied as a potential antihypertensive drug (*22*). Thus potential human tragedies were averted by use of the primate as a model system.

Research on Alzheimer's disease does not have the advantage of a primate model that replicates the disorder as well as the MPTP-primate model of Parkinson's disease. Aged rhesus monkeys—those older than 23 years of age—have cognitive and memory deficits and develop senile plaques with neurites derived from cholinergic and other transmitter systems. These aged macaques, while they do not have Alzheimer's disease, nonetheless provide a system for studying the relations between age-associated cognitive deficits and pathological changes that occur in certain transmitter systems of primates and humans (20). Alzheimer's disease is primarily a disease of cortical derangement and cognitive impairment. The well-developed cerebral cortex of primates makes these animals extremely valuable for research on Alzheimer's disease. One of the neuropathological changes that has received intensive evaluation by scientists is the cholinergic deficit in the neocortex and the forebrain limbic system. Alzheimer's patients lose 75 to 85% of their neurons in the nucleus basalis of Meynert (nbM). The memory impairments of monkeys in which lesions have been made in the nbM and medial septum are similar to the memory impairments of humans with Alzheimer's disease. The same type of lesion in the rat, which does not have a well-defined nbM, does not produce the type of neuropathologic or behavioral impairments seen in primates (17).

There are other damaged areas in the brains of patients with Alzheimer's disease and an alteration in a number of neurotransmitters (23). It is unlikely that a single lesion can reproduce the spectrum of neuropathological changes. A better model of Alzheimer's disease may be produced by lesions in the nbM and locus ceruleus (17) in the lower brainstem.

Aging

In 1981, Dr. Edward Brandt, then Assistant Secretary for Health, said at a meeting marking the 20th anniversary of the NIH regional primate research centers: "As we begin to delve more deeply into the health problems of aging humans, we will be turning more frequently to nonhuman primates for clues" (24). Like humans, primates have life-spans of multiple decades. Captive rhesus monkeys can live into the fourth decade. Monkeys older than 20 years are the equivalent of humans 60 to 70 years old

(13). Chimpanzees can survive well into their sixth decade.

As they grow older, humans and primates experience many of the same age-related changes in anatomy, physiology, mental function, and behavior (25). Monkeys, like humans but unlike rodents, undergo a significant reduction in total brain weight between early and late adulthood. As macaque monkeys age, their brain weight reduces at a rate equal to or greater than the brain weight changes that occur in humans as they age. The losses in brain weight in monkeys and humans occurs in the forebrain, brainstem, and cerebellum, and at the same rate (13).

Rhesus monkeys are often used in studies of the aging visual system, because these monkeys have an ocular aging process that is similar to humans, in both the time course of development of presbyopia and in the frequency of the occurrence of senile cataracts and glaucoma (12). In aging humans and rhesus monkeys, reductions occur in visual acuity as measured by amplitude of electroretinographic and evoked potential responses to light stimuli. The vitreous body of the aging rhesus eye, like the human eye, undergoes gradual multifocal liquefaction which may increase risk of retinal detachment. Lens opacification also occurs in aging rhesus and humans. The degenerative changes of the maculae of macques older than 20 years of age are similar to the loss of pigmentation and vascular lesions of senile macular degeneration of humans. However, it must be noted that there are some differences in the way that aging changes the visual systems of primates and humans. From the research viewpoint, one of the most notable is that intraocular pressure does not increase with aging in primates as it does in humans (13).

As they age, primates also have lower levels of certain neurotransmitters. The capillary walls become thinner in the cerebral cortex with age, suggesting changes in the blood-brain barrier. In aging primates, the coronary vessels thicken, a common antecedent of atherosclerosis in humans that is often related to the behavioral deficits observed in both aged humans and primates (see below) (26).

Indeed, the similarities of the learning and

memory deficits that accompany aging in primates and humans make these animals excellent models for studying intellectual and social aspects of aging (27), correlating the cognitive/memory changes to neurochemical and neuropathological processes (22), and devising and attempting experimental interventions and treatments (27). Primates, as they grow older, have deficits in memory for recent, but not immediate, events, an increased sensitivity to interfering stimuli, and decreased behavioral flexibility (27). Other memory changes that occur in these animals as they age include slowed reversal learning, increased stereotyping of spontaneous behavior, increased reaction time, changes in sensory processing, and reduced long-term memory. Additional research is required for an understanding of these changes and possible ways of prevention and treatment in humans (13, 27).

The use of primates in research on aging is limited somewhat by the availability of macaques and other monkeys that can be characterized as living to an "old age." However, several of the NIH regional primate research centers, which have been documenting and studying the physical and behavioral changes that accompany aging in primates, have colonies which include rhesus monkeys over 20 years old. One center has a group of chimpanzees 45 to 55 years of age. This represents a considerable achievement, because in the wild these apes rarely live more than 35 years.

Reproduction

Primates provide scientists with the closest models of all aspects of reproduction. Similarities exist in the prenatal development of sexual phenotype, in the endocrine control of the reproductive cycle, and in complications of reproductive processes during mature life and during aging. Investigation of the endocrine mechanisms in macaques that underlie the determination of the sexual phenotype have shown the role of prenatal hormone exposure (28). Such experiments would be ethically unacceptable in the human, but have great significance with regard to the development of reproductive competence in children, with direct relation to the psychological problems associated with misdiagnosis of neonatal sex. Similarly, investigations using primates are shedding light on the role played by the elevated levels of testosterone, follicle-stimulating hormone (FSH), and luteinizing hormone (LH) detected in the early months of life. Reversible elimination of these elevated levels by administration of agonists of gonadotropin-releasing hormone (29) show that the perinatal elevations of steroid hormones may influence subsequent normal sexual development, at least in the male (30).

Primates pass through puberty during development, as do humans. The endocrine control of this change is still unclear, but components of the change (known as adrenarche), reflected in alteration of secretions by the adrenal cortex, are susceptible to study only in certain nonhuman primates, including the chimpanzee and possibly the baboon (31). Such investigations have been directed to determination of the possible existence of a novel pituitary hormone specifically associated with the maturation process. The endocrine control of spermatogenesis in the male and ovarian function in the female is similar to that of the human in those primate species which exhibit a menstrual, as opposed to an estrous, cycle. The hypothalamo-hypophyseal control of the menstrual cycle in primates and humans is quite different from that of the estrous cycle in other animal species (32). Using monoclonal antibodies against components of the sperm surface, and sophisticated analyses of sperm surface composition as the sperm traverses the epididymis, investigators are for the first time gaining an understanding of changes at the molecular level associated with the acquisition of fertilizing capacity by primate spermatozoa. This understanding will help in development of specific methods for control of male fertility at the cellular level (33).

Information on the pituitary control of the menstrual cycle, including the understanding that it is dependent upon intermittent hormonal stimuli, and that similarly intermittent stimuli exist in the male, has been derived from study of the primate (34). Understanding of this control mechanism has opened the way for identification of abnormalities in the pulsatile secretion of reproductive hormones in the human female which have been implicated in precocious and delayed puberty, anovulation, and inadequate luteal phase. The last two clinical problems have a direct bearing on individual fertility. Macaques have been used for comparison of potential hormonal treatments for endometriosis (*35*). The similarities between human and primate females extend beyond endocrine control and into the behavior patterns resulting from the presence of hormones associated with reproduction. These similarities, plus the difference in social influences and patterns between human and primate females, make it practical and valuable to investigate the effects on behavior of continuous alteration of normal circulating hormone levels, as occurs when women are exposed to contraceptive hormones (*36*).

The anatomical similarities between the species, together with the physiologically similar response to endocrine stimuli, frequently make the primate suitable for the evaluation and development of novel methods for fertility control as well as for investigations of pregnancy. Evaluation of, for example, Gossypol as a male contraceptive was conducted with primate males. Those studies permitted the determination of a site of action for the compound and provided evidence of potential side effects. Hormonal methods for male contraception are also being tested in primates. Analogs of gonadotropin-releasing hormone effectively block pituitary hormone support for spermatogenesis, resulting in reversible sterility. Use of primate models has also shown the presence of undesirable effects of certain of these analogs on secondary sexual characteristics, and further experiments are being undertaken in an effort to identify a mixture of hormones that will avoid the problems (*37*). The anatomical, physiological, and endocrinological similarity of human and primate females makes the latter suitable for evaluation of novel contraceptive modalities that cannot be tested initially in the human because of the possibility of unacceptable pregnancy rates during the test period. Such methods include

nonhormonal modification of cervical secretions to prevent sperm penetration (*38*).

An early breakthrough in the area of the immunology of pregnancy resulted from the use of primates—the identification of the Rh (Rhesus) factor. Today, species as diverse as the marmoset and baboon are being used to study the immunology of implantation and pregnancy, with the goal of developing contraceptive vaccines and determining the mechanism of early pregnancy loss associated with immunological deficits (*39*). The application of techniques for gamete recovery and in vitro fertilization, very similar to those used in the human, make it feasible to investigate the teratogenic effect of drugs used in the ovulatory and early pregnancy period as well as to evaluate further any risks associated with the process of in vitro maturation and fertilization of gametes. The value of primates in testing for teratogenicity was tragically illustrated by the demonstration of adverse effects of thalidomide on the human fetus in the 1950s and early 1960s. While thalidomide did not reveal its teratogenicity in rodents, its potential for causing fetal abnormalities was readily demonstrated by testing in primates (*40*).

Research in this area is not a one-way street in which animal research is beneficial solely to humans. For example, methods for monitoring fetal development with amniocentesis and ultrasound are important in the management of the captive primate colony. Techniques for collection of gametes developed for domestic species have been modified to permit collection and storage of sperm and ova from the primate. These techniques will have an important role to play in the conservation and management of endangered primate species, such as the gorilla, pygmy chimpanzee (bonobo), and golden lion tamarin (*41*).

As in the human, nonhuman primates exhibit considerable longevity. The chimpanzee and rhesus macaque, in particular, may be suitable models for study of the menopause and its associated clinical and physiological changes such as osteoporosis. The underlying mechanism for the endocrine changes that occur at the time of menopause in the human are not known, but similar alterations occur over a more prolonged period in the chimpanzee, thus permitting a more detailed longitudinal analysis of the changes as they occur (*42, 43*). Such studies complement those conducted in the clinical setting, and extend the evaluation of treatment methods into areas not ethically possible with humans.

Behavior

The primate order is composed of species which, while diverse with respect to some elements of social organization, ecology, and behavioral function, share a marked gregariousness, a large brain, and a relatively long period of development. As a consequence of these shared features, primate species exhibit prolonged dependence on others after birth. In addition, they display highly complex behavior which is modified by learning and by social, environmental, and experiential factors. Primates, therefore, provide excellent models for the study of behavioral phenomena ranging from basic social structure and function to analysis of cognitive capacity, including learning and communication. Of particular importance, primate studies provide the opportunity to examine the development, expression, and biological etiology of complex behaviors in a system that is not modified by cultural influences. While each species has its own behavioral proclivities, based on the interaction of biology and experience (*2*), primates in general develop socially and relate to each other and their environments in ways that are more similar to humans than to other animals (*44*).

Behavioral investigations of monkeys and apes have been conducted in a variety of settings including the natural habitat, captive social environments such as zoological parks, and the laboratory. Each study environment offers advantages and disadvantages which make it well suited for addressing certain research issues but not others. For example, studies in the field, which have expanded greatly in the past two decades, have focused on a wide range of species and provided a wealth of information on the basic structure of social systems, behavioral patterns, and the interplay between behavior and ecology (*45*). The disadvantages of field studies include the inability to control variables, the need to monitor complex environments, and, increasingly, the destruction of natural habitats (*44*). On the other hand, the typical laboratory environment, in which animals are housed either singly or in small groups, offers great control of extraneous variables and allows the investigation to focus on a particular behavior in a minimal social context. In such a setting the animal is far removed from the natural environment, and isolated from many variables that normally exert an influence. While the range of relevant research issues is consequently narrowed, the laboratory environment is highly suited for focus on a particular behavior, such as cognition, or to measuring or manipulating biological variables to examine their behavioral influences. An intermediate environment may be found in the captive social setting, such as the large colony which has been maintained on a Puerto Rican island for several decades and permits systematic study of a variety of behaviors in a population of established genetic identity and history (*46*). In addition, socially housed animals living in outdoor corrals may be trained to routine capture and handling, permitting the assessment of the influence of biological variables on behavior in a social context; this kind of setting combines some of the advantages of the field and the laboratory (*47*).

Studies of primates have focused on a wide variety of behavioral phenomena with emphasis on such areas as social organization, mother-infant interactions, aggression, growth and development, puberty, communication, learning, and memory (*48*). Some topics that address questions of fundamental theoretical interest have been extensively studied in multiple contexts. For example, reproductive behavior has been examined from such diverse viewpoints as mating strategies and sexual competition to the role of hormones in the regulation of sexual and reproductive behavior (*48*). Because of the similarities in human and primate endocrinology, primates provide scientists with an ideal system for detailing the role of hormones in behavior. Scientists interested in reproduction not only must select a primate species appropriate to the research question, but also a research environment suitable for the study, because the sexual interactions of some primates vary with the conditions under which the animals are studied (*49*). Rhesus macaques, for example, are seasonal breeders when exposed to an outdoor environment, but reproduce year-round when living indoors (*48*).

Another area of behavioral research is communication. Monkeys can transmit information through their vocalizations (*50*), and apes are able to learn and communicate with American Sign Language (*51*), plastic chips that represent English words (*52*), and a computer-operated keyboard of word-symbols that represent objects, actions, events, and people (*53*) (Fig. 1). In addition to defining the requisites for language acquisition in the human species, language studies with apes have enabled scientists to develop and evaluate language systems for handicapped humans (*8*).

Also directly applicable to understanding human behavioral problems are studies on the effects of infant separation from mothers, mother surrogates, and peers. The separation studies were based on observations of the protest-despair-detachment sequence of reactions by human children separated from their parents (*54*). Signs of distress in young monkeys separated from their mothers are, along with accompanying physiologic changes, very similar to the symptoms that occur in depressed humans (*55*). Numerous biological alterations occur in activity, sleep, heart rate, temperature, endocrine function, immune function, and monomine systems (*56*). Separation studies with primates allow investigators to distinguish between the neurobiological mechanisms that mediate and the social factors that modulate the separation response (*57*) and apply the findings to treatment of human situations.

It has been theorized that separation or object loss underlies the development of

Fig. 1. Studies of the ability of chimpanzees and other primates to learn and use a symbolic language are among the investigations of primate cognition. At the Language Research Center, a joint endeavor of the Yerkes Primate Research Center of Emory University and Georgia State University, bonobo chimpanzees (*Pan paniscus*) such as the one shown here have proven especially adept at learning the symbolic language, originally called Yerkish, on language boards.

depression in humans, although not all studies concur. Object loss or separation seems to lead to the development of a grief reaction but not to severe depression except in otherwise vulnerable individuals. Research to define the vulnerability in humans has been limited, and indeed, this is an area in which animal studies may define the conditions which, in humans, lead to depressive-type responses (58). Prospective studies with primates in which environmental, social, and individual physiologic factors are systematically manipulated will be instructive. Studies also can focus on individual predisposition (possibly genetically determined), which may interact synergistically with the other variables to cause a particular type of behavior and/or physiological response.

Recent research has strengthened the theory that genetic-environmental interactions are worthy of additional study (59) in both primates and humans. Recent studies of human toddlers and preschool children have revealed the existence of developmentally stable individual differences in personality or behavioral characteristics. Some of the children consistently showed fearfulness, anxiety, and cautious withdrawal in response to novelty or challenge, and also had individual differences in psychophysiological and adrenocorticoid reactivity that seem to closely parallel the results with monkeys (59, 60). A related and newly emergent area of research, psychoneuroimmunology, combines several disciplines to focus on the psychological events (including environmental, social, and behavioral stress) that can alter immune responses and consequently susceptibility to disease. Psychologists, neuroscientists, immunologists, and endocrinologists are utilizing the many similarities in behavioral and immunologic function of primates and humans to address questions relating to psychosocial influences on immune competence (61).

Atherosclerosis

Until the 1960s, the laboratory animals primarily used in cardiovascular research included rabbits, chickens, dogs, rats, and swine.

Today, however, primates are also used extensively for research on human atherosclerosis. One reason for this is that the plaques that develop in monkeys and humans are virtually identical in microscopic and biochemical appearance (62). These similarities provide scientists with several unique opportunities in atherosclerosis research. In some primate species, dietary manipulation can produce hyperlipidemia which resembles that seen in humans; the males of some primate species have a greater susceptibility to clinical disease, a phenomenon that occurs in some human populations (63). Primates are appropriate for studies to identify the mechanisms of atherosclerotic destruction at cellular and molecular levels (64); to determine the course and progress of atherosclerotic disease; and to define the relative influence of such risk factors as hypertension, diabetes, tobacco, alcohol, gender, fats and other nutrients, obesity, and heredity (65). Primates are also ideal models for studying the extent to which psychological and social phenomena influence the development of atherosclerotic lesions (66).

Old World monkeys are preferred to New World monkeys for studies of atherosclerosis. However, significant species differences exist among Old World monkeys that influence the selection of a primate model for atherosclerosis research. For example, stumptail macaques tend to exhibit obesity with increasing age, a characteristic that limits their use. Baboons have a high prevalence of naturally occurring arterial lesions, but are not prone to developing diet-induced atherosclerosis (63). In cynomolgus and rhesus monkeys, atherosclerosis also occurs naturally (67), but in the rhesus, lesions develop to a lesser degree (63). Among the New World primates only the squirrel monkey, which is susceptible to both spontaneous and diet-induced atherosclerosis, is frequently used in atherosclerosis research. In addition, genetic strains exist that are either hyper- or hyporesponsive to dietary cholesterol (63).

Stumptails fed atherogenic diets have a high prevalence of hypertension and high-fat diets generate coronary artery atherosclerosis in both male and female stumptail macaques. In contrast, rhesus and cynomolgus macaques more closely mirror the human species in that males are more prone to develop diet-induced disease than are females (*67*).

Rhesus monkeys also can undergo diet-induced regression of atherosclerosis to a much greater extent than do cynomolgus monkeys (*67*). Various therapies against atherosclerosis are being developed and tested in primates. Indeed, the effectiveness of calcium-blocking drugs in modifying the atherosclerotic process was demonstrated in monkeys after positive studies with rabbits (*68*). Research on high blood pressure can be conducted with monkeys, because the natural hormones that control blood pressure in humans and primates are identical. Monkeys also are models for research on the genetic transmission of high blood pressure (*69*).

Infectious Diseases and Vaccine Development

During the past 25 years, primates have been extensively studied as models for a variety of naturally occurring and experimentally induced bacterial, viral, parasitic, and fungal infections that cause disease in humans (*7, 70, 71*). In many instances, primates are the only animal species in which these diseases occur spontaneously or can be experimentally induced (*70*).

The understanding and control of infectious diseases such as poliomyelitis, yellow fever, measles, and rubella depended on research with primates (*7*). Unusual infectious agents such as the etiologic agent of kuru, a slow virus, have been identified through primate research (*72*). Scientists did not have an animal model for hepatitis B until it was discovered that the chimpanzee could be infected with the virus. Subsequently chimpanzees and marmosets were used in the development of a vaccine against this disease (*73*).

For some infectious diseases, species other than primates are more suitable research models. The primates are truly important in infectious disease studies that are unique to primates and humans, or in which the immunological responses under consideration closely resemble those seen in humans.

Probably no better example of the need for animal models for the intervention of human disease can be found than in the current epidemic of the acquired immunodeficiency syndrome (AIDS). Animal models are needed to test antiviral therapy and to aid in vaccine development. Currently, only chimpanzees can be infected with strains of the human immunodeficiency virus (HIV) and approximately 90 chimpanzees have been infected to date. Of these, none has yet developed an AIDS-like disease. This is not surprising since only a small number of chimpanzees have been infected for more than 4 years and the incubation period for AIDS may extend up to 10 years in humans. It is beyond the scope of this article to provide an exhaustive review due to the extremely large amount of work and the rapid pace of new developments in this area. Recent reviews of animal models of retroviral infections and acquired immunodeficiency diseases have included various animal and nonhuman primate infections (*74, 75*).

Of considerable interest are the occurrence of simian immunodeficiency viruses (SIV), which are closely related morphologically and antigenically to HIV and other lentiviruses. The original SIV viruses (initially termed STLV-III) were isolated from captive rhesus macaques. Preliminary studies revealed that the isolates induced an immunosuppressive disease in juvenile rhesus monkeys in a relatively short time, with many characteristics similar to those of AIDS in humans (*75*). Since that time, similar STLV-III–like viruses were isolated from African Green monkeys and sooty mangabey monkeys (Fig. 2). The viruses appear to cause little or no pathogenicity in their natural hosts: however, the sooty mangabey isolate has been shown to induce immunosuppressive disease in macaques (*76*). Monkeys infected with SIV provide models for development and testing of drugs and vaccines and for studies of the influences of various suspected cofactors in the pathogenesis of AIDS. Even though the SIV studies in monkeys cannot at this time replace safety and efficacy testing in chimpanzees, they will greatly complement studies for which suffi-

Fig. 2. Sooty mangabey monkeys at the Yerkes Primate Research Center of Emory University.

Cancer

Primates are susceptible to many oncogenic viruses: type C viruses are related to cancers in owl monkeys and gibbons (77), and foamy agents (a subgroup of the leukoviruses) appear to be responsible for lymphomas in rhesus macaques (78). A variety of herpes viruses are oncogenic in cotton top marmosets, owl monkeys, and other primates (79). Further studies in primates may help to establish the viral etiology of certain tumors. Primates are rarely used in studies of chemical carcinogenesis, because of the cost and limited supply of primates, the time period needed to complete life studies (80), and the primates' extended latency period of tumor development (which can be as long as 10 to 12 years). Although it mirrors the human situation, the long latency period restricts the use of primates for rapid testing of chemicals (81). While primates cannot replace rodents for screening chemical carcinogens, the animals can be used to screen chemicals for which data from studies of rodents are ambiguous and conflicting, or chemicals to which large numbers of humans are exposed (80).

Primates have been used in the development and testing of monoclonal antibodies against certain forms of cancer, primarily those that arise from solid tumors. Because of the immunological similarities between humans and primates, particularly chimpanzees, monoclonal antibodies derived from primate material should be highly specific when used in laboratory diagnostic assays. For example, human melanoma and leukemia-associated antigens have been defined by antisera from primates (82). The immunological similarities should increase the safety of monoclonal antibodies when used in human treatment.

Because of their immunological and physiological similarities to humans, primates are used in evaluating certain cancer treatments, such as the effectiveness of recombinant granulocyte-macrophage colony-stimulating factor (GM-CSF) on hematopoietic reconstruction after autologous bone marrow transplantation (*83*).

Limitations and Constraints in the Use of Primates

Systematic scientific study of primates is constrained by a number of factors including progressive destruction of the natural habitat of many species, the high cost of producing and housing monkeys and apes, problems attendant to the large size and aggressive disposition of some primates, the potential for disease transmission between humans and primates, and the objection of anti–animal research organizations which target dogs, cats, and primates in particular.

Expense is a major limitation in the conduct of primate research. For example, a rhesus monkey costs from $600 to $2000 to purchase, and several hundred dollars annually to maintain. Consequently, studies requiring large numbers of primates are rarely feasible. In addition to cost factors, the supply of many species of primates is limited. Indeed, most of the commonly used monkeys and apes are obtained from domestic breeding programs. Nonetheless, the low reproduction rates (as compared to rodents, for example) and long developmental period limit supply and elevate costs. Space for housing and care of the animals is also a consideration, with substantial space and expensive materials required.

The size and natural aggressiveness of some primate species pose potential problems for researchers that require special precautions in the handling, management, and care of primates and in training personnel. In particular, investigators must employ procedures designed to eliminate stress in the animal and protect personnel and animals from injury and disease. For example, herpes B, a neurotropic virus, occurs naturally in macaque monkeys; all macaques should be presumed to carry the virus, which may cause fatal infections in humans and certain other primate species (*84*). Similarly, the Marburg virus was first discovered because it produced a fatal hemorrhagic fever in laboratory workers and personnel who were in contact with infected African Green monkeys (*85*). While there is presently no evidence that the SIVs can be transmitted to or cause illness in humans, research and animal care personnel must be especially cautious about handling primates known to be infected with or likely to harbor SIV. The incidence of tuberculosis in captive primate colonies has decreased in recent years because few primates are imported from the wild and because surveillance programs have been instituted in both animal colonies and research personnel. Surveillance eliminates infected animals and decreases the spread of disease in domestic colonies (*86*).

Another impediment to primate research is the animal rights movement, which holds the view that research with animals, particularly primates, is unnecessary, inhumane, and unethical. The campaign against primate research is actually based on the scientists' rationale for studying primates: the biological and behavioral similarities of primates to humans. As an example, antiresearch groups in 1983 singled out the NIH regional primate research centers as a target, and opposition to primate research has remained as a major focus of the animal rights agenda. One consequence of this activity has been new legislation, including a directive to the Department of Agriculture to develop standards for physical environments that promote the "psychological well-being" of laboratory primates. The U.S. Department of Agriculture's interpretation of that stipulation has been slow to be articulated chiefly because neither the regulators nor the scientists who study primates can define a concept as vague as "psychological well-being." Ideally, the interpretation will ultimately be based on scientific studies, some of which are under way, and on factors that influence the development and expression of normal behavior in laboratory primates. Such studies, by qualified scientists, are needed before primate research laboratories are required by federal regulation to build costly new facilities and adopt new labor-intensive husbandry routines (*87*) that may turn out to be irrelevant to the "psychological well-being" of primates. This illustrates how well intentioned legislation based on emotional rather than empirical arguments can negatively impact scientific inquiry, with as yet, unmeasured consequences for behavioral and biomedical research directed toward the improvement of both human and animal health.

References and Notes

1. C. G. Sibley and J. E. Ahlguist, *J. Mol. Evol.* **26**, 99 (1987).
2. D. M. Rumbaugh, in *G. Stanley Hall Lecture Series*, A. M. Rogers and C. J. Scheirer, Eds. (American Psychological Association, Washington, DC, 1985), vol. 5, pp. 7–53.
3. T. Ruch, in *Biology and Pathology of Monkeys; Studies of Human Diseases in Experiments on Monkeys, Proceedings of the International Symposium in Sukhumi, 17 to 22 October 1966* (Tbilisi, Moscow, 1966), p. 104; R. A. Mittermeier and A. F. Coimbra-Filho, in *Reproduction in New World Primates*, J. P. Hearn, Ed. (MTP, Lancaster, U.K., 1982), pp. 3–37; W. I. Gay, in *Primates, the Road to Self-Sustaining Populations*, K. Benirschke, Ed. (Springer-Verlag, New York, 1986), pp. 513–520.
4. U.S. Congress, Office of Technology Assessment, *Alternatives to Animal Research Testing and Education* (Government Printing Office, Washington, DC, 1986).
5. D. R. Mann, D. C. Collins, M. M. Smith, M. J. Kessler, K. G. Gould, *J. Clin. Endocrinol. Metab.* **63**, 1277 (1986).
6. W. R. Dukelow, in *Nonhuman Primate Models for Human Diseases*, W. R. Dukelow, Ed. (CRC, Boca Raton, FL, 1983), preface; D. O. Johnsen and L. A. Whitehair, in *Primates, the Road to Self-Sustaining Populations*, K. Benirschke, Ed. (Springer-Verlag, New York, 1986), pp. 499–511.
7. F. A. King and C. J. Yarbrough, *Physiologist* **28**, 75 (1985).
8. M. A. Romski, R. A. White, C. E. Millen, D. M. Rumbaugh, *Psychol. Rec.* **34**, 39 (1984).
9. F. A. King, testimony presented before the Committee on the Use of Laboratory Animals in Biomedical and Behavior Research of the Commission on Life Sciences and Institute of Medicine, 11 February 1986, Washington, DC; J. H. R. Maunsell and W. T. Newsome, *Annu. Rev. Neurosci.* **10**, 363 (1987); J. Moran and R. Desimone, *Science* **229**, 782 (1985); H. Spitzer, R. Desimone, J. Moran, *ibid.* **240**, 338

(1988).
10. F. A. King, in *Using Psychological Science, Making the Public Case,* F. Farely and C. Null, Eds. (Federation of Behavioral, Psychological, and Cognitive Sciences, Washington, DC, 1987), pp 5–12.
11. E. Satinoff and P. Teitelbaum, in *Handbook of Behavioral Neurobiology: Motivation*, F. A. King, Ed. (Plenum, New York, 1983), p. 6.
12. L. Z. Bito, *Exp. Eye Res.* **39**, 807 (1984).
13. D. M. Bowden and D. D. Williams, *Adv. Vet. Sci. Comp. Med.* **28**, 305 (1984).
14. T. N. Wiesel, *Nature* **299**, 583 (1982).
15. R. G. Boothe, V. Dobson, D. Y. Teller, *Annu. Rev. Neurosci.* **8**, 495 (1985).
16. J. A. Gammon, R. G. Boothe, C. V. Chandler, M. Tigges, J. R. Wilson, *Invest. Ophthalmol. Vis. Sci.* **26**, 1636 (1985); R. G. Boothe, L. Kiorpes, M. R. Carlson, *J. Pediatr. Ophthalmol. Strabismus* **22**, 206 (1985).
17. J. H. Kordower and D. M. Gash, *Integr. Psychiatry* **4**, 64 (1986).
18. E. D. Louis, P. D. Williamson, T. D. Darcey, *Yale J. Biol. Med.* **60**, 255 (1987).
19. R. S. Burns *et al., Proc. Natl. Acad. Sci. U.S.A.* **80**, 4546 (1983); R. A. E. Bakay, D. L. Barrow, M. S., Fiandaca, A. Schiff, D. C. Collins, *Ann. N.Y. Acad. Sci.* **495**, 623 (1987); R. A. E. Bakay, D. Barrow, A. Schiff, M. Fiandaca, *Congr. Neurol. Surg. Sci. Program* **35**, 210 (1985); D. E. Redmond, Jr., *et al., Lancet* i, 1125 (1986).
20. L. C. Cork, C. A. Kitt, R. G. Struble, J. W. Griffin, D. L. Price, *Prog. Clin. Biol. Res.* **229**, 241 (1987).
21. R. Lewin, *Science* **240**, 390 (1988).
22. J. W. Langston, *Integr. Psychiatry* **78**, 64 (1986).
23. J. Hardy *et al., Neurochem. Int.* **7**, 545 (1985).
24. E. N. Brandt, Jr., speech for the 20th anniversary of the regional primate research centers program, 19 November 1981, Atlanta, GA.
25. J. M. Ordy, in *Neurobiology of Aging*, J. M. Ordy, Ed. (Plenum, New York, 1975), vol. 16, pp. 575–594.
26. D. Bowden, Ed., *Aging in Nonhuman Primates* (Van Nostrand Reinhold, New York, 1979).
27. R. T. Bartus, R. L. Dean, B. Beer, *Psychopharmacol. Bull.* **19**, 168 (1983); L. D. Byrd *et al., Primate Rep.* **14**, 132 (1986).
28. R. W. Goy and J. A. Resko, *Recent Prog. Horm. Res.* **28**, 707 (1972).
29. Gonadotropin-releasing hormone agonists are molecules modified from the basic decapeptide gonadotropin-releasing hormone, which is a hypothalamic hormone responsible for stimulation of release of follicle-stimulating hormone and luteinizing hormone from the

anterior pituitary gland.

30. D. R. Mann *et al., J. Clin. Endrocrinol.* **59**, 207 (1984); D. R. Mann, K. Wallen, D. C. Collins, K. G. Gould, unpublished data.

31. G. B. Cutler *et al., Endocrinology* **103**, 2112 (1978).

32. D. L. Healy and G. D. Hodgen, *Excerpta Med. Int. Congr. Ser.* **658**, 117 (1985); R. Johnson, *Res. Resour. Rep.* **4** (no. 6), (1986).

33. L. G. Young, B. T. Hinton, K. G. Gould, *Biol. Reprod.* **32**, 399 (1985).

34. E. Knobil, *Recent Prog. Horm. Res.* **30**, 1 (1974).

35. D. R. Mann, D. C. Collins, M. M. Smith, M. J. Kessler, K. G. Gould, *J. Clin. Endocrinol. Metab.* **63**, 1277 (1986).

36. R. D. Nadler, J. F. Dahl, D. C. Collins, K. G. Gould, M. E. Wilson, in *Comparative Reproduction in Mammals and Man* (National Museums of Kenya, Nairobi, 1989) pp. 30-33.

37. F. B. Akhtar, G. R. Marshall, E. Nieschlag, *Int. J. Androl.* **6**, 461 (1983); D. R. Mann, M. M. Smith, K. G. Gould, D. C. Collins, *Fertil. Steril.* **43**, 115 (1985).

38. K. G. Gould and A. H. Ansari, *Am. J. Obstet. Gynecol.* **145**, 92 (1983).

39. J. P. Hearn, *J. Reprod. Fertil.* **76**, 809 (1986); G. D. Hodgen, in *Animal Models for Research on Contraception and Fertility,* N. J. H. Alexander, Ed. (Harper and Row, New York, 1985), pp. 425–436.

40. C. S. Delahunt and L. J. Lassen, *Science* **146**, 1300 (1964); A. G. Hendrickx, L. R. Axelrod, L. D. Claborn, *Nature* **201**, 958 (1966); G. G. Hendrickx and M. A. Cukierski, *Prog. Clin. Biol. Res.* **235**, 73 (1987).

41. K. G. Gould and D. E. Martin, in *Primates, the Road to Self-Sustaining Populations,* K. Benirschke, Ed. (Springer-Verlag, New York, 1986), pp. 425–443.

42. W. V. Holt, *ibid.,* pp. 413–424.

43. K. G. Gould, M. Flint, C. E. Graham, *Maturitas* **3**, 157 (1981).

44. D. L. Cheney, R. M. Seyfarth, B. B. Smuts, R. W. Wrangham, in *Primate Societies*, B. B. Smuts *et al.*, Eds. (Univ. of Chicago Press, Chicago, 1987), pp. 491–497.

45. J. van-Lawick Goodall in *Animal Behavior Monographs*, J. M. Cullen and C. G. Beers, Eds. (Baillier, Tindall and Cassell, London, 1968), vol. 1, part 3, pp. 161–311; S. A. Altmann, *Social Communication Among Primates* (Univ. of Chicago Press, Chicago, 1967); I. Devore, Ed., *Primate Behavior* (Rinehart and Winston, Boulder, CO, 1965).

46. R. G. Rawlins and M. J. Kessler, Eds., *The Cayo Santiago Macaques: History, Behavior and Biology* (State University of New York, Albany, 1986).

47. M. L. Walker, T. P. Gordon, M. E. Wilson, *J. Med. Primatol.* **11**, 291 (1982).

48. R. D. Nadler, J. G. Herndon, J. Wallis, in *Comparative Primate Biology*, vol. 2A, *Behavior, Conservation and Ecology*, G. Mitchell and J. Irwin, Eds. (Liss, New York, 1986), pp. 363–407; M. E. Wilson, *Endocrinology*, **119**, 666 (1986); I. S. Bernstein, *Behav. Brain Sci.* **4**, 419 (1981); I. S. Bernstein and T. P. Gordon, *Am. Sci.* **63**, 304 (1974); T. P. Gordon, *Am. Zool.* **21**, 185 (1981); K. Wallen and L. A. Winston, *Physiol. Behav.* **32**, 629 (1984); I. S. Bernstein and L. E. Williams, in *Comparative Primate Biology*, vol. 2A, *Behavior, Conservation and Ecology*, G. Mitchell and J. Irwin, Eds. (Liss, New York, 1986), pp. 195–213; H. D. Steklis and M. J. Raleigh, *Neurobiology of Social Communication in Primates* (Academic Press, New York, 1980); R. A. Hinde and M. J. A. Simpson, *Ciba Found. Symp. 33* (Elsevier/North-Holland, New York, 1975), pp. 39–67.

49. K. Wallen, *Science* **217**, 375 (1982).

50. R. M. Seyfarth, D. L. Cheney, P. Marler, *ibid.* **210**, 801 (1980).

51. R. A. Gardner and B. T. Gardner, *ibid.* **165**, 664 (1969).

52. D. Premack, *Intelligence in Ape Language*, in *Ape and Man* (Erlbaum, Hillsdale, NJ, 1976).

53. E. S. Savage-Rumbaugh, in *Conditioned Response to Symbol*, H. S. Terrace, Ed. (Columbia Univ. Press, New York, 1986), p. 433; D. M. Rumbaugh, T. V. Gill, von E. C. Glaserfeld, *Science* **182**, 731 (1973); E. S. Savage-Rumbaugh, D. Rumbaugh, S. T. Smith, J. Lawson, *ibid.* **210**, 922 (1980).

54. J. Bowlby, *Psychoanal. Study Child* **15**, 9 (1960); J. Bowlby, *Int. J. Psycho-Anal.* **41**, 8 (1960).

55. L. A. Rosenblum and G. S. Paully, *Psychiatr. Clin. N. Am.* **10**, 437 (1987).

56. N. H. Kalin and M. Carnes, *Progr. Neuro-Psychopharmacol. Biol. Psychiatry* **8**, 459 (1984).

57. M. Reite and R. A. Short, *Arch. Gen. Psychiatry*, **35**, 1247 (1978); _____, I. C. Kaufman, A. J. Stynes, J. D. Pauley, *Biol. Psychiatry* **13**, 91 (1978); M. R. Gunnar, C. A. Gonzalez, B. L. Goodlin, S. Levine, *Psychoneuroendocrinology* **6**, 65 (1981); M. L. Laudenslayer, M. Reite, R. J. Harbeck, *Behav. Neural Biol.* **36**, 408 (1982).

58. W. T. McKinney, E. C. Moran, G. W. Kraemer, in *Frontiers of Clinical Neuroscience*, R. M. Post and J. C. Ballenger, Eds. (Williams and Wilkins, Baltimore, 1984), pp. 393–406.

59. S. J. Suomi, in *Anxiety-Like Disorders in Young Nonhuman Primates, Anxiety Disorders of Children*, R. Gittelman, Ed. (Guilford, New York, 1986), pp. 1–23.

60. J. Kagan, in *Measuring Emotions in Infants and*

Children, C. E. Izard, Ed. (Cambridge Univ. Press, New York, 1982), pp. 38–66.

61. R. Ader, *Psychoneuroimmunology* (Academic Press, New York, 1981).

62. T. B. Clarkson, in *The Use of Nonhuman Primates in Cardiovascular Diseases*, S. S. Kalter, Ed. (Univ. of Texas Press, Austin, 1980), p. 452; T. B. Clarkson *et al.*, *Ann. N.Y. Acad. Sci.* 162, 103 (1969); C. B. Taylor, G. E. Cox, P. Manolo-Estrella, J. Southworth, *Arch. Pathol.* 74, 16 (1962); C. B. Taylor, P. Manolo-Estrella, G. E. Cox, *ibid.* 76, 239 (1963); C. B. Taylor *et al.*, *ibid.*, p. 404.

63. M. P. Jokinen, T. B. Clarkson, R. W. Prichard, *Exp. Mol. Pathol.* 42, 1 (1985).

64. R. Ross, *J. Am. Geriatr. Soc.* 31, 213, 1983).

65. R. W. Wissler, in *The Use of Nonhuman Primates in Cardiovascular Diseases*, S. S. Kalter, Ed. (Univ. of Texas Press, Austin, 1980), pp. 15–32.

66. T. B. Clarkson, K. W. Weingard, J. R. Kaplan, M. R. Adams, *Circulation* 76 (Suppl. I), 120 (1987).

67. T. B. Clarkson, M. S. Anthony, R. W. Prichard in *The Comparative Pathology of Nonhuman Primate Atherosclerosis, Regression of Atherosclerotic Lesions*, M. R. Malinow and V. H. Blatow, Eds. (Plenum, New York, 1984), pp. 61–78.

68. W. W. Parmley, S. Blumlein, R. Sievers, *Am. J. Cardiol.* 55, 165B (1985).

69. W. R. Hendee, J. M. Loeb, M. R. Schwarz, S. J. Smith, *Use of Animals in Biomedical Research, the Challenge and Responsibility*, AMA White Paper (American Medical Association, Chicago, 1988).

70. H. M. McClure, *Vet. Sci. Comp. Med.* 28, 267 (1984).

71. K. F. Soike, S. R. S. Rangan, P. J. Gerone, *ibid.*, p. 151.

72. D. C. Gajdusek, C. J. Gibbs, Jr., M. Alpers, *Nature* 209, 794 (1966).

73. W. I. Gay, in *Primates, the Road to Self-Sustaining Populations*, K. Benirschke, Ed. (Springer-Verlag, New York, 1986), pp. 513–520.

74. N. W. King, *Vet. Pathol.* 23, 345 (1986).

75. R. C. Desrosiers and N. L. Letvin, *Rev. Infect. Dis.* 9, 438 (1987).

76. J. M. Ward, *Am. J. Pathol.* 127, 199 (1987); H. M. McClure *et al.*, *Proceedings of the Third International Conference on AIDS*, June 1987, Washington, DC, p. 212.

77. H. Rabin, in *Recent Advances in Primatology*, D. J. Chivers and E. H. R. Ford, Eds. (Academic Press, New York, 1978), vol. 4, pp. 101–115.

78. J. L. VandeBerg, *Genetics of Nonhuman Primates in Relation to Viral Diseases in Viral and Immunological Diseases in Nonhuman Primates* (Liss, New York, 1983), pp. 39–66.

79. M. D. Daniel, N. W. King, R. D. Hunt, in *Nonhuman Primate Models for Human Diseases*, W. R. Dukelow, Ed. (CRC Press, Boca Raton, FL, 1983), pp. 55–56; R. D. Hunt, in *Recent Advances in Primatology*, D. J. Chivers and E. H. R. Ford, Eds. (Academic Press, New York, 1978), vol. 4, pp. 117–124.

80. R. H. Adamson and S. M. Sieber, *Basic Life Sci.* 24, 129 (1983).

81. B. A. Lapin, *J. Med. Primatol.* 11, 327 (1982).

82. J. R. Held, in *Viral and Immunological Diseases in Nonhuman Primates*, S. S. Kalter, Ed. (Liss, New York, 1983), pp. 3–16; H. F. Siegler, R. S. Metzgar, T. Mohoakumar, G. M. Stuhlmiller, *Fed. Proc.* 34, 1642 (1983).

83. A. W. Nienhais *et al.*, *J. Clin. Invest.* 80, 573 (1987).

84. A. Hellman, M. N. Oxman, R. Pollack, Eds., *Biohazards in Biological Research* (Cold Spring Harbor Laboratory, Cold Spring Harbor, NY, 1973).

85. R. Siegert and W. Slenczka, in *Laboratory Diagnosis and Pathogenesis in Marburg Virus Disease*, G. A. Martini and R. Siegert, Eds. (Springer-Verlag, Berlin, 1971), pp. 155–160.

86. H. M. McClure *et al.*, in *Primates, the Road to Self-Sustaining Populations*, K. Benirschke, Ed. (Springer-Verlag, New York, 1986), pp. 531–556.

87. D. M. Bowden, *Neurosci. Newslett.* 19 (no. 2), 1 (1988).

88. This article and several of the studies cited in it were supported by base grant RR-00165 from the Animal Resources Branch, Division of Research Resources, National Institutes of Health. The Yerkes Center is fully accredited by the American Association for the Accreditation of Laboratory Animal Care. We thank R. A. E. Bakay for assistance with revisions of portions of this article and C. Y. Powell, M. A. Milne, E. J. Lovell, and N. Johns for editorial and secretarial contributions.

18

The Human as an Experimental System in Molecular Genetics

Ray White and C. Thomas Caskey

The human species holds a special fascination for scientific investigation. As a consequence, skilled investigators have been drawn to direct study of the human, independently of specific experimental problems or advantages. Earlier researchers had selected *Drosophila* for its favored genetic traits and because of its short generation time, ease of culture and other experimentally useful characteristics. Additional, crucially important experimental advantages are embedded in the enormous background of knowledge that has accrued from decades of detailed study and that continues to be expanded by the current cohort of skilled investigators. Techniques that have been developed in *Drosophila* are now being applied to the study of human genetics. However, the human system has extrinsic advantages that are peculiar to itself.

Funding is one of the requisite tools of research. The human system has fared well in this competition, because a primary motive of society in the funding of research is the hope of discoveries that will ease the human condition. Proposals to clarify mechanisms of specific diseases have had a high priority. The human social context has created a critical mass of knowledge and of investigators committed to the human as the system of choice for study of many fundamental problems in molecular biology and genetics.

A Complex Surveillance System Exists for Human Populations

No other species is observed so closely for variation as *Homo sapiens*. Physicians examine hundreds of millions of individuals, sometimes detecting even the most subtle variations. Geneticists are complemented by clinicians and researchers in hematology, oncology, immunology, cardiology, and neurology. Each group has made significant contributions toward the identification of genetic defects that may be associated with complex conditions such as leukemia, arthritis, hematologic disorders, vascular disease, and psychiatric and neurological anomalies. Because this screening takes place on such a large scale, examples of even quite rare events are often documented. As a consequence, the catalog of human genetic variants now describes more than 4000 distinct Mendelian conditions (1); by comparison, only 700 genetic loci have been identified in the mouse (2).

R. White is an investigator at the Howard Hughes Medical Institute and professor in the Department of Human Genetics, University of Utah School of Medicine, Salt Lake City, UT 84132. C. T. Caskey is an investigator at the Howard Hughes Medical Institute as well as a professor and director of the Institute of Molecular Genetics, Baylor College of Medicine, Houston, TX 77030. This chapter is reprinted from *Science* 240, 1483 (1988).

The human system, therefore, comes equipped with an astonishingly broad array of characterized mutant phenotypes; thousands of human genes have been identified by virtue of the phenotype they confer when inherited in mutant form. The mutations range from straightforward deficiencies in enzymes that are important in basic metabolic pathways to subtle alterations that affect our psychology. The number of individual subjects examined in detail by phenotypic experts, the physicians, is far greater than for any other system.

Clinical surveillance uncovers rare metabolic mutations

The clinical description of a patient may identify a genetic variation, but rarely determines the etiology of the defect. Further characterization depends on a judicious choice of appropriate testing methods. For example, molecular and biochemical study of human heritable metabolic diseases has been a highly productive means of identifying genes. In this arena, the human has an advantage over the mouse as an experimental system, since metabolic disorders frequently bring the patient, of his or her own volition, to the physician. Table 1 details some of the disorders for which biochemical analysis has provided insight into the deleterious mutation and has suggested a logical strategy for cloning the normal gene.

For each genetic anomaly listed, identification of a defective protein or enzyme, or of a compound produced in excess or accumulated in postmortem tissue in affected individuals, permitted investigators to accurately predict the nature of the genetic deficiency. A wide spectrum of molecular approaches was then applied for successfully cloning the putative gene as Table 1 shows; cloning strategies are outlined in Fig. 1. Automation and other improvements in procedures for sequencing and synthesizing amino acids and nucleic acids have greatly simplified these alternatives (3). Once a gene is cloned, its chromosomal "home" can be readily identified by somatic cell hybridization.

Table 1. Mendelian disorders in which identified biochemical abnormalities have led to gene cloning. Abbreviations: HPRT, hypoxanthine-guanine phosphoribosyltransferase; ASA, arginosuccinase synthetase; PH, phenylalanine hydroxylase; LDL, low-density lipoprotein.

Disorder	Biochemical and genetic abnormality	Deficiency	Method for cloning gene
Lesch-Nyhan syndrome	X-linked hyperuricemia	HPRT	1. Overexpression: isolation of cDNA from cells in which the gene is amplified (33) 2. Isolation of genomic fragment after nuclear DNA gene transfer (34)
Citrullinemia	Autosomal recessive; serum hyperammonemia and elevated serum/urinary citrulline	ASA	Overexpression: isolation of cDNA from cells with abnormal gene regulation (35)
Phenylketonuria	Autosomal recessive; elevated serum/urinary phenylalanine	PH	Antibody enrichment of polysomes containing PH mRNA (36)
Type IIa hypercholesterolemia	Autosomal dominant; hypercholesterolemia and elevated plasma LDL	LDL receptor	Antibody identification of a cDNA in a prokaryotic expression vector (37)
Tay-Sachs	Autosomal recessive; excess GM_2 ganglioside in postmortem brain tissue of children	Hexosaminidase A, α subunit	Antibody identification of a cDNA in a prokaryotic expression vector (38)
Hemophilia A	X-linked deficiency in intrinsic clotting pathway	Factor VIII	Synthetic oligonucleotide screening of genomic libraries (39)

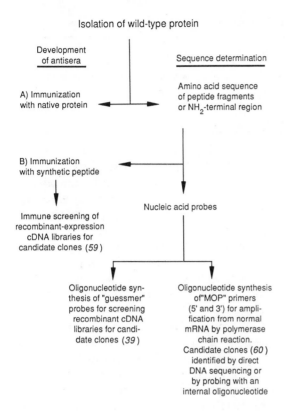

Isolation of wild-type protein

Development of antisera

Sequence determination

A) Immunization with native protein

Amino acid sequence of peptide fragments or NH$_2$-terminal region

B) Immunization with synthetic peptide

Immune screening of recombinant-expression cDNA libraries for candidate clones (59)

Nucleic acid probes

Oligonucleotide synthesis of "guessmer" probes for screening recombinant cDNA libraries for candidate clones (39)

Oligonucleotide synthesis of "MOP" primers (5' and 3') for amplification from normal mRNA by polymerase chain reaction. Candidate clones (60) identified by direct DNA sequencing or by probing with an internal oligonucleotide

Fig. 1. Common strategies for isolation of disease genes.

We estimate that more than 200 heritable genetic defects have been isolated and mapped by this approach.

Because multiple examples even of rare mutations are often available, a cloned gene provides access to a number of distinct mutations that may occur at the same locus. A variety of different mutations observable in a specific gene can provide valuable information on normal gene function and regulation, and insights into the natural history of the disorder in the population.

For example, the large number of mutations that have been examined at the hemoglobin loci in thalassemia patients have specified a wide spectrum of mutational lesions capable of reducing gene activity (4). Many of these have been analyzed, providing insights into messenger RNA metabolism as well as the specific molecular mechanism underlying the disease state for each patient. Examples of defects in transcription initiation, loss of proper

splice sites, and emergence of new, improper splice sites have documented the belief that "anything that can go wrong, will go wrong" in human genetic systems.

Furthermore, mutational variants at the low-density lipoprotein (LDL) receptor locus (*LDLR*) yielded insights into the mechanism of receptor-mediated endocytosis (5). Variants were identified through an extensive population survey that detected many families who were segregating hypercholesterolemia. Mutant alleles that specifically block transport of the LDL receptor molecule to the membrane, or prevent association of the receptor with coated pits, or interfere with binding to the ligand, were each characterized; in turn, each mutation so identified served to characterize a specific step in lipid metabolism.

Nevertheless, for many inborn errors of metabolism the enzyme deficiency is unknown, and significant research opportunities remain. For example, the nature of the defect (or defects) resulting in X-linked Menkes' syndrome (6) and the autosomal-recessive Wilson's disease (7), both related to defective copper metabolism, is yet unresolved.

Medical observation uncovers rare but significant cytogenetic events

High-resolution cytogenetic studies often provide vital clues to the chromosomal location of an inherited defect or a somatic mutation. Again, even though the diseases and the presence of cytogenetic abnormalities may be rare (with a combined incidence in the population of about 0.05%), the large size of the population surveyed permits us to observe association between specific cytogenetic events and disease phenotypes frequently enough to make correlations. Several examples are listed in Table 2.

Because of this surveillance system, a few individuals have been detected who are phenotypically males but have two X chromosomes. Analysis has revealed that a small portion of the Y chromosome has been translocated to the X, and has recently led to the identification of a putative testis-determin-

Table 2. Diseases in which observation of cytogenetic abnormality has led to cloning of the implicated genes.

Disease	Cytogenetic findings	Genes implicated
Burkitt lymphoma	Translocation (8;14)	*myc*; μ heavy-chain enhancer (*9*)
Chronic myeloid leukemia (CML)	Translocation (9q;22q)	*abl*; *bcr* (*10*)
Duchenne muscular dystrophy	Deletion Xp21	Dystrophin (*34*)
Retinoblastoma	Deletion 13q14	*Rb* (*35*)

ing factor gene (*8*).

When chromosomal translocations are associated with specific tumor types, specific oncogenes can be implicated, as they have been in Burkitt lymphoma and chronic myeloid leukemia (CML). In these cases the translocation breakpoints are near the known location of an oncogene that was originally identified by its ability to cause cellular transformation to malignancy. In Burkitt lymphoma, both implicated genes involved in the translocation were cloned prior to their being associated with the disease (*9*); in CML, the *abl* oncogene facilitated identification of its junction partner, *bcr*, which was previously unknown (*10*). Apparently, translocation causes juxtaposition of new regulatory regions, displaces normal regulatory elements, and alters the expression of genes involved in regulation of cell growth and differentiation. The ability to compare the location of translocation breakpoints with the known locations of oncogenes greatly enhances our ability to identify specific genes that are activated by a translocation; the large number of loci already mapped in the human genome facilitates such comparisons. This approach is complemented by the application of pulsed-field gel analysis (*11*), which can detect translocation junction fragments over long stretches of DNA. These methods may prove particularly useful for identifying the gene (or genes) involved in acute promyelocytic leukemia, where 95 percent of patients show a (15;17) translocation (*12*), and in acquired hematopoietic defects commonly associated with three deletions in chromosome 5 (*13*).

One form of X-linked mental retardation, which affects approximately 1 in 2000 males, is associated with another type of chromosomal abnormality, a fragile site at Xq27; it is the object of intensive cloning efforts (*14*). This kind of aberration, in which gaps or breaks occur more frequently when cells from affected individuals or carriers are grown in vitro, was first observed in the human system. The location of other fragile sites on human chromosomes, in some cases, appears to coincide with the location of proto-oncogenes (*15*).

Chromosomal deletions associated with such heritable diseases as Duchenne muscular dystrophy (DMD) and retinoblastoma (RB) have provided critical information leading to cloning of the gene involved (*16, 17*). The large size of the DMD gene (2000 kb) could not have been predicted beforehand. In both DMD and RB, cytogenetic alterations in patients' cells were the first clue to the molecular basis of the disease. Subsequently, they helped significantly in the identification of the genes; the deletions enabled investigators to sort quickly through libraries of cloned DNA segments for sequences specific to the region of interest, and even to enrich for them (*18*). The availability of region-specific clones then allowed the investigators to identify new deletions that had escaped cytogenetic detection because of their very small size. Cloned DNA sequences corresponding to the segment missing in the smallest deletions became excellent probes for investigating complementary DNA libraries derived from tissues where the putative genes were likely to be expressed. The large number of independent mutations harbored by the population as a whole at the loci for DMD and RB played an important role in these stepwise protocols.

Deletions within chromosome 13 found in surveys of patient populations not only provided the basis for the mapping of the gene for retinoblastoma to chromosome 13 but also opened the door for the significant and unexpected discovery of a new class of oncogenes, the "recessive oncogenes." The association of deletions with the disease suggested that hemizygosity for a mutant allele might be important in the disease; studies with DNA markers confirmed that indeed, loss of a chromosome 13 homolog was associated with tumorigenesis (19). This observation strongly supported the idea that a recessive, mutant allele of the retinoblastoma gene was located on the homolog that was retained, with its phenotype revealed by loss of the normal allele.

Subsequently, chromosome loss or deletion has been characterized in a number of tumor types, both inherited and sporadic (20). Loss of a characteristic chromosome has recently served as a key to the mapping of the loci for two important inherited tumor types, acoustic neurofibromatosis (21), and multiple endocrine neoplasia, type I (22).

A substantial number of disorders that includes Prader-Willi syndrome (23), Miller-Dieker syndrome (24), and the myeloid and acute promyelocytic leukemias mentioned earlier (10, 12) have been associated with specific cytogenetic abnormalities, thus localizing important genes and giving molecular biologists a starting point for eventual identification of the molecular defect. It is possible to predict that the application of cytogenetic studies to diagnosis of birth defects will contribute to the localization and, ultimately, identification of human developmental genes as well.

Mapping and Identification of Genes

Often the gene in which a human mutation occurs has no biochemical or cytogenetic correlate and is recognized only by its phenotypic effect on people who inherit the mutant allele. The biochemical and cytogenetic approaches outlined above are not applicable then. However, classical linkage analysis in families segregating the disease can still permit localization and identification of a gene whose product is unknown.

The logic underlying this approach is simple in concept and has been used in many other systems but its implementation in human systems is intricate. The basic premise of linkage analysis is that the defective gene must be located at some specific place on a specific chromosome. By correlating the inheritance of specific chromosomal regions (defined by DNA markers) with inheritance of the putative gene (defined by its phenotypic effect on family members) one can map the mutation to a specific region of a particular chromosome. The disease locus will show co-inheritance, or genetic linkage, with a DNA marker segment in its immediate vicinity on the chromosome. The linked marker may in turn serve to diagnose the inheritance of the mutant gene by family members in whom the phenotypic effect is not yet apparent.

The linkage approach depends completely, however, on our having genetic markers that will allow us to trace the inheritance of each part of every chromosome. A system of markers to cover the entire genome, based on direct detection of common variations in DNA sequence within the population, was considered several years ago (25); restriction enzymes would detect the variations in DNA sequence and DNA probes would define the fragments derived from specific genetic loci. The method is applicable both to arbitrary fragments of human DNA and to DNA segments associated with cloned genes. Large-scale screening is required to identify the combinations of DNA probes and restriction enzymes that detect common population variants in DNA sequence. The importance of identifying human genes that cause disease has stimulated development of hundreds of genetic markers in recent years, making the linkage approach now generally feasible. A growing number of mutations associated with human disease have been localized to specific chromosomal regions by linkage to DNA markers in affected families; Table 3 is a partial list.

For each of the disorders listed, major efforts are now under way to identify and clone the gene that harbors the mutation. Once localized to a particular chromosomal region, the gene must be more precisely mapped to a segment that contains only 1 to 3 million base pairs of DNA; a chromosomal segment of this size will contain only a few genes that are likely

Table 3. Chromosomal localization of defective genes by linkage anlysis in affected families.

Disease	Chromosomal location
Chronic granulomatous disease*	Xp21 (40)
Huntington's chorea	4p16 (41)
Cystic fibrosis	7q21–22 (42)
Familial adenomatous polyposis	5q21–22 (27)
Polycystic kidney disease	16p13 (43)
Multiple endocrine neoplasia, type IIA	10q4 (44)
Type 1 neurofibromatosis	17cen-q21 (45)

*Cytochrome oxidase deficiency has been identified as the CGD defect.

to be expressed in specific tissues affected by the disease in question. Chromosomal landmarks such as translocation breakpoints or small deletions can help to implicate a "candidate" gene within such a region, as described above. A biological assay for the candidate gene based on its activity would be helpful at this point, but usually none is available; direct sequencing may be the only way to correlate the disease phenotype with specific mutations that are capable of altering the product of a candidate gene.

Sometimes, DNA sequence variants associated with the locus of a candidate gene can be detected by probes based on cloned DNA segments from the gene itself. Genetic markers constructed this way allow investigators to confirm or eliminate candidate genes by testing them for linkage to specific diseases. For example, reason once suggested that familial polyposis, a genetic disorder characterized by large numbers of adenomatous colonic polyps that can become malignant, might be a consequence of inherited mutations in one of the *ras* oncogenes, either Kirsten or Harvey, because mutant K-*ras* genes in tumor cells are characteristic of human carcinoma of the colon. However, variant DNA sequences in the vicinity of each of these candidate oncogenes were shown to be unlinked to the polyposis locus in families segregating the disease (26); linkage tests subsequently located the gene on chromosome 5 in a region that had been implicated by a rare deletion (27). Linkage tests with candidate genes for cystic fibrosis (28) and mytonic dystrophy (29) are being undertaken now, since the regions containing these two

defects have been narrowed sufficiently that candidates can be identified. Even failure to find linkage is valuable, because quick elimination of suspected genes from candidacy frees investigators to pursue other leads.

Similarly, familial hypercholesterolemia (FH) often results from inheritance of a mutant *LDLR*. However, lesions in other genes involved in LDL metabolism, for example apolipoprotein B, could presumably confer a similar phenotype. The question of genetic etiology can be resolved for individual FH pedigrees by linkage studies with genetic markers that are associated with the LDL receptor and the apolipoprotein B genes (30). As more human genes are identified and cloned, and speculation arises as to what phenotype a mutant gene might confer on its host, it seems likely that this approach will become even more effective.

In vitro techniques complement genetic studies

Human cells grow well in culture. In consequence, the intrinsic interest in human molecular biology has led to development of cell lines derived from various human tissues, on a scale well exceeding that for any other organism. To provide investigators free access to genetically valuable human cell lines, a Human Genetic Mutant Cell Repository has been established by the National Institute of General Medical Sciences (NIGMS). The 1986/87 catalog of the repository lists almost 4000 human cell lines carrying a wide variety of mutations, ranging from metabolic variants to translocations.

Moreoever, human genes can be cloned readily in *Escherichia coli*; again, the number of cloned human genes greatly exceeds the total for any other organism. The American Type Culture Collection (ATCC) now carries over 1000 human DNA clones and is adding more than 300 new ones every year. These in vitro resources allow investigators flexibility to pursue basic questions in human biology, untrammeled by problems of experimentation with the intact organism.

Advantages and Disadvantages of the Human System: A Challenging Mixture

The human system at times clearly lacks the intrinsic advantages one might hope for in an experimental organism. From a geneticist's point of view, the generation time is overlong and social proscription prevents the investigator from carrying out controlled crosses. This inability to construct strains of known genotype denies access to one of the most powerful approaches geneticists have always used to critically test hypotheses. Can we overcome these liabilities of the human experimental system and recover the power of classical genetics?

Recent advances in DNA technology—specifically, the development of many markers for the genome—offer some relief by making it possible to define inherited genotypes much more accurately than before. Genotypes in the offspring of natural human matings can now be determined in enhanced detail, so that individuals of a specific genotype can be examined to test hypotheses. The large set of linkage markers now available is a strong advantage, having grown out of the need to answer specific questions in the human system.

The need to protect individual rights and to obtain informed consent from the participant places serious constraints on the range of experimental possibilities, although recent concern for the rights of laboratory animals is

narrowing this experimental gap. On the positive side, human subjects are long-lived relative to the experimental lifetime of the investigator; he or she need not establish a strain or maintain a colony in order to complete a study. Furthermore, clinical research centers provide an avenue for direct experimentation in humans; the rapid progress stemming from in vitro technologies will increase the demand for direct testing of hypotheses in the intact human organism.

It is important to bear in mind that the human is not the only system available for testing hypotheses that have been formulated in the human. Construction of individuals of a specific genotype (the construction of transgenotes through in vitro gene implantation, for example) will be banned from the human system for the foreseeable future. For many purposes, however, all mammals are virtually identical; gene regulation during early development should be the same in mouse and human and pig. When information is available about human genes responsible for aberrant phenotypes, one can transfer genetically altered homologous genes to another mammalian species, the mouse for example, to test hypotheses related to human disease. Examples of such models have already been reported and the related technologies required to establish them are summarized in Table 4. The field is rapidly advancing, aided by the capacity to carry out legitimate recombination in embryonic stem cells of mice (*31*) and possibly to engineer mouse mutants. The challenge with the human, then, is to evaluate the experimental context carefully and to make choices on animal systems that are appropriate to the needs and questions at issue for man.

The genetics of complex diseases

Complex, common diseases subtend an area of genetics almost unique to the human. Many such disorders are reflected in relatively subtle phenotypes—psychiatric diseases, for example—that would not be recognized easily in other genetic systems. The human sensitivity to

Table 4. Human disease equivalent derived from genetic alterations in the mouse. HGH, human growth hormone; MBP, myelin basic protein; HPRT, hypoxanthine guanine phosphoribosyl transferase; MDX, X-linked muscular dystrophy.

Genetic alteration	Method	Phenotype
Introduction of HGH into GH-deficient mice	Nuclear microinjection of HGH minigene	Growth deficiency corrected (46)
Introduction of mouse MBP gene into MBP-deficient mice	Nuclear microinjection of MBP nuclear gene	Shiverer phenotype corrected (47)
Introduction of mutant collagen gene into wild-type mice	Nuclear microinjection of mutant minigene	Osteogenesis imperfecta (48)
Introduction of activated human *ras* and *c-myc*	Nuclear microinjection of inducible minigene	Induction of malignancy (48)
Inactivation of mouse HPRT gene	Insertion of retrovirus into HPRT locus in embryonic stem cells	HPRT deficiency in F_3 generation (49)
Mutation at locus for MDX	Male mutagenesis followed by identification of female carriers	MDX phenotype and genotype (50, 51)

complex variations presents an opportunity for research into interactions between genotype and environment, research that may reveal genetic bases for behavioral traits of fundamental interest. For example, the study of genes that confer predisposition to schizophrenia may shed light on mechanisms of psychological homeostasis. Providing discrete windows into complex metabolic and biochemical systems has been the traditional contribution of those who analyze mutation; for complex disorders, we take advantage of detailed and large-scale human surveys to identify intriguing variants.

The apparent familiality of cardiovascular disease suggests that significant genetic components participate in a wide range of phenotypes, from hypertension to premature atherosclerosis. Genetic analysis promises eventual identification of the individual genes involved, with the concomitant expectation that clear relationships between individual alleles and specific environmental components subsequently can be identified. If we can clarify the genetics, the environmental components may become recognizable on their own. In the case of cardiovascular disease, the phenotypic associations between clinical disease and lipid profiles or apolipoprotein concentrations in blood have come from epidemiological studies whose large scale and expense reflect our special concern with human disease. Without biochemical correlates to study as genetic traits,

however, it would be much harder to investigate the genetics underlying the familial nature of cardiovascular diseases.

Genetic diagnosis

A particularly rewarding characteristic of human genetic research is that new findings frequently can be translated directly into clinical interventions. Notable in this category are the ability to determine preclinically, or even prenatally, genotypes of individuals at risk for a disease, or to discriminate among individuals for forensic purposes (32).

Information derived from molecular approaches to human heritable disease is already being applied to the prevention of severe genetic disease, the identification of patients at high risk for neoplasia and atherosclerosis, and the detection of carriers for common recessive traits. As the methods rapidly improve in speed, simplicity and cost, many are becoming "standard of practice"; a selective summary of DNA tests that are being applied with increasing frequency is given in Table 5. These successes are a direct outgrowth of the study of humans as a genetic system.

The challenge for DNA-based diagnostics will expand greatly as the genetic components of common, complex diseases emerge in detail. In general, disorders such as cardiovascular dis-

Table 5. Determination of disease risk by DNA-based methods.

Disease	DNA method	Heterozygotes identified	Prenatal diagnosis available?
Duchenne muscular dystrophy*	Linkage association and detection of deletions	Asymptomatic female carriers (52)	Yes (53)
Cystic fibrosis*	Linkage association	Asymptomatic carriers in index families (54)	Yes (54)
Lesch-Nyhan syndrome	Point mutation; deletions (55)	Asymptomatic female carriers	Yes (56)
Retinoblastoma	Linkage association (57)	Presymptomatic heterozygotes in index families	Yes (57)
Adult polycystic kidney disease*	Linkage association (43)	Presymptomatic heterozygotes in index families (43)	Yes (35)
Huntington's chorea	Linkage association (41)	Presymptomatic heterozygotes in index families (58)	Yes

*DNA-based detection is now standard of practice.

ease, cancer, and diabetes become manifest only in adult life. An individual likely will benefit from knowing, well in advance of symptoms, whether he or she is at special risk because of genetic predisposition to an adult-onset disease whose medical management may forestall major complications of the disorder.

Progress in our ability to diagnose before the development of symptoms will have to be matched with programs of therapeutic intervention; retinoblastoma and adult polycystic kidney disease provide cases in point. Finally, and particularly for adult-onset neurodegenerative disorders like Huntington's chorea, for which no therapeutic option presently exists, the predominant benefits will be related to halting disease in subsequent generations, at the psychological expense of those who are identified as presymptomatic heterozygotes.

Ethical challenges

Challenges to ethics and values will come with our increased ability to diagnose the presence of predisposing genes for common disorders well in advance of symptoms. One of the earliest is likely to arise from the problem of insurability of individuals found to be carrying a predisposing allele. Insurance companies will want access to such information and may mandate testing

for the presence of the predisposition. Is an unlucky draw in the genetic lottery an appropriate basis for deciding insurability? Many think not, while others maintain that persons who do not carry predisposition to a given illness should not be required to pay higher insurance rates to support those who do. It is important to bear in mind that predisposition does not automatically result in the development of the disease. The dialogue will intensify as more and more predisposing alleles are characterized and we begin to discover that most of us are predisposed to *something*. This issue may well add fuel to the debate over national health insurance.

Sometimes, knowing that he or she harbors a strongly predisposing disease allele can severely damage a person's quality of life; we mentioned Huntington disease earlier. This dilemma is not a new one to medicine, since diagnoses of chronic disease are frequently made (for example, mental retardation, cancer, hypertension, renal failure, coronary artery disease, and multiple sclerosis).

Medical treatment and genetic counseling in the United States function on the principle of disclosure to the patient. Undoubtedly the new era of genetics will evoke difficult challenges in patient management, but they hopefully should be no worse than those that derive from lack of information.

References and Notes

1. V. A. McKusick, *Mendelian Inheritance in Man* (Johns Hopkins Univ. Press, Baltimore, ed. 7, 1986).
2. M. C. Green, in *Genetic Variants and Strains of the Laboratory Mouse*, M. C. Green, Ed. (Fischer, Stuttgart, 1981), p. 8.
3. M. Hunkapiller *et al.*, *Nature* 310, 105 (1984).
4. S. H. Orkin and H. Kazazian Jr., *Annu. Rev. Genet.* 19, 131 (1984).
5. M. S. Brown and J. L. Goldstein, *Science* 232, 34 (1986).
6. D. M. Danks *et al.*, *Lancet* i, 1100 (1972).
7. D. W. Cox, F. C. Fraser, Sass-Kovtsak, *Am. J. Hum. Genet.* 24, 646 (1972).
8. D. C. Page *et al.*, *Cell* 51, 1091 (1987).
9. P. Leder, *IARC Sci. Publ.* 60, 475 (1985).
10. C. M. Croce *et al.*, *Proc. Natl. Acad. Sci. U.S.A.* 84, 7174 (1987).
11. D. C. Schwartz and C. R. Cantor, *Cell* 37, 67 (1984).
12. A. M. Bitter *et al.*, *Human Path.* 18, 211 (1987); J. D. Rowley and J. R. Testa, *Adv. Cancer Res.* 36, 103 (1982).
13. M. M. LeBeau *et al.*, *Science* 231, 984 (1986).
14. R. L. Nussbaum and D. H. Ledbetter, *Annu. Rev. Genet.* 20, 109 (1986).
15. J. J. Yunis and A. L. Soreng, *Science* 226, 1199 (1984).
16. A. Monaco *et al.*, *Nature* 323, 646 (1986).
17. S. Friend *et al.*, *ibid.*, p. 643.
18. L. M. Kunkel, A. P. Monaco, W. Middlesworth, H. D. Ochs, S. A. Latt, *Proc. Natl. Acad. Sci. U.S.A.* 82, 4778 (1985).
19. W. Cavanee, T. Dryja, R. Phillips, R. Benedict, B. Godbout, B. Gallie, A. L. Murphree, R. L. White, *Nature* 305, 5937 (1983).
20. M. Hansen and W. Cavanee, *Trends Genet.* 4, 125 (1988).
21. G. A. Rouleau *et al.*, *Nature* 329, 246 (1987).
22. C. Larsson, B. Skogseid, K. Oberg, Y. Nakamura, M. Nordenskjold, *ibid.* 332, 85 (1988).
23. D. H. Ledbetter *et al.*, *N. Engl. J. Med.* 304, 325 (1981).
24. W. B. Dobyns *et al.*, *J. Pediatr.* 102, 552 (1983).
25. D. Botstein, R. White, M. Skolnick, R. Davis, *Am. J. Hum. Genet.* 32, 314 (1980).
26. D. Barker *et al.*, *Mol. Biol. Med.* 1, 199 (1983).
27. W. F. Bodmer *et al.*, *Nature* 328, 614 (1987); M. F. Leppert *et al.*, *Science* 238, 1411 (1987).
28. R. Williamson, personal communication.
29. A. Roses, personal communication.
30. M. Leppert *et al.*, *Am. J. Hum. Genet.* 39, 300 (1986).
31. K. R. Thomas and M. R. Capecchi, *Cell* 51, 503 (1987).
32. A. Jeffreys, V. Wilson, S. Thein, *Nature* 314, 67 (1985).
33. J. Brennand, A. C. Chinault, D. S. Konecki, D. W. Melton, C. T. Caskey, *Proc. Natl. Acad. Sci. U.S.A.* 79, 1950 (1982).
34. D. J. Jolly, A. C. Esty, H. V. Bernard, T. Freidmann, *ibid.*, p. 5038.
35. T. S. Su, H. G. O. Bock, W. E. O'Brien, A. L. Beaudet, *J. Biol. Chem.* 256, 11826 (1981).
36. K. Robson, T. Chandra, R. MacGillivray, S. L. C. Woo, *Proc. Natl. Acad. Sci. U.S.A.* 79, 4701 (1982).
37. D. W. Russell *et al.*, *ibid.* 80, 7501 (1983).
38. R. Myerowitz and R. L. Proia, *ibid.* 81, 5394 (1984).
39. J. J. Toole *et al.*, *Nature* 328, 616 (1987); J. Gitschier *et al.*, *ibid.* 312, 326 (1984).
40. B. Royer-Pokora *et al.*, *ibid.* 322, 32 (1986).
41. J. F. Gusella *et al.*, *ibid.* 306, 234 (1983).
42. L.-C. Tsui *et al.*, *Science* 230, 1054 (1985); B. Wainwright *et al.*, *Nature* 318, 384 (1985); R. White *et al.*, *ibid.*, p. 382.
43. S. T. Reeders *et al.*, *Nature* 317, 542 (1985).
44. C. Mathew *et al.*, *ibid.* 328, 527 (1987); Simpson *et al.*, *ibid.*, p. 538.
45. D. Barker *et al.*, *Science* 236, 1100 (1987); B. R. Seizinger *et al.*, *Cell* 49, 589 (1987).
46. R. E. Hammer, R. D. Palmiter, R. L. Brinster, *Nature* 311, 65 (1984).
47. C. Redhead *et al.*, *Cell* 48, 703 (1987).
48. E. Sinn *et al.*, *ibid.* 49, 465 (1987).
49. M. R. Hoehn, A. Bradley, E. Robertson, M. J. Evans, *Nature* 326, 295 (1987).
50. V. M. Chapman, D. R. Miller, D. Armstrong, C. T. Caskey, *Proc. Natl. Acad. Sci. U.S.A.* 86, 1292 (1989).
51. J. S. Chamberlain *et al.*, *Science* 239, 1416 (1988).
52. C. T. Caskey, P. Ward, F. Hejtmancik, *Adv. Neurol.* 48, 83 (1988).
53. J. F. Hejtmancik *et al.*, in *Nucleic Acid Probes in Diagnosis of Human Genetic Diseases*, A. M. Willey, Ed. (Liss, New York, 1988), pp. 83–100.
54. J. E. Spence *et al.*, *Hum. Genet.* 76, 5 (1987).
55. R. A. Gibbs and C. T. Caskey, *Science* 236, 303 (1987).
56. J. T. Stout, L. G. Jackson, C. T. Caskey, *Prenatal Diagnosis* 5, 183 (1985).
57. W. K. Cavenee *et al.*, *N. Engl. J. Med.* 314, 1201 (1986).
58. G. J. Meissen *et al.*, *ibid.* 318, 535 (1988).
59. R. A. Young and R. W. Davis, *Proc. Natl. Acad. Sci. U.S.A.* 80, 1194 (1983).
60. C. C. Lee *et al.*, *Science* 239, 1288 (1988).
61. We would like to thank R. Foltz for her assistance.

19

Fetal Research

John T. Hansen and John R. Sladek, Jr.

Human development occurs in two entirely different environments, one prenatal and the other postnatal. Prenatal development encompasses the embryonic and fetal periods, whereas postnatal development involves the passage through infancy, childhood, and adolescence to adulthood. These two environments could not be more different. The safe and nutritive environment of the womb predictably yields to the more hostile existence of life after birth. Nevertheless, the relatively short prenatal existence has always held a fascination for us as we marvel at the apparent recapitulation of our developmental history. Advances in scientific understanding now are at the point where the homunculus of our ancestors' imaginations has given way to an appreciation of the intricate patterning faithfully reproduced by our genetic blueprint. Our ability to intervene prenatally when nature's course deviates has long been limited to the physician's crude palpations and auscultations, methods woefully inadequate to diagnose, let alone treat, fetal problems. Only through persistent scientific inquiry, driven by our inherent curiosity about our development, have we now reached the threshold of prenatal diagnosis and treatment necessary to ensure the mother's safety or save an endangered life. The fetus, once a captive of its own environment, an enigma to be protected but left untreated, finally has gained the status of patient. Accordingly, fetal research itself enters an important new era.

In this chapter, we review some of the significant contributions of fetal research and fetal tissue research over the past 20 years. It is important to draw a distinction between fetal research, that is, research performed on the living fetus in utero, versus fetal tissue research that focuses on tissues or cells derived from the dead fetus, obtained as a result of spontaneous or induced abortion (*1*). By its very nature, scientific inquiry that involves fetal research or the use of fetal tissues often is obscured in the larger ethical, moral, and legal questions surrounding the use of fetuses, especially human fetuses, in research of any kind. These concerns are not trivial, for they strike at the heart of our moral dilemma regarding abortion, or the use of invasive procedures on a patient (the fetus) who can neither be informed nor grant consent. The resolution of these concerns and the answers to the ethical and legal questions will require honest, open dialogue from all aspects of society before, and if, a consensus is ever forthcoming. Our intent is not to debate whether fetal research should continue; rather, our focus will be on why fetal research and fetal tissue research are done at all, what procedures are feasible, and how this research benefits mankind.

The authors are professors in the Department of Neurobiology and Anatomy, University of Rochester School of Medicine and Dentistry, Rochester, NY 14642. This chapter is reprinted from *Science* **246**, 775 (1989).

Prenatal Diagnosis

Fetal research plays a vital role in the continued ability to diagnose a variety of fetal disorders, from genetic inborn errors in metabolism to congenital malformations (Table 1). Approximately 150,000 children in the United States alone, representing 3 to 5% of all live births each year, are born with congenital abnormalities (2). Ultrasonography, a noninvasive procedure that permits visualization of the fetus without apparent risk to fetus or mother, is one of the most important diagnostic advances available to the physician (3) and is used as an aid for the accurate guidance of instruments. Ultrasonography is also used to assess fetal movements and gross fetal malformations. For example, neurological defects such as anencephaly, spina bifida, and hydrocephalus can be diagnosed with ultrasonography. Heart defects, which occur on the order of 1% of all live births (4), and various obstructive disorders of the gastrointestinal or urinary tracts also may be visualized with this noninvasive approach.

In contrast, early diagnosis of inherited chromosomal abnormalities, fetal disease, and metabolic deficiencies require invasive intervention. Amniocentesis, the withdrawal of amniotic fluid, has dramatically changed the physician's ability to diagnose, counsel, and implement treatment (5). The assessment of chromosomal abnormalities, amniotic infections, fetal lung maturation, and the severity of hemolytic disease related to Rhesus (Rh) factors is now possible; however, most amniocentesis is used for cytogenetic studies (6). In addition to direct chromosomal analysis, recombinant DNA technology now makes it possible to diagnose a large number of genetic disorders. Presently, more than 4000 disorders in man are known or suspected of being due to a single gene mutation, and as many as 300 gene mutations in humans may be X-linked (7, 8). By means of recombinant technology, many gene mutations may be identified either directly or with the use of restriction fragment length polymorphisms (RFLP). Disorders such as Huntington's disease, Duchenne muscular dystrophy, sickle-cell anemia, hemophilia, and cystic fibrosis have been diagnosed by the use of RFLP. Inborn errors in metabolism also may be assessed by culturing fetal cells suspended in the amniotic fluid sample and subjecting their resulting gene products to enzyme analysis assays. Although prenatal diagnosis of most inborn errors of metabolism are made by analyzing the gene product, several direct determinations of unique metabolites in the amniotic fluid sample are also possible (9). For example, hexosaminidase A, the deficient enzyme of the autosomal recessive disorder Tay-Sachs disease, can be diagnosed directly (10). Endocrine disorders such as adrenogenital syndrome are diagnosed during the prenatal period by direct assay for the elevated levels of 17α-hydroxyprogesterone in amniotic fluid (5). Amniocentesis also is a vital diagnostic procedure for the detection of neural tube defects, such as spina bifida, encephalocele, and anencephaly. These neural tube defects, for example, affect about 1 to 2 in 1000 live-born

Table 1. Examples of noninvasive and invasive procedures used to diagnose or treat fetal disorders. Details are provided in (3, 6, 9, 47).

Noninvasive	Invasive
Patient history	Amniocentesis
Uterine size	Chorionic villus sampling
Fetal activity	Percutaneous umbilical blood sampling
Fetal heart rate	Fetoscopy
Ultrasonography	Blood or tissue biopsies
Estimate age	Structural abnormalities
Evaluate growth	Fetal therapy
Detect gross malformations	Blood transfusions
Determine multiple gestation	Drug administration
Determine sex	Surgical intervention
	Fetal cell transplants

infants in the United States and Canada (*6*). These defects result from the failure of the embryonic neural tube to close, and their diagnosis relies on the determination of elevated levels of α-fetoprotein, a glycoprotein normally found in fetal serum (*9*). The α-fetoprotein leaks through the membrane covering such neural tube defects and accumulates in the amniotic fluid and maternal serum (*6*).

One significant drawback of amniocentesis is that it usually is not performed before 15 or 16 weeks gestation, and any final diagnosis dependent on cell culture must be delayed an additional 2 to 3 weeks (*3, 6, 9*). Moreover, the rate of pregnancy loss relating to amniocentesis is approximately 0.5% in the United States (*11*). Earlier diagnosis of chromosomal abnormalities is possible by using ultrasound-guided chorionic villus sampling, which may be performed as early as 8 weeks gestation. Chorionic villus sampling, although valuable for gathering karyotyping data at earlier gestational ages, does pose a slightly higher risk of fetal loss than amniocentesis (*12*).

Fetoscopy, that is, percutaneous transabdominal uterine endoscopy, provides additional advantages for prenatal diagnosis. Anatomical malformations may be directly visualized, and fetoscopy may be used to obtain blood or tissue biopsy samples (*6, 9*). Since 1983, a newer sampling procedure for obtaining fetal blood samples, called percutaneous umbilical blood sampling (PUBS), has proved valuable for diagnosing fetal hemolytic disease and a number of genetic disorders (*13*). During PUBS, an ultrasonographically guided needle is inserted directly into an umbilical vessel to withdraw a fetal blood sample. The procedure may be performed on an outpatient basis, does not require maternal sedation, and is safer for the fetus than fetoscopy (*13*). Nevertheless, PUBS is still considered an experimental procedure and should only be performed at selected medical centers (*13*). Procedures that involve collecting amniotic fluid, blood, urine, or other body fluids are used to diagnose almost 100 genetic diseases that result from single gene mutations (*8*). Tissue biopsies are especially valuable in prenatal diagnoses when chorionic villus sampling or amniocentesis results are equivocal, and for gathering information about multifactorial inherited congenital anomalies not easily or readily diagnosed by chromosomal or biochemical abnormalities present in the amniotic or other fetal fluid samples (Table 2).

Diagnostic procedures such as those described above are possible because of technical advances developed from fetal research. Refinements of these procedures are first developed in suitable lamb or nonhuman primate animal models and then judiciously introduced into the clinical setting (*9*). Additionally, a number of biopsy procedures are being developed and perfected. For example, blood, skin, or liver may be biopsied by the use of fetoscopy. About 100 enzyme deficiency disorders can be diagnosed from cultured fibroblasts, and another 100 deficiencies are diagnosed from specific cell types obtained from fetal tissue biopsies (*8*). However, before these invasive procedures become standard clinical practice, they must be carefully tested for their safety and effectiveness in clinical volunteers. To illustrate this point, some en-

Table 2. Diagnostic and therapeutic benefits and applications of fetal research and fetal tissue research. A more complete listing of specific benefits and applications may be found in (*1, 3, 47, 48*).

Fetal research	Fetal tissue research
Amniocentesis	Cell growth—normal and abnormal
Blood transfusions	Cell line development
Chorionic villus sampling	Cell plasticity
Drug therapy	Drug testing
Fetoscopy	Fetal cell transplantation
Percutaneous umbilical artery sampling	Immunology
Pregnancy management	Karyotyping studies
Ultrasonography	Vaccine development
Ventriculoamniotic shunts	
Vesicoamniotic shunts	

zyme deficiencies can only be diagnosed from fetal liver cells. Needle biopsies of the fetal liver are possible, but questions concerning liver damage, intraperitoneal bleeding, or fetal injury surround this procedure. The answers to these questions were obtained by experimenting with fetal liver biopsy procedures on fetuses of patients undergoing second-trimester abortions (9). The biopsy procedures were successful. Consequently, enzyme deficiencies such as glucose-6-phosphate deficiency, which occurs in von Gierke's disease and is related to the liver's ability to store glycogen, may now be diagnosed (9). Similarly, several rare enzyme deficiencies of the urea cycle, for example, carbamyl-phosphate synthetase and ornithine transcarbamylase, may be diagnosed from fetal liver biopsies (14).

Research on the fetus is essential before diagnosed disorders can be treated. The efficacy of vaccines, such as the rubella vaccine for the prevention of German measles, or the titration of drugs can only be tested in pregnant women. The fetus is not an innocent bystander if maternal treatment necessitates medical intervention. Virtually all commonly used drugs with the possible exception of insulin, heparin, dextrose, and thyroxine pass through the placenta to varying degrees (15). Therefore, the safety of medications such as hormones, diuretics, anticonvulsants, anesthetics, and analgesics must be tested first in utero to determine their effect on the fetus. Moreover, fetal disorders such as cardiac arrhythmias are responsive to antiarrhythmic drugs such as digitalis and may be treated directly while in utero (9, 16). In instances where substances do not cross the placenta, or do so poorly and at low levels, medications or nutritional supplements may be administered directly into the amniotic fluid where oral ingestion and gastrointestinal absorption by the fetus can occur.

Surgical Intervention

For those disorders affecting a single organ system or resulting from an isolated congenital malformation, unencumbered by multifactorially inherited abnormalities, surgical intervention may provide the most promising prognosis. Obstructive hydrocephalus and urethral obstruction are among several anatomical malformations amenable to surgical intervention in utero.

Obstructive hydrocephalus, a condition that occurs with an incidence of about 5 to 25 per 10,000 births and is characterized by dilation of the brain's ventricular system due to the obstruction of the normal cerebrospinal fluid (CSF) pathways, leads to significant brain compression and neurologic dysfunction. The surgical insertion of a ventriculoamniotic shunt with a one-way valve that permits the release of CSF into the amniotic fluid offers one possibility for decompressing the brain (17). Obstructive uropathy and the resulting damage to the developing kidney also may be corrected by the surgical placement of a suprapubic drainage catheter. The catheter is guided into the distended fetal urinary bladder by the aid of sonography, and the accumulated urine is drained into the amniotic fluid (18). These surgical procedures, and others still under development, were made possible because suitable animal models were available (19). This experimentation is difficult because only larger animal species such as rabbits, lambs, or nonhuman primates can be used. The animal models for obstructive disorders such as those discussed above mimic the clinical condition and replicate closely the human pathophysiology. The monkey is particularly useful for these studies because, like humans, the pregnant monkey uterus is susceptible to premature labor and late gestational miscarriage (9). However, suitable animal models for most human genetic and metabolic disorders do not exist.

The Future of Fetal Research

The benefits of fetal research include (i) the development of vaccines, (ii) advances in prenatal diagnosis, (iii) detection of anatomical malformations, (iv) assessment of safe and effective medications, and (v) the development and refinement of in utero surgical therapies. Animal models have been essential in the advancement of most of these applications and are vital for determining potential risks before clinical applications. Frequently, appropriate animal models are not available or are inadequate for risk assessment. In these instances clinical fetal research becomes important; many diseases and malformations occur during fetal development and if the problems can be addressed early, often before birth, the neonate stands a much better chance of living a normal life. Current federal regulations limit fetal research to only those procedures that pose "minimal risk" to the fetus or that can be of direct therapeutic benefit to an endangered fetus (1). Perhaps, until we fully explore the ethical issues surrounding fetal research, this is an appropriate standard. Nevertheless, advances continue to be made in laboratory animal experiments and in countries where the potential benefits of clinical fetal research are regarded as outweighing the potential erosion of ethical standards (20). Clearly, there has been a decline in the number of investigators willing to face criticism consequent to conducting clinical fetal research in spite of the potential benefits generated by such studies (1). Thus, fetal research at present appears to have plateaued and may advance only slowly until the larger questions affecting social responsibility are addressed.

Fetal Tissue Research

Fetal tissue research differs from fetal research in that it involves studies on fetal cells rather than on living fetuses. Fetal tissue research has benefited a number of biomedical areas by providing cell lines to study gene regulation, pattern formation during embryogenesis, and model systems for cell interaction and function. Vaccines, such as the polio vaccine, have been developed in fetal tissues, and a variety of studies on cell growth and regulation have led to an understanding of chromosomal abnormalities, cancer and tumorigenesis, and cellular immunology (Table 2). These advances are possible because of some of the unique characteristics of fetal cells. They have the ability to rapidly divide and grow in culture, are

pluripotent with respect to their developmental lineage, may be cryopreserved and subsequently reanimated, have lower antigenicity, and will survive and grow if transplanted into a supportive host environment.

Fetal cells are used to establish cell lines that provide model systems with which to study events in cell differentiation and growth. In vitro and in vivo analyses of stem cell lineages are instrumental in helping researchers better understand complex cell interactions during normal and abnormal fetal development (21). The process of culturing and growing fetal cells has been used by molecular biologists to understand gene regulation, protein synthesis, and other cellular mechanisms. Genetic engineering experiments have advanced to the point where investigators can now immortalize cells and develop gene constructs that can be used to design cells that express specific functional or secretory activities (22). Additionally, fetal cells are used to replicate human viruses that may be used to develop and test vaccines (23). The rapidly dividing fetal cells of the central nervous system are used to test their susceptibility to the acquired immunodeficiency syndrome (AIDS) virus, studies crucial for determining the rate of infection of fetal cells to maternally transmitted AIDS (24). Finally, fetal cells are used to screen new pharmaceutical agents to determine their risk as teratogens or carcinogens. These experiments are essential before clinical trials may be undertaken. One need only recall the thalidomide episode of the 1960s as a grim reminder of the value of careful fetal screening before patient use. Maternal intake of the sedative thalidomide early in pregnancy, as reported in Germany and England, led to an unusually high incidence of limb-reduction deformities (25). Once thalidomide was recognized as the causative agent, it was withdrawn from the market, but not before an estimated 3000 malformed infants were born (26).

One of the more exciting and promising applications of fetal tissue research has been the use of fetal cells as therapeutic tools to treat clinical disorders. In one such instance, fetal cells were used to treat another fetus in utero. In June 1988, French physicians Jean-Louis

Touraine, an immunologist, and obstetrician Daniel Raudrant treated a 30-week-old fetus diagnosed with a rare, and nearly always fatal, immune deficiency disease (bare lymphocyte syndrome) by injecting immune cells from the thymus and liver of two aborted fetuses into the umbilical cord of the deficient fetus (27). This daring clinical experiment was based on the results of animal studies that demonstrated that second trimester fetal liver contains a rich source of hematopoietic stem cells (28). At this stage, the donor fetus immune system is not yet developed, so normally histoincompatible stem cells may be transplanted into the immunodeficient host fetus to establish a viable population of reconstituted T cells (29). After the birth of this infant, a second injection of cells was given; subsequent blood tests suggest that some of the cells have seeded and multiplied in the infant's spleen, liver, and bone marrow. Although there is hope that this infant will develop a normal immune system, the prognosis is still guarded.

Fetal cells are also being examined in animal studies and clinical trials for their potential to reverse insulin-deficient diabetes mellitus, a disease that affects millions of people worldwide, including an estimated 11 million in the United States (30). Animal experiments demonstrate that if fetal pancreas is transplanted before the differentiation of the problematic exocrine cells (which produce the lytic digestive enzymes), subsequent development of the islet cells necessary for insulin production will occur (31). However, one limitation for the success of fetal pancreatic transplants has been the presence of significant quantities of immunogenetic lymphoid tissue in the pancreas. Selective cell culturing before transplantation of the pancreas has been successful in removing most of the immunogenetic cells (32). If this approach ultimately proves successful, it will be an important advance in this field of transplantation because immunosuppression to avoid graft rejection is undesirable in diabetics who already are at increased risk for infection. Cell transplant efforts such as this highlight the benefits of using fetal cells: their ability to survive and multiply necessitates the grafting of only small numbers of cells; their lower or absent antigenicity

eliminates the requirement of tissue matching and immunosuppression; and they are adaptable to the host environment. Although fetal islet cell transplantation is still experimental, initial clinical trials suggest that some patients who have received grafts can produce their own insulin, thus decreasing their requirement of daily exogenous insulin (33). However, these results are not universal and a number of questions remain (34).

Nowhere else is the revolutionary idea of fetal cell transplantation as a therapeutic tool more evident than as a treatment for neurodegenerative disorders such as parkinsonism or Alzheimer's disease. Because these diseases are progressive and affect millions of people in the United States alone, medical researchers have explored the feasibility of grafted fetal nerve cells to restore damaged neural circuits. The economic impact of these neurodegenerative diseases is significant when one considers the lost productivity, forced early retirement, and costs of therapy and nursing care.

The field of neural transplantation has a long and varied history, but its full potential perhaps was revealed as a result of pioneering studies by Olson and colleagues (35). These investigators demonstrated that the anterior eye chamber provided a nutritive and immunologically "privileged site" for the transplantation of cells. Moreover, subsequent experiments showed that grafted neurons could reinnervate a previously denervated target (36). However, adult neurons do not divide and, once damaged or lost, cannot be replenished from endogenous sources. Because of this, grafts of fetal neurons were investigated and found to be capable of partially reestablishing damaged neural circuits (37). Not only could fetal cells survive and establish neural contacts, but they also had the ability to synthesize and release appropriate transmitter substances (38). Additionally, implanted fetal nerve cells could ameliorate specific cognitive (39), neuroendocrine (40), and motor deficits (41). Virtually every region of the central nervous system, from the olfactory neuroepithelium to the spinal cord, can be grafted, with minimal immunological consequence (35).

Although a variety of animal models of neurodegenerative disorders are used by transplant neurobiologists, the nonhuman primate model of Parkinson's disease has provided the greatest incentive for fetal grafting experiments. The selective neurotoxin *N*-methyl-4-phenyl-1,2,3,6-tetrahydropyridine (MPTP) administered to monkeys produces a parkinsonian-like syndrome characterized by rigidity, resting tremor, progressive akinesia, flexed posture, and episodes of freezing during movement (*42*). Moreover, MPTP rather selectively affects the dopaminergic cells of a brainstem region known as the substantia nigra, causing their degeneration in a manner that anatomically and physiologically mimics human parkinsonism (*43*). The loss of the neurotransmitter dopamine from these neurons, which project to target neurons in a large subcortical region termed the striatum, appears responsible for the movement disorders of parkinsonism. Many Parkinson's disease patients, therefore, benefit from the administration of the drug L-dopa, a dopamine precursor, which is converted to dopamine in the brain to replenish the dopamine deficient striatum.

However, L-dopa therapy is ineffective in a large percentage of patients, and other patients progressively become refractory to the drug over a period of 5 to 10 years. Because few pharmacologic options remain for these patients, neural grafting of dopamine-producing cells has been considered as one alternative. Initial studies in rodents with grafted fetal tissue from the mesencephalon brain region giving rise to dopaminergic neurons of the substantia nigra were promising (*41*). The use of fetal dopaminergic neurons appeared obvious because one could replace degenerating cells in the host with cells of like origin that presumably carry the correct genetic programs for dopamine synthesis, cell growth, connectivity, transmitter release, and receptivity.

Subsequent experiments by Redmond and colleagues (*43*) in African Green monkeys that were rendered parkinsonian by MPTP administration confirmed earlier rodent studies and demonstrated the efficacy of fetal nerve cell transplantation in nonhuman primates. Seven and one-half months after transplantation of fetal nigral neurons, these investigators observed significant behavioral improvement, as well as dopamine neuron survival and increased levels of dopamine in the host striatum. The implanted neurons appeared integrated with the host, extending numerous small fibers into the adjacent neural parenchyma. Long-term studies and experiments to determine the specificity of fiber sprouting from the implanted neurons are not yet completed, so caution is warranted (*44*). Nevertheless, current scientific wisdom suggests that fetal dopaminergic neurons presently may be the best tissue source to graft in parkinsonian patients, usurping the use of the adrenal medullary autografts, which exhibit very poor survival in monkeys (*45*) and have minimal effects as used presently in humans (*46*). Several centers around the world, including two in the United States, have already performed human fetal nigral grafts in patients with Parkinson's disease. It is still too early to objectively assess the results.

The Future of Fetal Tissue Research

The benefits of studying fetal cells are many, and the clinical potential for their use as therapeutic tools is just now being realized (Table 2). Vaccine development, study of human viruses and the development of specific therapies for the treatment of infections such as AIDS, the assessment of risk factors and toxicity levels in drug production, and the initiation of transplantation trials are important and necessary contributions of fetal cell research to biomedical science (*1*). Ongoing animal experiments and a source of human fetal cells are critical for studying fatal blood diseases (sickle-cell anemia, aplastic anemia, and leukemia), or for addressing nervous system disorders including optic nerve damage, degenerative disorders of the brain, and spinal cord damage. On the horizon lies the potential to reverse insulin-deficient diabetes and immunodeficiency disorders and to address cognitive dysfunctions. Current federal and state regulations permit the use of fetal tissues and cells obtained from dead fetuses, and all 50 states have adopted the Uniform Anatomical Gift Act, which sustains this essential need for continued research to advance our scientific knowledge and biomedical applications (*1*). Such advances have brought us to the point where we no longer stand by helplessly in the face of fetal malformations, nor are we left impotent to respond to treatable disorders. With a growing ability to diagnose and treat, with a new-found knowledge to shape and direct developmental events, and with an awareness of how to replace and restore that which is old, we must remain cognizant of the delicate interplay between responsible moral behavior and the desire to maintain and improve the quality of human life.

References and Notes

1. D. Lehrman, *Summary: Fetal Research and Fetal Tissue Research* (Association of American Medical Colleges, Washington, DC, 1988).
2. L. M. Hill, R. Breckle, W. C. Gehrking, *Mayo Clin. Proc.* **58**, 805 (1983).
3. P. W. Soothill, K. H. Nicolaides, C. H. Rodeck, *J. Perinat. Med.* **15**, 117 (1987).
4. K. H. Nicolaides and S. Campbell, in *Reviews in Perinatal Medicine*, E. M. Scarpelli and E. V. Cosmi, Eds. (Liss, New York, 1989), pp. 1–30.
5. W. L. Nyhan, *Clin. Symposia* **32**, 31 (1980).
6. A. A. Lemke, in *Current Perinatology*, M. Rathi, Ed. (Springer-Verlag, New York, 1989), pp. 247–256.
7. V. A. McKusick, *Mendelian Inheritance in Man* (Johns Hopkins Univ. Press, Baltimore, MD, ed. 8, 1988).
8. H. Galjaard, in *Reviews on Perinatal Medicine*, E. M. Scarpelli and E. V. Cosmi, Eds. (Liss, New York, 1989), pp. 133–172.
9. M. R. Harrison, M. S. Golbus, R. A. Filly, *The Unborn Patient* (Grune & Stratton, Orlando, FL, 1984).
10. J. Friedland, G. Perle, A. Saifer, L. Schneck., B. W. Volk, *Proc. Soc. Exp. Biol. Med.* **136**, 1297 (1971).
11. J. L. Simpson, M. S. Golbus, A. O. Martin, G. E. Sarto, *Genetics in Obstetrics and Gynecology* (Grune & Stratton, New York, 1982) p. 110.
12. G. G. Rhoads *et al.*, *N. Engl. J. Med.* **320**, 609 (1989).
13. A. Ludomirski and S. Weiner, in *Clinical Obstetrics and Gynecology*, R. M. Pitken, Ed. (Harper & Row, Hagerstown, MD, 1988), vol. 31, pp. 19–26.
14. C. H. Rodeck, A. D. Patrick, M. E. Pembrey, C. Tzannatos, A. E. Whitefield, *Lancet* ii, 297 (1982).
15. D. F. Hawkins, in *Reviews in Perinatal Medicine*, E. M. Scarpelli and E. V. Cosmi, Eds. (Liss, New York, 1989), pp. 91–131.
16. J. W. Wladimiroff and P. A. Stewart, *Br. J. Hosp. Med.* **34**, 134 (1985).
17. W. H. Clewell, *ibid.*, p. 149; W. H. Clewell *et al.*, *N. Engl. J. Med.* **306**, 1320 (1982).
18. M. R. Harrison, D. K. Nakayama, R. Noall, *J. Pediatr. Surg.* **17**, 965 (1982).
19. M. R. Harrison, J. Anderson, M. A. Rosen, N. A. Ross, A. G. Hendrickx, *ibid.*, p. 115; M. R. Harrison, N. Ross, R. Noall, A. A. de Lorimier, *ibid.* **18**, 247 (1983).
20. B. J. Culliton, *Science* **241**, 1423 (1988); *ibid.* **242**, 1625 (1988); D. E. Koshland, Jr., *ibid.* **241**, 1733 (1988).
21. C. S. Goodman and N. C. Spitzer, *Nature* **280**, 208 (1979); N. LeDouarin, *The Neural Crest* (Cambridge Univ. Press, Cambridge, 1982).
22. E. Nichols, *Human Gene Therapy* (Harvard Univ. Press. Cambridge, MA, 1988); H. Varmus, *Science* **240**, 1427 (1988).
23. J. F. Enders, T. H. Weller, F. C. Robbins, *Science* **109**, 85 (1949).
24. H. Minkoff, D. Nanda, R. Menez, S. Fikrig, *Obstet. Gynecol.* **69**, 285 (1987); S. Sprecher, G. Soumenkoff, F. Puissant, M. Deguedre, *Lancet* i, 288 (1986).
25. W. G. McBride, *Lancet* ii, 1358 (1961); W. Leng, *ibid.* i, 45 (1962).
26. C. S. Delahunt and L. J. Lassen, *Science* **146**, 1300 (1964); A. G. Hendrickx, L. R. Axelrod, L. D. Claborn, *Nature* **201**, 958 (1966); E. Rubin and J. L. Farber, *Pathology* (Lippincott, Philadelphia, PA, 1988).
27. J. Langnone, *Time* **133**, 71 (3 April 1989); at the time of this writing we are unaware of any peer-reviewed report of this clinical trial.
28. R. Namikawa *et al.*, *Immunology* **57**, 61 (1986).
29. J.-L. Touraine, *Immunol. Rev.* **71**, 103 (1983).
30. *Diabetes Facts and Figures* (American Diabetes Association, Alexandria, VA, 1986).
31. J. Brown, J. A. Danilovs, W. R. Clark, Y. S. Mullen, *World J. Surg.* **8**, 152 (1984); T. E. Mandel, *ibid.*, p. 158; C. J. Simeonovic, K. M. Bowen, I. Kotlarski, K. J. Lafferty, *Transplantation* **30**, 174 (1980).
32. D. A. Hullett *et al.*, *Transplantation* **43**, 18 (1987).
33. G. Farkas, S. Karacsonyi, M. Szabo, G. Kaiser,

Transplant Proc. 19, 2352 (1987).

34. B. E. Tuch, A. G. R. Sheil, A. B. P. Ng, R. J. Trent, J. R. Turtle, *Transplantation* 46, 865 (1988).

35. L. Olson, *Histochemie* 22, 1 (1970); L. Olson, A. Björklund, B. J. Hoffer, in *Neural Transplants: Development and Function*, J. R. Sladek, Jr., and D. M. Gash, Eds. (Plenum, New York, 1984), pp. 125–165.

36. B. K. Beebe, K. Mollgard, A. Björklund, U. Stenevi, *Brain Res.* 167, 391 (1979); A. Björklund and U. Stenevi, *Cell Tissue Res.* 185, 289 (1977).

37. R. D. Lund and S. D. Hauschka, *Science* 193, 582 (1976); U. Stenevi, A. Björklund, N. A. Svendgaard, *Brain Res.* 114, 1 (1976).

38. R. H. Schmidt, A. Björklund, U. Stenevi, S. B. Dunnett, F. H. Gage, *Acta Physiol. Scand.* 522, 19 (1983).

39. K. A. Crutcher, R. P. Kesner, J. M. Novak, *Brain Res.* 262, 91 (1983); S. B. Dunnett, W. C. Low, S. D. Iversen, U. Stenevi, A. Björklund, *ibid.* 251, 335 (1982); F. H. Gage, A. Björklund, U. Stenevi, S. B. Dunnett, P. A. T. Kelley, *Science* 225, 533 (1984); W. C. Low *et al.*, *Nature* 300, 260 (1982).

40. D. Gash, C. Sladek, J. R. Sladek, Jr., *Peptides* 1, 125 (1980); D. Gash, J. R. Sladek, Jr., C. D. Sladek, *Science* 210, 1367 (1980); M. J. Gibson *et al.*, *ibid.* 225, 949 (1984); A. J. Silverman *et al.*, *Neuroscience* 16, 69 (1985).

41. W. J. Freed *et al.*, *Ann. Neurol.* 8, 510 (1980); M. J. Perlow *et al.*, *Science* 204, 643 (1979).

42. R. Burns, S. Markey, J. Phillips, C. Chiueh, *Can. J. Neurol. Sci.* 11, 166 (1984); J. Langston, L. S. Forno, C. S. Rebert, I. Irwin, *Brain Res.* 292, 390 (1984).

43. D. E. Redmond, Jr. *et al.*, *Lancet* i, 1125 (1986); J. R. Sladek, Jr., T. J. Collier, S. N. Haber, R. H. Roth, D. E. Redmond, Jr., *Brain Res. Bull.* 17, 809 (1986).

44. R. Lewin, *Science* 237, 245 (1987); J. R. Sladek, Jr., and I. Shoulson, *ibid.* 240, 1386 (1988).

45. J. T. Hansen *et al.*, *Exp. Neurol.* 102, 65 (1988).

46. C. G. Goetz *et al.*, *N. Engl. J. Med.* 320, 337 (1989).

47. W. A. Hogge and M. S. Golbus, in *Prenatal and Perinatal Biology and Medicine*, N. Kretchmer, E. J. Quilligan, J. D. Johnson, Eds. (Harwood Academic Publishers, Chur, Switzerland, 1987), pp. 29–67.

48. E. M. Scarpelli and E. V. Cosimi, Eds., *Reviews in Perinatal Medicine* (Liss, New York, 1989).

49. We thank all of our colleagues involved with our transplant programs, especially D. E. Redmond, D. M. Gash, T. J. Collier, R. H. Roth, and S. N. Haber. Several of the studies cited in the article were supported by USPHS grants PO1-NS24032 and PO1-NS25778, and a grant from the Pew Charitable Trust. We also thank D. Baram for helpful suggestions.

20

How Many Species
Are There on Earth?

Robert M. May

Over a century ago, Darwin and others provided the broad outline of an answer to the question of how life has evolved on Earth and how species originate. The next question would seem to be how we use this basic understanding to estimate—from first principles—how many species are likely to be found in a given region or, indeed, on Earth as a whole.

Surprisingly, this question of "how many species?" has received relatively little systematic attention, from Darwin's time to our own. At the purely factual level, we do not know to within an order of magnitude how many species of plants and animals we share the globe with: fewer than 2 million are currently classified, and estimates of the total number range from under 5 million to more than 50 million. At the theoretical level, things are even worse: we cannot explain from first principles why the global total is of the general order of 10^7 rather than 10^4 or 10^{10}.

This article first surveys various kinds of empirical and theoretical studies that are helping to give us a better idea of how many species, or how many individual organisms, we might expect to find in a given environment. Such studies include the structure of food webs, patterns in the relative abundance of species, patterns in the number of species or number of individuals in different categories of physical size, and general observations about trends in the commonness or rarity of organisms. The article then reviews current evidence about the total number of species on Earth, indicating lines of research that could sharpen the estimates. We do not end up with a list of answers, but rather with a list of more sharply focused questions.

The Structure of Food Webs

Cohen and Briand (*1, 2*) have compiled and analyzed a catalog that now includes 113 food webs, embracing a wide variety of natural environments (55 food webs from continental settings—23 terrestrial and 32 aquatic—along with 45 coastal and 13 oceanic webs). The data for these food webs are of uneven quality, with the most notable problem being that some studies identify individual species ("blue jays") whereas others deal with aggregates ("spiders," "copepods," or even "zooplankton"); some studies articulate individual species of predators at upper levels but aggregate coarsely at lower trophic levels (*3*). Even so, some remarkable regularities emerge from Cohen and Briand's analysis of these data (*1, 2*).

For one thing, the average number of other species with which any one species inter-

R. M. May is in the Department of Zoology, Oxford University, Oxford, OX1 3PS, England. This chapter is reprinted from *Science* **241**, 1441 (1988).

acts directly is consistently around 3 to 5 (*4*). The number is consistently higher (average, 4.6) in relatively constant environments than in fluctuating ones (average, 3.2). There are also consistent and quantitative patterns in the proportions of basal, intermediate, and top predator species (those whose links reach only upward, both ways, and only downward, respectively); the ratios are 0.19 : 0.53 : 0.29, respectively (*5*). A similar pattern of "link-scaling" invariance is found for the ratio of links among the four categories of basal-intermediate, basal-top, intermediate-intermediate, and intermediate-top (0.27 : 0.08 : 0.30 : 0.35, respectively). Most interestingly, these two quantitative patterns in the proportions of species in different trophic categories, and in "link-scaling," can be deduced from the empirical observation that each species is directly connected to roughly four others, along with the assumption that the species are ordered in a cascade or hierarchy, such that a given species can prey on only those below it and can be preyed on only by those species above it in the hierarchy [an assumption that several authors (*2, 6*) have independently suggested may follow from body-size considerations between predators and their prey].

Other patterns in the ratios between numbers of interacting species in different trophic levels are the subject of continuing investigation. Hawkins and Lawton (*7*) have observed that food chains comprising green plants, insect herbivores, and insect parasitoids include over half of all known species of metazoans, so that understanding what determines the richness of parasitoid species could be a major step toward understanding the diversity of terrestrial communities. They analyzed data for 285 species of herbivorous insects, from 42 families in Britain, and found the typical such species to be attacked by 5 to 10 species of parasitoids; the number depends significantly on the geographical range of the host insect, on the architecture of the host plant, and to a lesser extent on a variety of other factors (*7*). Preliminary data suggest that the tropics are roughly similar to Britain, in that herbivorous insects are hosts to around five to ten species of parasitoids (*8*). Other studies document systematic patterns in the number of

phytophagous insect species associated with different plant hosts (*9*) and in the ratios between numbers of species of prey and predators of various kinds (*10*).

It could be that many of these apparent patterns tell us more about the workings of the human mind, and about how we tend to collect and categorize data, than they do about the natural world (*11*). Moreover, the populations in real food webs can have extremely complex dynamical behavior, with nonlinearities in density-dependent factors producing cyclic or chaotic changes in abundance and with unpredictable environmental fluctuations adding further complications; it seems unlikely that the salient features of such dynamical systems can be captured in static analyses of food web graphs (*12*). These caveats and complications notwithstanding, the patterns discussed above are intriguing. If they stand up to further study, they could simplify the task of understanding diversity. It could be, for example, that one need only understand what determines the number of plant species, and then the total faunal diversity could be deduced from appropriate rules.

Relative Abundance of Species

Real understanding of food webs in particular, and of diversity in general, must go beyond the mere presence or absence of species to an understanding of relative abundance. In early successional communities, and in environments disturbed by toxins or "enriched" by pollution, steeply graded distributions of species relative abundance (SRA) are commonly seen, with a handful of dominant species accounting for most of the individuals present. Conversely, in relatively undisturbed "climax" communities consisting of many species, relatively even distributions of relative abundance are typical; very often, such SRAs are distributed according to a "canonical lognormal" distribution, as illustrated in Fig. 1.

Such trends in SRAs show up in studies of old field succession (*13*). The effects of pollution or other systematic disturbances reveal the same trend, except that time effectively runs

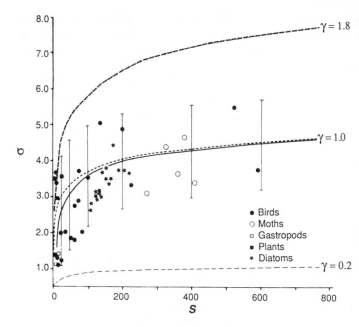

Fig. 1. A plot of S, the number of species, versus σ, the standard deviation of the logarithms of the relative abundances, for various communities of birds, moths, gastropods, plants, and diatoms. The dashed line labeled $\gamma = 1.0$ shows the relation between S and σ for Preston's (15) "canonical" lognormal distribution; the lines labeled $\gamma = 0.2$ and $\gamma = 1.8$ are the bounds to the range of S-σ relations that might be expected from general mathematical properties of the lognormal distribution, for large S and reasonable ranges of values for the total number of individuals, N. The solid line is the mean relation predicted by Sugihara's (17) model, and the error bars represent ± 2 standard deviations about this mean.

backward, so that the progression is from evenness to dominance (14).

It is not surprising that the relative abundances within a fairly large and relatively undisturbed group of species will be disturbed lognormally. The relative abundances are likely to be governed by the interplay of many more or less independent factors. It is in the nature of the equations of population dynamics that these several factors should compound multiplicatively, and the statistical central limit theorem applied to such a product of factors implies a lognormal distribution. This general observation, however, tells us nothing about the relation between σ (the standard deviation of the logarithms of the relative abundances) and S (the total number of species present). The puzzling fact is that very many assemblies have SRAs that obey the canonical lognormal distribution, that is, that have the unique relation between σ and S illustrated by the curve labeled $\gamma = 1.0$ in Fig. 1, although this curve represents just one of an infinite family of possible lognormal distributions (15).

It has been conjectured that the canonical property may be merely an approximate mathematical property of all lognormal distributions for large S; the dashed curves in Fig. 1 labeled $\gamma = 1.8$ and $\gamma = 0.2$ represent plausible boun-

daries to the σ-S relation on this basis (16). The data put together by Sugihara (17) in Fig. 1 make it clear, however, that real SRAs obey the canonical relation more closely than can be explained by these mathematical generalities alone.

Sugihara has also suggested a biological mechanism that will produce the observed patterns. He imagines the multidimensional "niche space" of the community as being a hypervolume broken up sequentially by the component species (with any fragment being equally likely to be chosen for the next breakage, regardless of size), such that each of the S fragments denotes the relative abundance of a species. Although the biological status of this assumption is debatable, it generates patterns of SRA in accord with a large number of data (the solid line in Fig. 1 shows the mean relation between S and σ predicted by the model, and the error bars show the range of ± 2 standard deviations about the mean). Such a fit does not, of course, validate the model; it is possible that other biological assumptions could produce similar distributions of SRA.

One problem with essentially all the data that have been compiled for SRA is that they focus on particular taxonomic groups ("birds," "moths"). To understand how communities are assembled, it may be more relevant to inquire about the relative abundance within ecologically similar groups (putting birds together with bats and some large insects, for instance).

Number of Species Versus Physical Size

A variety of other patterns in the distribution and abundance of organisms have received little attention. For example, how many species do we expect to find in different categories of physical size, within a given region?

The meager amount of available information bearing on this question is reviewed elsewhere (18). Figure 2 gives one representative study, showing the way in which all 3000 or so mammalian species, excluding bats and marine mammals, are apportioned among mass classes (19). A corresponding analysis, but restricted to the mammal species of Britain, again excluding bats and marine mammals, is also shown in Fig. 2 (18). Although Britain's mammals appear to obey the global pattern of species versus size, appropriately scaled down, this may not be true in general; there is no a priori reason to expect the species-size patterns for faunal assemblies from relatively small areas to be the same as those from large (and correspondingly more environmentally diverse) areas.

Figure 2 and similar analyses represent rough assessments of the facts. Very few ideas have been advanced in explanation of these facts about species-size distributions. Hutchinson and MacArthur (20) have advanced arguments for expecting an L^{-2} relation between the number of species and the characteristic length of constituent individuals, L. The argument is essentially that, for terrestrial organisms, the world is seen as two-dimensional, and therefore the possibility of finding new roles (and thence new species) may scale as L^{-2}. This conjectured L^{-2}, or $M^{-2/3}$, relation, where M is mass, is illustrated by the dashed straight line in Fig. 2.

Number of Individuals Versus Physical Size

Other patterns can be sought in the relation between numbers of individuals and their physical size (mass or characteristic length). For example, in a particular region, how is the number of individual animals in the size class from 0.1 to 1 cm related to the number in the class from 1 to 10 cm (21, 22)?

In particular, Morse et al. (21) and Brown and Maurer (23) have collated data about populations of phytophagous insects and of birds, respectively, and have advanced qualitative explanations for these data. Morse et al. began with the assumption that roughly equal amounts of energy flow through each size category; although very unlikely to be true in

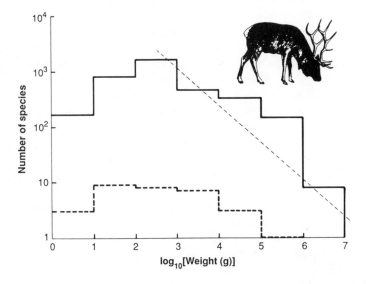

Fig. 2. The numbers of species, S, of all terrestrial mammals (solid histogram) and of British mammals (dashed histogram), excluding bats, are shown distributed according to mass categories (mass expressed in grams) (18, 19). Note the doubly logarithmic scale. The thin dashed line illustrates the shape of the relation $S \sim L^{-2}$, where L is the characteristic length (20).

general, this assumption is supported by some evidence from organisms ranging widely in size (*24*). Given this assumption, along with the usual manner in which metabolic costs become relatively larger at smaller sizes, the total number of individuals, N, in the size class with characteristic mass M and length L may be expected to scale as $N \sim M^{-0.75} \sim L^{-2.25}$ (*25*). That is, for a 10-fold decrease in characteristic length we would, on this basis, expect a roughly 180-fold increase in the total number of individuals.

Recent insights into the fractal geometry of nature suggest, however, that the structure of the habitat—and hence the number of possible ways of making a living—is unlikely to scale linearly with L (*26*). Consider, for example, the circumference of a large tree, or any other "one-dimensional" object. If we measure it on a 10-cm scale, we get one answer. On a 1-cm scale, we will often get another, larger answer. A yet larger answer would be obtained on a 1-mm scale, and so on. The circumference of the tree is thus not simply one dimensional but has a "fractal dimension," D, such that the perceived length, $l(\lambda)$, depends on the step-length of measurement, λ, as λ^{1-D} : $l(\lambda) = c\lambda^{1-D}$, where c is a constant. If $D = 1.5$, for example, a 10-fold reduction in the measurement scale (from, say 10 cm to 1 cm) will result in the apparent length increasing by a factor

$10^{0.5} \simeq 3$. Morse *et al.* applied these notions to measure the profiles of various kinds of vegetation at different scales, concluding that D for such habitats ranged from around 1.3 to around 1.8, with an average around 1.5 (Fig. 3). That is, for herbivorous insects that exploit their surroundings in an essentially one-dimensional way (using the edges of leaves, or the like) a 10-fold decrease in physical size produces a roughly 3-fold increase in the apparently available habitat; for creatures that exploit their environment in an essentially two-dimensional way (using surfaces rather than edges), the effect must be squared, so that a 10-fold decrease in physical size produces an effectively 10-fold increase in apparent habitat. These two factors (the one-dimensional factor 3 and the two-dimensional factor 10) are likely to bound the range of possibilities found in actual assemblies of insects.

Combining these fractal aspects of habitat perception with the metabolic considerations discussed above, Morse *et al.* concluded that a 10-fold decrease in characteristic length, L, is likely to produce an increase in N that lies between 500 and 2000 (that is, roughly between 3 and 10 times 180). As shown in Fig. 4, this very rough expectation is borne out surprisingly well by data for the number of individual arthropods of different body lengths found on

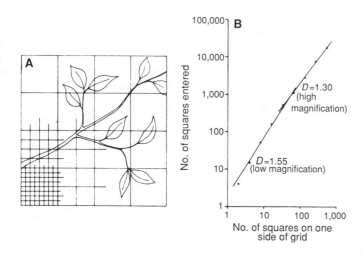

Fig. 3. (A) Photographs of plants at various magnifications were placed under a grid by Morse *et al.* (*21*). The number of squares entered by the outline of the plant were counted, starting with a coarse grid of two large squares on one side, then 2^n squares, with n varying from 2 to 6 or 7, depending on the grid size. For ease of representation, the plant's leaves in this figure are drawn flat; in reality they are oriented at all angles with respect to the grid. Also for clarity, the progressively finer divisions are only illustrated in one corner of the figure. The logarithm of the number of squares entered by the outline of the plant is then plotted against the logarithm of the number of squares along one side of the grid, as shown in (B). The slope of the line equals the fractal dimension, D. (B) Data gathered in this way for Virginia creeper, photographed without leaves in early spring. The twigs were photographed at one scale, then parts of the same twigs were rephotographed at a higher magnification, permitting D to be estimated at two levels of resolution.

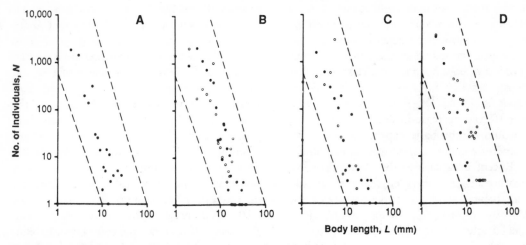

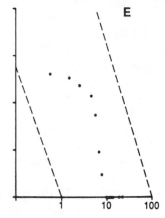

Fig. 4. Data plotted by Morse *et al.* (*21*) on the number of individual arthropods (mainly insects) of different body lengths, collected from vegetation: (**A**) understory foliage in primary forest in Costa Rica; (**B**) Osa secondary vegetation (solid dots) and Kansas secondary vegetation (open dots); (**C**) Tobago primary riparian vegetation (solid dots) and Icacos vegetation (open dots); (**D**) understory foliage in cacao plantations in Dominica (solid dots) and in Costa Rica (open dots); and (**E**) birch trees at Skipwith Common, North Yorkshire. The lower bound prediction that, for an order of magnitude decrease in body length, there should be a roughly 500-fold increase in the number of individuals, is indicated by the lower dashed line on each graph. The upper bound prediction—roughly 2000-fold increase for an order of magnitude decrease in body length—is shown by the upper dashed line.

vegetation in places ranging from primary forests, primary riparian vegetation, and secondary vegetation in the New World Tropics to temperate habits, for example, birch trees on Skipwith Common in Yorkshire.

The study by Morse *et al.* is a frankly speculative one. I have chosen to highlight it because it provides an explicit example where our thinking about aspects of population abundance and diversity needs to acknowledge that nature is often not Euclidean but rather may have fractal geometry, with organisms existing in spatial and temporal frameworks that are, as it were, jagged on every scale (*27*). This is an example where new mathematical concepts interact with biological ideas in potentially surprising ways (the chaotic behavior of many simple population models is another example).

Species Numbers, Species Abundance, and Body Length

Still other studies have focused on empirical relations between the abundance of individual species and the body size of constituent individuals (*25, 28*). I think any eventual understanding of the total number of species in a given environment, and thence ultimately of the diversity of life on Earth, will need to be based on a clear understanding of the interplay among all the factors discussed above. Yet most of the few existing studies have singled out one or other aspect (species size, species abundance) from the interwoven mosaic.

Exceptions are the work on birds by Brown and Maurer (*23*) and the recent study by Morse *et al.* of the relations among species number, species abundance, and body length for 859 species of arboreal beetles in lowland rain-forest trees in Borneo (*29*). Figure 5A summarizes the results, showing the total number of species in different categories of population abundance and physical size (both plotted logarithmically); Fig. 5, B through E, correspondingly shows the number of species in different trophic categories. Although Fig. 5 does have some interesting structural details [for discussion, see (*29*)], it is essentially simple. It is encouraging that Fig. 5 has the basic features one would have guessed from the separate studies of species abundance, species size, and abundance size in different groups, as discussed above.

Commonness and Rarity

In the discussion above, some of the species found in a given region are confined to that region, whereas others (which are part of the species-size and other distributions in the region) are distributed much more widely.

Partly for this reason, and partly for its intrinsic interest, it would be nice to know more about the distribution of geographical ranges within different taxonomic groups of species. What fraction of all bird species, for example, range globally over 10% of the globe, over 1%, and so on? Hanski (*30*), Brown (*31*), Root (*32*), Rapoport (*33*), and others have made a start toward answering this question, for diverse collections of organisms including vascular plants, intertidal invertebrates, terrestrial arthropods, planktonic crustaceans, and terrestrial vertebrates (especially birds), but much remains to be learned.

Intuitive ideas about commonness and rarity usually make reference both to geographical distribution and to local abundance. Such considerations often swirl together in ways that make it difficult to define exactly what constitutes a rare species. One type of rareness is, for example, exhibited by the silver sword, *Argyroxyphium macrocephalum*, that grows only in the crater of the Haleakala volcano on Maui. Although there are around 50,000 individuals in the large, local popula-

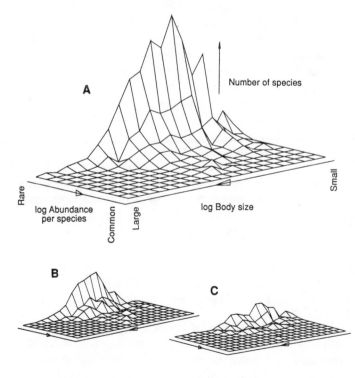

Fig. 5. (A) The height of each intersection is proportional to the number of beetle species that have a particular combination of body length [plotted logarithmically on a scale that extends from 0.5 mm, "small," to 30 mm, "large"] and abundance [plotted as logarithms to the base 2, on the conventional "octave" scale of Preston (*15*)]; for details, see (*29*). (B through E) The same information for the separate beetle guilds of herbivores, predators, fungivores, and scavengers, respectively (*29*).

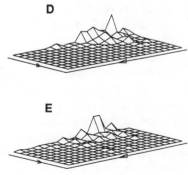

tion of this plant, its restriction to the one volcanic crater would make it very rare by most definitions. Another type of rareness is exhibited by the grass *Setaria geniculata*, which is found from Massachusetts to California and on down through tropical South America to Argentina and Chile but which is not abundant anywhere. This grass is rare in the sense that its populations are "chronically sparse" (*34*) everywhere in its broad range.

There have been a variety of proposals for codifying ideas about commonness and rarity. In particular, Rabinowitz *et al.* (*34*) have considered three different kinds of questions that arise in thinking about rarity: (i) is the species distributed over a broad geographical area, or is it endemic to some restricted location; (ii) whatever its range, is the species found in a wide variety of habitats, or is it specialized to one kind of site; and (iii) is the species abundant somewhere in its range, or are its numbers everywhere small. These three considerations combine to give eight categories, only one of which (broad distribution, unspecialized habitat, large populations) ordinarily corresponds to the species being called "common." Rabinowitz *et al.* noted that the archetypal "rare" species, with narrow distribution, specialized habitat, and small numbers, represents only one of several different kinds of rarity. These investigators pursued their ideas by applying them to the plants surveyed in the *Biological Flora of the British Isles* (which gives detailed distribution maps and notes about the habitat and population of 177 of the 1822 native British plants). Rabinowitz *et al.* asked 15 colleagues to classify each of the 177 species according to the eightfold category scheme described above (*35*). This process gave clear consensus for 160 of the 177 species, and the results are summarized in Table 1.

Most species (149 versus 11) are abundant somewhere, and most species (137 versus 23) have a wide geographical range (Table 1). A narrower majority (94 versus 66) have restricted habitat specificity.

Of the eight categories, species with wide ranges and large population sizes, but restricted habitat specificities, predominate (71 of 160 species, or 44%). Most of these "rare habitat"

species are specialists of marsh, sand dunes, bogs, or forest floors; wherever their habitat exists, they are predictably present (*36*). The category that is conventionally called "common" comes a close second, exemplified by species such as heather, *Calluna vulgaris*, or English oak, *Quercus robur* (58 of 160 species, or 36%). The remaining six categories are all less well represented, collectively accounting for only 20% of the total. The most frequent of these six are what are usually called "endemic rarities," specializing in one type of habitat but abundant in that habitat (14 of 160 species, or 9%); the Lady Orchid, *Orchis purpurea*, in Kent is an example. Other categories are uncommon, and one is unrepresented in the British flora: Rabinowitz *et al.* found no species with small populations in a variety of habitats but with a narrow geographic distribution. The absence of this category may reflect the small sample size, or it may reflect ecological mechanisms that are not yet fully understood.

We need more of these kinds of empirical studies of the multifactorial determinants of commonness and rarity (*37*). Not only do such studies illuminate fundamental questions about diversity, but they have practical implications for conservation biology. For instance, Table 1 helps justify the attention traditionally given by conservationists to "endemic rarities": not only are these species, with their narrow ranges and restricted habitat specificities, easily destroyed, but they are also numerically the most prevalent category of rare plants. Rabinowitz *et al.* speculated, moreover, that a better understanding of how endangered and extinct species are apportioned among their eight categories (or among other, alternative categories) may offer "clues about the causes of the endangered state" (*34*, p. 200).

How Many Living Species Have Been Recorded?

So far, this article has dealt with issues that must be resolved if we are ever to estimate the number of species in a given region, or on Earth, from basic principles. The second part of the article now reviews our current ignorance

Table 1. The distribution of 160 plant species from the *Biological Flora of the British Isles,* classifed into eight categories according to geographic distribution (wide or narrow), habitat specificity (broad or restricted), and local abundance (somewhere large or everywhere small) (*34*).

Local population size	Geographic distribution			
	Wide habitat specificity		Narrow habitat specificity	
	Broad	Restricted	Broad	Restricted
Somewhere large	58	71	6	14
Everywhere small	2	6	0	3

about the simple facts of how many species there actually are.

Living things may be divided into five kingdoms, distinguished by different levels of cellular organization and modes of nutrition. Two of these kingdoms, the prokaryotic monerans and the eukaryotic protists, comprise microscopic unicellular organisms, and together they account for something like 5% of recorded living species. The fungal and plant kingdoms represent roughly another 22% of species. The animal kingdom thus comprises the majority (more than 70%) of all recorded living species (*38*). Table 2 gives a rough account of how the species in the different animal phyla are apportioned according to the habitat of the adult creatures; each phylum represents a distinct body plan, with fundamental differences that distinguish it from all the others (*39*).

Table 2 shows that most phyla are found in the sea, and more particularly in benthic environments; many phyla are found only in benthic habitats. On the other hand, by far the most abundant category of recorded living species is terrestrial insects. To a rough approximation and setting aside vertebrate chauvinism, it can be said that essentially all organisms are insects. Hutchinson (*40*, p. 149) has suggested that "the extraordinary diversity of the terrestrial fauna, which is much greater than that of the marine fauna, is clearly due to the diversity provided by terrestrial plants." Although it is true that in the sea vegetation does not form a structured environment (except close to shore) and that species generally have large geographical ranges (and the oceans are contiguous), closer examination suggests that there are subtle boundaries to dispersal in the

sea and that latitudinal zonation is often more marked in the sea than on land (*41*). Viewing these questions in another light, Ray (*42*) has observed that although the sea contains only 20% of all animal species, it contains systematically higher proportions of higher taxonomic units, culminating in 90% or more of all classes or phyla (largely because all phyla are found in the sea, and the bulk of classes are exclusively marine). These facts make it plain that the factors influencing how many species there are in any one place—food web structure, relative abundance, species-size patterns, and so on—can operate differently in different environments and on different spatial scales.

Any interpretation of information about diversity, such as that summarized in Table 2, is clouded by uncertainties about how different two groups of organisms have to be before we call them different species, and by the fact that some taxa (for example, vertebrates) have been studied in vastly more detail than others (for example, mites). Even within very well studied groups, some workers recognize many more species than others. This is especially the case for organisms that can reproduce asexually; thus some taxonomists see around 200 species of the parthenogenetic British blackberry, others see only around 20 (and a "lumping" invertebrate taxonomist may concede only 2 or 3). Some strongly inbreeding populations are almost as bad, with "splitters" seeing an order of magnitude more species than do "lumpers" (*43*). At a more fundamental level, Selander (*44*) observed that different strains of what is currently classified as a single bacterial species, *Legionella pneumophila*, have nucleotide sequence homologies (as revealed by DNA

Table 2. The number of species (to within an order of magnitude) in the different animal phyla, classified according to the habitat of adult animals. Most phyla are predominantly marine and benthic, some exclusively so. The numbers 1 through 5 indicate the approximate number of recorded living species: 1 means 1 to 10^2; 2 means 10^2 to 10^3; 3 means 10^3 to 10^4; 4 means 10^4 to 10^5; and 5 means 10^5 or more. After (38). Abbreviations: B, benthic, P, pelagic; M, moist; X, xeric; Ec, ecto; and En, endo.

Phylum / Subphylum	Marine B	Marine P	Freshwater B	Freshwater P	Terrestrial M	Terrestrial X	Symbiotic Ec	Symbiotic En
Porifera	3		1				1	
Placozoa	1							
Orthonectida								1
Dicyemida								1
Cnidaria	3	2	1	1			1	
Ctenophora	1	1						
Platyhelminthes	3	1	3		2		1	4
Gnathostomulida	2							
Nemertea	2	1	1		1		1	
Nematoda	3	1	3	1	3	1	3	3
Nematomorpha								2
Acanthocephala								2
Rotifera	1	1	2	2	1		1	1
Gastrotricha	2	2						
Kinorhyncha	2							
Loricifera	1							
Tardigrada	1		2		1			
Priapula	1							
Mollusca	4	2	3		4	1	2	1
Kamptozoa	1	1					1	
Pogonophora	2							
Sipuncula	2				1			
Echiura	2							
Annelida	4	1	2		3		2	
Onychophora					1			
Arthropoda								
Crustacea	4	3	3	2	2		2	2
Chelicerata	2	1	2	2	4	3	2	1
Uniramia	1	1	3	2	5	3	2	2
Chaetognatha	1	1						
Phoronida	1							
Brachiopoda	2							
Bryozoa	3		1					
Echinodermata	3	1						
Hemichordata	1							
Chordata								
Urochordata	3	1						
Cephalochordata	1							
Vertebrata	3	3	2	3	3	3	1	1

hybridization) of less than 50%; this is as large as the characteristic genetic distance between mammals and fishes. Relatively easy exchange of genetic material among different "species" of microorganisms could mean that basic notions about what constitutes a species are necessarily different for vertebrates than for bacteria. But I think there are likely also to be systematic trends toward greater lumping of species of small and relatively less-studied organisms, and toward greater splitting as we approach the furries and featheries.

In Table 3, I attempt to give a rough impression of how the efforts of professional taxonomists and systematists are currently distributed among the major groups of organisms. Obviously the vertebrates, which comprise only 3% of all animal species, receive a disproportionate amount of attention. One result is that new birds continue to be found at the rate of about three species per year (against a total of around 8000 species), and new mammals at the rate of around one genus per year (against a total of around 600 genera), which contrasts with the possibility that there may be more than ten insect species for every one yet classified (45).

Setting all these reservations and biases aside, the total number of living organisms that have received Latin binomial names is currently around 1.5 million or so (46). Amazingly, there is as yet no centralized computer index of these recorded species. It says a lot about intellectual fashions, and about our values, that we have a computerized catalog entry, along with many details, for each of several million books in the Library of Congress but no such catalog for the living species we share our world with. Such a catalog, with appropriately coded information about the habitat, geographical distribution, and characteristic abundance of the species in question (no matter how rough or impressionistic), would cost orders of magnitude less money than sequencing the human genome; I do not believe such a project is orders of magnitude less important. Without such a factual catalog, it is hard to unravel the patterns and processes that determine the biotic diversity of our planet.

How Many Living Species Are There?

Until recently, the total number of species was thought to be around 3 million to 5 million. This estimate was obtained roughly as follows (46). For the species of mammals, birds, and other larger animals that are relatively well enumerated, there are roughly twice as many species in tropical regions as in temperate ones. The total number of species actually named and recorded is around 1.5 million, and two-thirds of these are found in temperate regions. Most of these are insects. But most insects that have actually been named and taxonomically classified are from temperate zones. Thus, if the ratio of numbers of tropical to temperate species is the same for insects as for mammals and birds, we may expect there to be something like two yet-unnamed species of tropical insects for every one named temperate species. Hence the overall crude estimate of a total of roughly three times the number currently classified, or around 3 million to 5 million.

This estimate is open to several questions. For one thing, the total includes relatively few species of bacterial, protozoan, and helminth parasites, largely because such parasites are usually studied in connection with economically important animal hosts. But it could be that essentially every animal species is host to at least one specialized such parasitic species (47), which would immediately double the estimated total. For another thing, the Acarina (mites), both tropical and temperate, are even less well studied than tropical insects; it was largely tropical insects that carried the estimate from the known 1.5 million to 3 million to 5 million, and mites could carry it significantly higher.

An indirect approach to the question of the number of species whose body size is small is through studies of species-size relations, such as that in Fig. 2. Figure 6 depicts a very crude estimate of the global totals of terrestrial animal species in different size categories (classified, on a logarithmic scale, according to characteristic body length); the data in Fig. 6 are the result of a multitude of rough and uncertain estimates (18). The dashed line indicates the scaling of numbers of species as L^{-2} (20); the fractal considerations reviewed in connection with Figs. 3 and 4 suggest the scaling might more appropriately be somewhere between $L^{-1.5}$ and L^{-3} (48). Whatever the detailed scaling relation at larger body sizes, it clearly breaks down for organisms whose characteristic body length is significantly below 1 cm. But these are exactly the same creatures—insects, mites, and the like—that have received relatively little atten-

Table 3. A rough indication of the relative effort devoted to animals from different taxonomic groups is given by the average number of papers listed in the *Zoological Record*, 1978 through 1987 (*54*).

Phylum Subphylum Class Order	Average number of publications per year (coefficient of variation, in percent)	Approximate number of recorded species	Papers per species per year
Protozoa	3,900 (10)	70,000	0.05
Porifera	190 (22)	10,000	0.02
Coelenterata	740 (12)	10,000	0.07
Echinoderma	710 (15)	6,000	0.12
Nematoda	1,900 (1)	1,000,000	0.002
Annelida	840 (9)	15,000	0.06
Brachiopoda	220 (14)	350	0.63
Bryozoa	160 (15)	4,000	0.04
Entoproctra	7 (53)	150	0.04
Mollusca	4,200 (8)	50,000	0.08
Arthropoda			
Crustacea	3,300 (9)	39,000	0.09
Chelicerata			
Arachnida	2,000 (6)	63,000	0.03
Uniramia			
Insecta	17,000 (7)	1,000,000	0.02
Coleoptera	2,900 (6)	300,000	0.01
Diptera	3,200 (7)	85,000	0.04
Lepidoptera	3,500 (9)	110,000	0.03
Hymenoptera	2,200 (9)	110,000	0.02
Hemiptera	1,700 (7)	40,000	0.04
Chordata			
Vertebrata			
Pisces	7,000 (13)	19,000	0.37
Amphibia	1,300 (12)	2,800	0.47
Reptilia	2,400 (7)	6,000	0.41
Aves	9,000 (10)	9,000	1.00
Mammalia	8,100 (12)	4,500	1.80
Mammalian orders			
Monotremata	20	3	6.8
Marsupialia	269	266	1.0
Insectivora	270	345	0.8
Dermoptera	2.2	2	1.1
Chiroptera	402	951	0.4
Primates	956	181	5.3
Edentata	38	29	1.3
Pholidota	5	7	0.7
Lagomorpha	173	58	3.0
Rodentia	1,538	1,702	0.9
Cetacea	360	76	4.8
Carnivora	1,157	231	5.0
Tubulidentata	2.7	1	2.7
Proboscidea	94	2	47
Hyracoidea	12	11	1.0
Sirenia	43	4	10.8
Perissodactyla	142	16	8.9
Artiodactyla	1,124	187	6.0
Pinnipedia	218	33	6.6

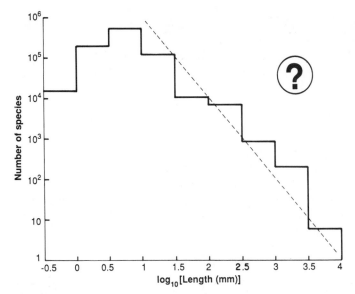

Fig. 6. A crude estimate of the distribution of number of species of all terrestrial animals, categorized according to characteristic length L. The dashed line indicates the relation $S \sim L^{-2}$, as in Fig. 2 (S is number of species) [after (*18*)]. The question mark emphasizes the crudity of these estimates and the inadequacy of the data for small size classes.

tion from taxonomists. Because we lack a fundamental understanding of the size-species relation itself, there is no reason to expect a simple extrapolation of the scaling law for large sizes to estimate accurately the number of unclassified smaller species. It is, however, interesting that the total number of species obtained by extrapolating down to around 1 mm or so is in the range 10 million to 50 million.

A sounder basis for an upward revision of the estimated number of species comes from Erwin's studies of the insect fauna in the canopy of tropical trees (*49*). Using an insecticidal fog to "knock down" the canopy insects, Erwin found that most tropical arthropod species appear to live in the tree tops. This is not so surprising, because this is where there is most sunshine as well as most green leaves, fruits, and flowers.

Erwin's original studies (*49*) were on canopy-dwelling beetles (including weevils) collected from *Luehea seemannii* trees in Panama over three seasons. He found more than 1100 species of such beetles, distributed among the categories of herbivore, predator, fungivore, and scavenger as shown in Table 4. To use this information as a basis for estimating the total number of insect species in the tropics, one needs to know what fraction of the fauna are specific to the particular tree species or genus under study; unfortunately, there are essentially no data bearing on this point. Erwin estimated 20% of the herbivorous beetles to be specific to *Luehea* (in the sense that they must use this tree species in some way for successful reproduction) (Table 4); the overall answer is more sensitive to this guess than to the corresponding figures of 5%, 10%, and 5% for predator, fungivore, and scavenger beetles, respectively. In this way, one arrives at the estimate of around 160 species of canopy beetles specific to a typical tropical tree.

Several other assumptions are needed to pyramid this estimate of 160 host-specific

Table 4. Estimated numbers of host-specific canopy beetles on *Luehea seemannii*, classifed into trophic groups (*49*).

Trophic group	Number of species	Estimated fraction host-specific (%)	Estimated number of host-specific species
Herbivores	682	20	140
Predators	296	5	15
Fungivores	69	10	7
Scavengers	96	5	5
Total	1100+	—	160

species of canopy beetles per tree to 30 million species in total. Slightly simplified, the argument runs as follows. First, Erwin noted that beetles represent 40% of all known arthropod species, leading to an estimate of around 400 canopy arthropod species per tree species. Next, Erwin suggested the canopy fauna is at least twice as rich as the forest-floor fauna and is composed mainly of different species; this increases the estimate to around 600 arthropod species that are specific to each species or genus-group of tropical tree. Finally, using the estimate of 50,000 species of tropical trees (*50*), Erwin arrived at the possibility that there are 30 million tropical arthropods in total. This estimate has been widely cited, often without full appreciation of the chain of argument underlying it.

Although it is easy to cavil at each step in Erwin's argument, the work is important in providing a new and focused approach to the problem of estimating how many species there are. Erwin does not so much answer the question as define an agenda of research.

First, the overall estimate depends almost linearly on the necessarily arbitrary assumption that 20% of the herbivorous beetles are found only on one species or genus-group of tree; changing this number to 10% would halve the global estimate to 15 million species. I think it likely that insects feeding in the canopies of rain-forest trees could be significantly less specialized in their use of food plants than are temperate insects, in order to help them deal with the sparse distribution of many tropical trees. Experiments that "knocked down" the canopy insect fauna from each of many neighboring trees of different species could shed light on these issues and provide a firmer basis for the estimates in the last column of Table 4 (*51*).

Second, the fact that 40% of taxonomically classified arthropod species are beetles is of doubtful relevance if, in truth, essentially all arthropods are unclassified tropical canopy dwellers. What we need to know is the fraction of the canopy fauna that are beetles. Again, this information could be obtained by systematic studies of the overall arthropod fauna in the canopies of a variety of tropical trees.

Third, the assumption that there are roughly two canopy species for each forest-floor species is also amenable to systematic study. Such studies should, in my view, reach below the forest floor into the soil, attempting to get a better idea of the species diversity of decomposing animals (including nematodes and other helminths) and other soil-dwellers.

More generally, I believe our ignorance of tropical mites—to name but one group—is at least as great as the ignorance about beetles and other arthropods that Erwin has exposed. These other groups may be similarly diverse. One proposal that is ambitious by ecological standards (although not by those in the physical sciences) is to assemble a team of taxonomists, with a comprehensive range of expertise, and then make a rough list of all the species found in one representative hectare in the tropical rain forest; it would be better to census several such sites (*52*). Until this is done, I will not trust any estimate of the global total of species.

Coda

For most of the history of life on Earth that is preserved in the fossil record, rates of extinction and rates of speciation have been roughly commensurate. If, however, we assume that something like half the extant species evolved in the last 50 million to 100 million years and that maybe half of all extant species will become extinct in the next 50 to 100 years if current rates of tropical deforestation continue, then contemporary rates of speciation are of order 1 million times slower than rates of extinction (*53*). Were speciation rates plotted as the *y*-axis on a graph 10 cm high, then on the same scale extinction rates would require an *x*-axis extending 100 km.

These circumstances give a special urgency to the kinds of studies called for above. Unlike essentially all other scientific disciplines, conservation biology is a science with a time limit, with the clock ticking faster as the human population continues to increase. We need to understand the world of living things for the same fundamental reasons that we need to understand the physics of the unimaginably small and of the unimaginably large. We also need such understanding to manage the biosphere in a sustainable way (which we do not appear to be doing at present) and to design rational strategies for preserving some habitats while exploiting others in ways that allow some fraction of the original flora and fauna to persist. I believe future generations will find it blankly incomprehensible that we are devoting so little money and effort to the study of these questions.

References and Notes

1. The details of 40 of these webs are fully documented in J. E. Cohen, *Food Webs and Niche Space* (Princeton Univ. Press, Princeton, NJ, 1978) and F. Briand, *Ecology* **64**, 253 (1983). The frequency distribution of chain lengths and other properties have been reported for all 113 webs in J. E. Cohen, F. Briand, C. M. Newman, *Proc. R. Soc. London Ser. B* **228**, 317 (1986).

2. F. Briand and J. E. Cohen, *Science* **238**, 956 (1987).

3. Other problems in the compilation of such catalogs of food webs include the following: one wants the "community" web, not a subset of species traced up from one resource or down from one top predator; one must decide, arbitrarily, which links are regarded as too weak or too unusual to list; different researchers may have followed different procedures or had different biases in tabulating individual webs; and so on. For further discussion, see R. M. May, *Nature* **301**, 566 (1983).

4. J. E. Cohen, C. M. Newman, F. E. Briand, *Proc. R. Soc. London Ser. B* **224**, 449 (1985).

5. J. E. Cohen and C. M. Newman, *ibid.*, p. 421. This study is based on 62 food webs only.

6. P. H. Warren and J. H. Lawton, *Oecologia (Berlin)* **74**, 231 (1987).

7. B. A. Hawkins and J. H. Lawton, *Nature* **326**, 788 (1987).

8. J. H. Lawton, personal communication.

9. These patterns depend significantly on the range of the host plant: D. R. Strong, J. H. Lawton, T. R. E. Southwood, *Insects on Plants* (Blackwell, Oxford, 1984); H. C. Godfray, *J. Econ. Entomol.* **9**, 163 (1984); P. W. Price, *Evolutionary Biology of Parasites* (Princeton Univ. Press, Princeton, NJ, 1980).

10. M. J. Jeffries and J. H. Lawton, *Freshwater Biol.* **15**, 105 (1985); F. Briand and J. E. Cohen, *Nature* **307**, 264 (1984).

11. This number of direct interactions with 3 to 5

12. R. T. Paine, *J. Anim. Ecol.* **49**, 667 (1980).

13. F. A. Bazzaz, *Ecology* **56**, 485 (1975).

14. D. Tilman, *Resource Competition and Community Structure* (Princeton Univ. Press, Princeton, NJ, 1982); R. M. May, Ed., *Theoretical Ecology: Principles and Applications* (Sinauer, Sunderland, MA, 1981).

15. F. W. Preston, *Ecology* **43**, 185 (1962); R. H. MacArthur and E. O. Wilson, *Island Biogeography* (Princeton Univ. Press, Princeton, NJ, 1967).

16. R. M. May, in *Ecology of Species and Communities*, M. Cody and J. M. Diamond, Eds. (Harvard Univ. Press, Cambridge, MA, 1975), pp. 81–120.

17. G. Sugihara, *Am. Nat.* **116**, 770 (1980).

18. R. M. May, in *Diversity of Insect Faunas*, L. A. Mound and N. Waloff, Eds. (Blackwell, Oxford, 1978), pp. 188–204.

19. L. Van Valen, *Evolution* **27**, 27 (1973). The pattern in Fig. 2 (and in Fig. 6) could be somewhat changed if the large mammals that have become extinct over the last million years or so are included.

20. G. E. Hutchinson and R. H. MacArthur, *Am. Nat.* **93**, 117 (1959).

21. D. R. Morse, J. H. Lawton, M. M. Dodson, M. H. Williamson, *Nature* **314**, 731 (1985).

22. D. Griffiths, *Am. Nat.* **127**, 140 (1986); D. Strayer, *Oecologia (Berlin)* **69**, 513 (1986).

23. J. H. Brown and B. A. Maurer, *Am. Nat.* **130**, 1 (1987).

24. E. D. Odum, *Fundamentals of Ecology* (Saunders, Philadelphia, 1953).

25. R. H. Peters, *The Ecological Implications of Body Size* (Cambridge Univ. Press, Cambridge, MA, 1983).

26. B. B. Mandelbrot, *Fractals: Form, Chance, and Dimension* (Freeman, San Francisco, 1977); C. Loehle, *Spec. Sci. Tech.* **6**, 131 (1983).

27. R. M. May, in *Ecological Concepts*, J. M. Cherrett, Ed. (Blackwell, Oxford, in press).

28. J. Damuth, *Nature* **230**, 699 (1981); R. H. Peters and J. V. Raelson, *Am. Nat.* **124**, 498 (1984); R. H. Peters and K. Wassenberg, *Oecologia (Berlin)* **60**, 89 (1983); J. H. Brown and B. A. Maurer, *Nature* **324**, 248 (1986).

29. D. R. Morse, N. E. Stork, J. H. Lawton, *J. Ecol. Entomol.* **13**, 25 (1988).

30. I. Hanski, *Oikos* **38**, 210 (1982).

31. J. H. Brown, *Am. Nat.* **124**, 255 (1984).

32. T. Root, *Atlas of Wintering North American Birds* (Univ. of Chicago Press, Chicago, 1988).

33. E. H. Rapoport, *Areogeography: Geographical*

Strategies of Species (Pergamon, Oxford, 1982).

34. D. Rabinowitz, S. Cairns, T. Dillon, in *Conservation Biology*, M. E. Soule, Ed. (Sinauer, Sunderland, MA, 1986), pp. 182–204.

35. The 177 plant species in the *British Flora* are clearly unlikely to be a random subset of the entire 1822 plant species in Britain. Understandably, very widespread species tend to be overrepresented, at the expense of middling species. Species with highly restricted distributions, however, form about 30% both of the *British Flora* and of the entire flora. For further details about possible biases, and about the consensual process among 15 ecologists and systematists, see (*34*).

36. The category of "rare habitat" may be overrepresented in the *British Flora* by virtue of the country's long coastline and many maritime species.

37. See, for example, the discussion of sparse and patchily distributed species in species-rich plant communities by P. J. Grubb [in *Community Ecology*, J. M. Diamond and T. J. Case, Eds. (Harper and Row, New York, 1986), pp. 207–226] and of experimental design in demographic studies of rare plants by J. Travis and R. Sutter [*Natural Areas J.* 6, 3 (1987)].

38. V. Pearse, *Living Invertebrates* (Blackwell, Oxford, 1987).

39. There are, of course, varying opinions about what constitutes a fundamentally different body plan, which is why different schemes recognize from 25 to 35 animal phyla.

40. G. E. Hutchinson, *Am. Nat.* 93, 145 (1959).

41. These ideas were developed further by E. C. Pielou, *Biogeography* (Wiley, New York, 1979).

42. G. C. Ray, *Am. Zool.* 25, 451 (1985).

43. Notable examples are provided by *Erophila* and *Arabidopsis*, British plants in the mustard family.

44. R. K. Selander, in *Population Genetics and Molecular Evolution*, T. Ohta and K. Aoki, Eds. (Springer-Verlag, Berlin, 1985), pp. 85–106.

45. J. M. Diamond, *Nature* 315, 538 (1985).

46. P. H. Raven, *Bull. Entomol. Soc. Am.* 29, 5 (1983).

47. C. A. Toft, in *Community Ecology*, J. M. Diamond and T. J. Case, Eds. (Harper and Row, New York, 1986), pp. 445–463.

48. In an essentially one-dimensional environment with fractal dimension D, Hutchinson and MacArthur's (*20*) argument suggests a scaling of $S \sim L^{-D}$; in an essentially two-dimensional environment, the scaling is $S \sim L^{-2D}$. When combined with the estimate $D \sim 1.5$ (*21*), this gives the result in the text.

49. T. L. Erwin and J. C. Scott, *Coleopt. Bull.* 34, 305 (1980); T. L. Erwin, *ibid.* 36, 74 (1982); *Bull. Entomol. Soc. Am.* 29, 14 (1983).

50. R. Howard, personal communication, cited in (*43*).

51. See, for example, N. E. Stork, *Ecol. Entomol.* 12, 69 (1987).

52. I suggest 1 ha as the smallest meaningful area for such a study, although not the least of our problems is that we do not understand just what the smallest scale need be to give sensible answers.

53. A. P. Dobson, personal communication.

54. Table 3 was compiled by A. P. Dobson, with help from A. M. Lyles and A. Merenlender. The number of species is taken from R. F. K. Barnes, *A Synoptic Classification of Living Organisms* (Blackwell, Oxford, 1983); most of the numbers are for recorded species, but some (especially nematodes and insects, as emphasized by the question mark) are Barnes's estimates. The numbers of publications come from *The Zoological Record*, averaged over the years 1978 through 1987 (with the coefficient of variation of the annual number shown in parentheses). *The Zoological Record* lists papers covering every aspect of the zoology of animals (scanning over 6000 journals and other documents), except for insects, where only publications dealing with systematics, taxonomy, or nomenclature are given; Table 3 therefore underestimates, by an unknown factor, work on insects relative to that on other animals. A further major complication is that, for some groups, many—if not most—of the papers cited in *The Zoological Record*, are paleontological, dealing with fossil species rather than extant ones. (This may apply, for example, to 80 to 90% of the papers on Brachiopods and roughly 30 to 40% of those on Molluscs, Poriferans, and Coelenterates).

55. This article has been shaped by helpful suggestions from R. K. Colwell, J. O. Corliss, P. J. DeVries, W. D. Hamilton, A. P. Dobson, H. S. Horn, A. R. Kabat, M. Kreitman, J. H. Lawton, J. May, G. Rosenberg, C. A. Toft, and two anonymous reviewers. Supported in part by NSF grant DMS87-03503.

Index

abl, 228
Acetylcholine (ACh), 137
ACh receptor, fish, 137
Actin, 42, 52
α-Actin genes, 181
α-Actinin, 51–53
Activated sludge process, 33
Acute promyelocytic leukemia, 228–229
Adaptation
  biochemical in fish, 142–143
  genetic in fish, 142–143
  to parasitism, 78, 80–82
  physiological in birds, 153
Adjuvants, 84–86
Adrenal medullary autografts, 241
Adrenal tissue implants, 212
Adrenarche, 214
Aging, primate models for, 213–214
Agrobacterium tumefaciens, 20
AIDS, 219
  HIV and, 11–12, 239
  retroviruses and, 1–2, 5
Algae, 107–108
Alleles
  expression studied in fish, 140
  in humans, 225–234
Allografting, 167–168
Allografts, 168–169
Allozyme diversity, survival and, 143
Altruism, 150–151
Alzheimer's disease, 240
  primate research and, 212–213
Ames test, 171
Amniocentesis, 236–237
Anabaena, 22–23
Animal rights movement, 221–222
Antibiotics, biosynthesis, 27
Antifreeze genes, 142
Antifreeze proteins, 142
Antigenic variation, trypanosome, 77–78, 81, 86
Antimorphs, 188
Anti-oncogenes, 165
Antisense transformation, 49–51
Apolipoprotein B, 230

Aquatic toxicology, 141–142
Arabidopsis, 117, 128
Ascaria suum, 87
Ascaris, 73, 86
Atherosclerosis
  genetic components, 232
  primate models, 213, 218–219
  in transplant, 167–168
Auditory space map, 158
Auxotrophic mutants, isolation of, 17
Avian leukemia/leukosis virus (ALV), 7–9, 180
Avirulence genes (*avr*), 27–28
Axon guidance, 73–74
Azacytidine, 183
Azidothymidine, 12

Baboons, 200, 212
Bacillus amyloliquefaciens, 31
Bacillus stearothermophilus, 31
Bacillus subtilis, 19, 22
Bacillus thuringiensis, 126
Bacteria
  genetic engineering of, 25–34
  research on, 17–23
Baleen whales, 193–195, 197–198, 201–202
Bare lymphocyte syndrome, 240
Barn owls, 157–158
Beaked whales, 196
Beetles, 251, 257–258
Behavior
  cetacean, 196–201
  primate models, 216–218
  study in rats, 173
Behavioral plasticity, 156
Benzo[*a*]pyrene, 141
Biochemical adaptation, fish, 142–143
Biogeography, avian, 148
Bioleaching, 31–32
Birds
  research on, 147–158
  species number, abundance, and body
    length, 250–251
Bird song, 151–152, 155–156, 195
Blubber, 193
Blue whales, 195, 197–198, 202

Bone marrow transplantation, 167, 221
Bordetella pertussis, 20
Bovine leukemia virus (BLV), 11–12
Bovine papilloma virus (BPV), 180
Bowhead whales, 195, 198, 202
Bradyrhizobium, 28
Brain
    aging and, 213
    avian, 156–158
    cetacean, 194
    development, avian studies, 156
    primate, 211–213
Bromosynil, 125
Brugia malayi, 85–87
Bryde's whale, 197–198, 202

Caenorhabditis elegans, 67–75, 86–87
Cancer
    primate research, 220–221
    retroviruses and, 1–2, 7–10
Carcinogenesis
    genetic mechanisms, 172
    multistep process, 170–171
    risk assessment in rats, 169–172
Carcinogens, chemical
    grc and susceptibility to, 165
    screening, 220
    study in rats, 169–172
    testing, 141–142
Cardiovascular disease
    genetic components, 232
    study in primates, 218
    study in rats, 172–173
Cascade systems, 20
Caulobacter crescentus, 22
Cell adhesion molecules (CAMs), *Dictyo-
    stelium*, 53
Cell-cell communication, development and,
    47–54, 67–68
Cell lines, human, 230–231
Cells
    differentiation, 239
    lineages, 140
    migration, 73–74
    movement, 140
Cellulomonas uda, 31
Cetaceans, 193–204
Chemotaxis, *Dictyostelium*, 47–48, 52–53
Chitin, 26
Chitinases, 26
Chlamydomonas, 108
Chlorinated compounds
    bacterial degradation, 30–31
Chorionic villus sampling, 236–237
Chromatin conformation, 187

Chromosomal markers, 186
Chromosome microdissection, 94–95
Chromosome walking, 128
Chronic granulomatous disease, 230
Chronic myeloid leukemia (CML), 228–229
Circumsporozoite (CS) protein, 84
Cis-acting elements, 181, 183
Citrullinemia, 226
Clathrin, 42
Clavibacter xyli, 26, 34
Clostridia, 22
Coat protein-mediated protection, 126
Codas, sperm whale, 196
Cognition, primate, 216–217
Collagen genes, 181
Color oncogene, 141
Communication
    avian, 151
    primate models, 216–218
Congenital abnormalities, 236–238
Contraception, 215
Contraceptive hormones, 215
Coronaviruses, 11
Crop improvement, genetic engineering and,
    125–126
Cryptosporidium, 77–78
γ-Crystallin genes, 181
Cyanobacteria, 22, 107
Cyclic AMP receptors, 49–52
Cyclosporine A, 168
Cystic fibrosis, 230, 233, 236
Cytogenetic analysis, 93–94
Cytogenetic events, 227–229
Cytogenetic mapping, 94

Daylength, 154
dct, 28–29
Delta endotoxins, 25–26
Density models, 79
Depression, primate models for, 218
Developmental biology
    C. elegans, 68–69
    cell-cell interactions in, 47–54, 67–68
    fish, 139–141
    plants, 108–116
    Xenopus laevis, 57–60
Diabetes mellitus
    fetal cells and, 240–241
    rat model, 172
Dictyostelium discoideum
    biochemistry and genetics, 47–51
    cell-cell communication, 53
    gene expression, 53
    morphogenesis, 54
    motility and chemotaxis, 52–53

signal transduction, 51–52
Differentiation, cellular, 21–22
Differentiation-inducing factor (DIF), 54
Discoidin, 49–51, 53
Disease resistance in plants, 126
Diving, cetacean, 194, 197, 203
DNA
  membrane activities and, 20–21
  probes, 229
  sequence, population variants in, 229
  supercoiling, 20
  synthesis, retroviral, 6
DNA tumor viruses, 184
Dolphins, 194–197, 199–204
Dominant negative mutations, 188
Dopamine, 241
Drosophila melanogaster
  genome, 91, 100n, 101n
  research on, 91–100, 225
  wingless, 9
Drug resistance, multiple, 78
Drugs, testing, 238
Duchenne muscular dystrophy, 228, 233, 236
Duffy blood group substance, 80

Echolocation, cetacean, 193–196, 199, 205n
Ecological divergence, 148
Ecology
  behavioral, 149–150
  community, 148–149
  evolutionary, 148
  population, 148–149
Economic defendability, theory of, 149–150
Effector-modulator systems, 20
Effectors, 19–20
Elastase genes, 181
Electrophorus electricus, 137, 139
Embryogenesis
  fish as models for, 140
  in plants, 112–113
  Xenopus laevis and, 57–64
Embryonic stem cells, recombination in, 231
Encephalization quotient (EQ), 194
Endangered species, 252
  cetacean, 203–204
  primate, 215
Endocrine disorders, 236
Endocrinology
  avian, 153–156
  fish as models in, 138–139
  human and primate, 216
Endometriosis, 215
Enhancers, 181–184

Environmental safety, recombinant bacteria and, 33–34
Enzyme deficiency disorders, 237–238
Enzymes, modified, 31
Epilepsy, 212
EPSPS, 125
Erwinia amylovora, 27
Escherichia coli, 17–18, 20–23, 28
Estrogen, 156
Ethology, 151–153
Eukaryotic cells, 18
  bacteria compared with, 39–40
  evolution of, 23
Evolution, 22–23
Extinction, rates of, 258
Extinct species, 252

F-actin, 52–53
Familial adenomatous polyposis, 230
Familial hypercholesterolemia, 230
Feeding efficiency, 149
Fertility, 215
Fetal biopsy, 237–238
Fetal cells, 239–240
Fetal research, 235–242
  regulations limiting, 239
Fetal therapy, 236
Fetal tissue research, 235, 239–241
  regulations affecting, 241–242
  transplantation, 240–241
α-Fetoprotein, 237
α-Fetoprotein genes, 181–182
Fetoscopy, 236–237
Fibroblast growth factor (FGF), 61
Filariasis, 85
Fin whales, 194–195, 198, 202
Fish
  evolution, 135
  freshwater and saltwater, 138
  interspecific hybrids, 140
  research on, 135–143
Florigens, 114
Follicle-stimulating hormone (FSH), 214
Food webs, 245–246, 253, 259n
Foraging strategies, cetacean, 197
fos, 184
Fractal geometry, 249–250, 255
Fragile site, chromosomal, 228
Frameshifts, retroviral, 10–11
Frankia, 28
Frost control, 26, 28
Fucus, 108, 111
Fuels, microbial production, 31

β-Galactosidase, regulation of synthesis, 18–
    19, 23
Gamete recovery, 215
Gametophytes, 106, 111–112
Gastrulation, amphibian, 58, 62, 64
Gelsolin, 52–53
Gene cloning strategies, 226–227
Gene expression
    Dictyostelium, 53
    immunoglobulin, 181, 184–185
    plants, 109–110, 112–117, 126–128, 130
    transgenic animals, 181–183
    regulation, 18–20, 109–110, 239
Genes
    cloned, uses of, 39
    developmental activation, 181
    5S RNA, 63–64
    mapping and identification, 229–231
    transfer in *Drosophila*, 96–98
    See also Human genes
Gene therapy, 13–14, 182
Genetic adaptation, in fish, 142–143
Genetic counseling, 233
Genetic defects, human population surveil-
    lance and, 225–229
Genetic diagnosis, 232–233
Genetic-environmental interactions, 218
Genetic markers, 229–231
Genomic imprinting, 169
β-Globin genes, 181, 183
    developmental activation, 182
Glucose-6-phosphate deficiency, 238
Gluphosinate, 125
Glycophorin A, 80–81
Glyphosate, 125
Gonadotropin-releasing hormone agonists,
    214, 222n
G proteins, 9, 51–52
Graft-versus-host reaction, 167
Granulocyte-macrophage colony-stimulating
    factor (GM–CSF), 221
    Gray whales, 195, 197–198, 202
Growth and reproduction complex (*grc*),
    165, 169, 171
Growth factor genes, 184
Growth hormone, effect in fish, 138–139

Heart grafting, 167
Heat shock genes, 142
Heat shock proteins, 19–20
Helminths, parasitic, 77–87
Hematopoietic stem cells, 13
Hemolytic disease, fetal, 237
Hemophilia, 226, 236
Hepatocyte cell lines, toxicity testing and, 171

Herpes B virus, 221
Heterochronic genes, 72
Heterochrony, 140, 143
Heterogenotic duplication, 41, 43
Histocompatibility (HLA) matching, 168
hok, 34
Homeostasis, maintenance of, 153
Homologous recombination, gene targeting
    by, 188
Hormone deficiencies, 182
Hormones
    behavior and, 216
    brain and, 155
    seasonal behavior and, 154–155
    social behavior and, 153–154
Host-versus-graft reaction, 167
hprt, 13
Human T cell leukemia virus (HTLV), 7, 10–
    12
Human genes
    cloning, 231
    identification, 226
Human growth hormone gene, 182–183
Human immunodeficiency virus (HIV), 11–
    12, 219, 239
Humpback whales, 194–195, 197–198, 202
Huntington's disease, 230, 233, 236
Hydrocephalus, obstructive, 238
Hypercholesterolemia, 172
Hyperlipidemia, 218
Hypersensitive response, 27
Hypertension, 218–219
    genetic components, 232
    rat model, 172
Hypothalamo-pituitary unit, 155
Hypoxanthine-guanine phos-
    phoribosyltransferase (HPRT), 187–188

ice gene, 26–27
Ice⁻ mutants, 26
Ice nucleation (Ice⁺) strains, 26
Immune competence, 218
Immune system, study with transgenic mice,
    184–185
Immunogenetics, studies with rats, 163–167
Immunoglobulin genes, expression of, 181,
    184–185
Inducing factors, 60–61
Induction, 57
    neural, 62
Industrial chemicals, microbial production, 31
Infectious diseases, primate research on, 219–
    220
Insect resistance, 126
Insecticides, *B. thuringiensis*, 26

Insertional mutagenesis, 117
Insertion mutations, 9
    retroviruses and, 5, 13–14
    in transgenic mice, 186–188
Insulin receptor, *Drosophila*, 99
Integration protein IN, 6–7
Integration, retroviral, 2, 6–7
Integrative disruption, 41, 43
int-1, 9–10
Introns, 182
In vitro fertilization, 215
Island biogeography, equilibrium theory, 148

jun, 9–10

Killer whales, 200, 202
Killifish (*Fundulus heteroclitus*), 136, 138,
    140, 142–143
Kingdoms, 253
Kinship, altruism and, 150–151
Klebsiella planticola, 31
Klebsiella pneumoniae, 28
Krill, 197

Laboratory animals, 210
Lactate dehydrogenase, 140
Lactobacillus casei, 31
Language acquisition, 217
Laser ablation, cell removal by, 185
Learning
    behavioral plasticity and, 156
    ontogeny and, 152
    primate models, 216–218
    rat models, 173
Legionella pneumophila, 253
Legumes, yield improvement, 28
Leishmania, 78, 80
Leishmania tropica, 79, 82
Lentiviruses, 219
Lesch-Nyhan syndrome, 187
Lethality, apparent, in yeast, 42, 44
Leukoencephalopathy, multifocal, 184
Light
    refractoriness to, 154–155
    timing cue, 154
Lineage markers, 185
Link scaling, 246
Linkage analysis, 229–231
Liver grafting, 167–168
Liver-focus induction, 171
Longevity, primate, 215
Luteinizing hormone (LH), 154, 214

Macroparasites, 78–79
Magnetic sensing, 203

Major histocompatibility complex (MHC),
    rat, 164–167
Malaria, 78, 80
    chemotherapy, 78
    vaccines, 83–85
Mapping, in brain, 157–158
Marburg virus, 221
Mating systems
    avian, 150, 158
    primate, 216
Maze experiments, 173
Medaka (*Oryzias latipes*), 136, 140–141
Meiotic recombination, 93
Melanomas, fish hybrid crosses and, 141
Menkes' syndrome, 227
Menopause, 215
Menstrual cycle, 214–215
Mental retardation, X–linked, 228
Meristem, 114
Merops bullockoides, 151
Merozoites, malaria, 79–81
Mesoderm induction, 60–62
Messenger RNA
    discovery, 19
    in plants, 108–109, 112–114, 116
Metabolic mutations, 226–227
Metabolism, inborn errors, 236
Metallothionein promoter, 182
Metallothioneins, 33
Metazoans, 246
MHC antigens
    polymorphism in rat, 164, 167, 169
    rat, mouse, and human, 165–166
    in transgenic mice, 185
    transplant rejection and, 167–169
Microdissection, cell removal by, 185
Microfilaments, 71
Microinjection, of DNA into pronucleus,
    179–183, 186–188
Microparasites, 78–79
Migration
    bird, 152–154
    terrestrial mammal, 202
    whale, 193, 201–203
Miller-Dieker syndrome, 229
Min proteins, 21
Mineral processing, bioleaching and, 31–32
Minke whale, 197–198, 202
Mitotic recombination, 93
Molecular genetics, human, 225–233
Moloney murine leukemia retrovirus (M–
    MuLV), 179
Monoclonal antibodies, from primate
    material, 220
Morphogen, 54

Morphogenesis
    Dictyostelium, 54
    in plants, 114
Mosaics, genetic, 69, 93
    fish, 136, 140
    lineage relations and, 185
mos, 184
Mosses, 107
Motility, *Dictyostelium*, 52–53
Mouse lung adenoma assay, 171
Mouse mammary tumor virus (MMTV), 182
MPTP primate model, 212
Multiple endocrine neoplasia, 229–230
Murine leukemia virus (MLV), 10–11, 13
Muscle
    assembly and function, 74
    induction, 61
Mutagenesis, 169, 171
Mutant complementation, 51
Mutants
    developmental, 140–141
    homeotic, 114
Mutations
    anucleolate, 63
    Dictyostelium, 48–50, 52
    Drosophila 91–97
    human gene, 226
    lethal, 97
    screens for, 92
    single gene, genetic disorders and, 236–237
c-*myc*, 9, 141
myc oncogenes, expression, 183–184
Myelination defect, 182, 184
Myosin, 52–53
Myosin heavy chain, 49–51, 53
Myotonic dystrophy, 230
Mysticetes, 193, 195, 197–201
Myxobacteria, 22

Na⁺,K⁺-ATPase, 138
Natural selection, 149–150
Navigation
    avian, 152–153
    cetacean, 203
Nematodes, 67, 86
Neocortex, evolution of, 156
Nervous system
    Ascaris, 73
    C. elegans, 72–74
    Drosophila, 98–99
    evolutionary states, 136–137
    primate models, 212
Neural plasticity, 156, 212
Neural transplantation, 168

Neural tube defects, 236–237
Neurobiology
    avian studies, 155–158
    fish as models in, 136–138
    primate, 211–213
Neurodegenerative diseases, 168
    fetal cell transplantation for, 240–241
Neuroendocrine secretion, 155
Neurofibromatosis, 184
    acoustic, 229
    type I, 230
Neurological disorders, primate models for, 212
Neuron grafts, fetal, 240–241
Neuropsychology, primate research in, 211–213
Neurotransmitters, 137
    aging effects on, 213
    C. elegans, 72–74
Niche space, 247
nif, 22, 28
nod, 28–29
Norway rat (*Rattus norvegicus*), 163
Nucleus laminaris, 157
Null mutations, 41–42

Odontocetes, 193, 195–196, 198–201
Oilseed rape, genetically engineered, 122, 131
Oncogenes, 2, 7–10, 14
    expression, 183–184
    fish, 142
    implication in disease, 228
    recessive, 229
Oncogenesis, models for, 183–184
Oncogenic viruses, 220
Ontogeny, behavioral, 152
Optimal foraging theory, 149
Orientation
    avian, 152–153
    cetacean, 203
Osteogenesis imperfecta II, 188
Osteoporosis, 215
Ovalbumin, 155

Parasitism, 22, 77–87
    adaptation to, 78, 80–82
Parkinsonism, 240–241
    primate models, 212
Particle gun method, 124
Patent protection, for genetically engineered plants, 34, 129–130
Pathogenicity genes, 27
Pattern formation
    Dictyostelium, 54
    embryonic, 71, 239

postembryonic, 71–72
P elements, 95–96, 101n
Percutaneous umbilical blood sampling
  (PUBS), 237
Phage T4, 18
Phenylketonuria, 226
Philadelphia chromosome, 9
Phyla, numbers of species in, 253–254
Physcomitrella patens, 107
Phytoalexins, 27
Phytochrome, 115–116
Pilot whales, 200, 203
Plant growth hormones, 27
Plant resistance genes (*R*), 28
Plant-microbe–pest interactions, 25–29
Plants
  bacterial pathogenesis and, 27
  developmental processes, 105–118
  dicotyledonous, 121
  disease-resistant, 126–127
  diversity, 106–108
  flowering, 106–108, 111–112, 114
  fractal dimension, 249
  genetic engineering, 105, 110–111, 117,
    121–131
  genomes, 116–117
  herbicide-resistant, 125–126
  insect-resistant, 126
  monocotyledonous, 121, 123
  nonflowering, 106
  nonvascular, 106–107
  transgenic, 121, 123–124, 126–130
  vascular, 106–107
Plasmids
  D. discoideum, 49
  xenobiotic catabolism and, 29
  yeast, 41
Plasmodium, 78, 80, 82
Plasmodium falciparum, 80–81, 84
Platyfish (*Platypoecilus maculatus*), 136, 141
Pneumocystis carinii, 77
Poeciliopsis, 143
Polarity, embryonic, 57–59, 61–62, 108
pol, 10–11
Pollutants
  accumulation in cetaceans, 204
  fish tumors and, 141
Polycyclic aromatic hydrocarbons, 141
Polycystic kidney disease, 230, 233
Polygyny, 150, 154, 198, 201
Polygyny threshold model, 150
Polymers, plant, 131
Polytene chromosomes, 93–94, 96
Population dynamics, 247
Population models, 250

Porpoises, 195, 199, 204
Prader-Willi syndrome, 229
Predator avoidance, 149
Pregnancy, immunology of, 215
Prenatal diagnosis, 236–238
Prevalence models, 79
Primate research proposals, criteria for evalua-
  tion, 210
Primates
  constraints in use of, 221–222
  research on, 209–222
Progenote, 23
Programmed cell death, 73
Prokaryotic cells, 18, 23
Prolactin, effect in fish, 138–139
Promoters, 182–184
Protamine genes, 181
Protein kinases, *src* gene products, 8–9
Protein polymorphisms, 142
Proteins
  eukaryotic, sequence conservation in, 42–
    44, 99
  expressed in *Dictyostelium*, 49–51
  genetic engineering and, 131, 183
  multimeric, 188
Proto-oncogenes, 7–10, 14, 141, 184
  fragile sites and, 228
Protozoans, parasitic, 77–87
Provirus, 2–7, 10, 13
  methylation, 183
Pseudomonas, 25, 27, 29–30
Pseudomonas fluorescens, 26, 34
Pseudomonas mendocina, 31
Pseudomonas putida, 29
Pseudomonas solanacearum, 28
Pseudomonas syringae, 26, 28, 34
Pseudotypes, 4–5
Psychiatric diseases, 231–232
Psychoneuroimmunology, 218
Puberty, 214–215

Radial glial cells, 156
Rainbow trout, 141–142
ras, 9, 141, 230
  expression, 183–184
RAS proteins, 42
Rat
  as substitute for humans, 170
  research on, 163–173
Receptor-mediated endocytosis, 2
Recessive lethal genes, 169
Recombinant DNA technology, 39–40
Recombinant bacteria, environmental safety
  of, 33–34
Recombinational switching, 22

Regeneration, of plants, 105, 107–108, 110–111, 123–124
Regulatory approval
  genetically engineered bacteria, 34
  genetically engineered plants, 128–129
Reporter genes, 27, 182
Repressor-effector hypothesis, 19
Reproduction
  avian, 153
  primate models, 214–216
Resource partitioning, 148
Restriction enzymes, 229
Restriction fragment length polymorphism (RFLP), 128
Retinoblastoma, 165, 228–229, 233
Retrotransposons, 2–3, 5–6, 11
Retroviruses, 1–14
  genetic manipulation with, 179–180, 183, 186–188
  genome, 2–4
  immune response to, 12
  immunodeficiency viruses (HIV and SIV), 7, 10–12, 219, 221, 239
  integration, 187
  life cycle, 2–4, 6, 11–12
  model systems, 5–11
  mutation induction by, 186–187
  oncogenic, 7–10
  vectors, 180
Reverse transcriptases, 6
Reverse transcription, retroviruses and, 2, 5–7
Rhesus (Rh) factor, 215
Rhesus monkeys, 212–213
Rhizobitoxin, 28
Rhizobium, 22, 28
Rhizobium japonicum, 28–29
Rhizobium leguminosarum, 29
Rhizobium meliloti, 28–29, 34
Ribosomal RNA genes, 63–64
Right whales, 195, 198
RNA polymerase, gene expression and, 19
Rorqual whales, 197–198
Rous sarcoma virus (RSV), 7–8, 11

Saccharomyces cerevisiae, 40, 42–44
Salmonella typhimurium, 20, 23
Schistosoma mansoni, 85
Schistosomes, 78
Schistosomiasis, 78–80, 85
Schistosomula, 81–83, 85
Schizophrenia, 232
Schizosaccharomyces pombe, 40, 43
Seals, 197
Seasonal behavior, hormones and, 154–155
SecA protein, 21

Seeds, plant, 112–113, 115
Selenocystein, 18
Selenopeptides, 18
Self-fertilization, 67, 69
Separation response, 217–218
Serratia marcescens, 26
Shigella flexneri, 20
Sickle-cell anemia, 236
Siderophores, 27
Signaling behavior, evolution of, 151
Signaling system, transmembrane, 47, 51–52
Signal transduction, 21
Simian immunodeficiency virus (SIV), 219, 221
Simian virus 40 (SV40) DNA, microinjection, 179
Site-directed mutagenesis, 188
Small bowel grafting, 167
Social behavior, hormones and, 153–154
Sodium channel
  Drosophila, 138–139
  fish, 137–139
Somatic recombination, 6
Song, in birds, 151–152, 155–156, 195
Sound production, cetacean, 194–196, 198
Soybean embryo development, 112–113
Speciation
  allopatric, theory of, 148
  avian, 148
  rates, 258
Species
  commonness and rarity, 251
  definition, 254
  number of living, 255–258
  number recorded, 252–255
  numbers, abundance, and body length, 250–251
  number versus physical size, 248
  relative abundance, 246–248
  total number on Earth, 245–259
  transgenic plant, 124
Speech acquisition, 152
Spemann-Mangold organizer, 57–58, 62
Sperm whales, 194–196, 199, 201–203
Splicing, retroviral, 10
Spontaneously hypertensive (SHR) strain, 172
Sporophytes, 106, 111–112
Sporozoites, malaria, 79–80, 83–84
src, 8, 10
Steroid hormones, 155–156, 210
Streptococcus pneumoniae, 17
Streptomyces, 22, 26
Sulfonylurea compounds, 125
Surgical therapies, in utero, 238

Swordtails (*Xiphophorus helleri*), 136, 140–142
Symbiosis, 22
Sym plasmids, 28

Tay-Sachs disease, 226, 236
Teleosts, 138
Teratogenesis, 169, 215
Territorial behavior, 149
Testis–determining factor gene, 227–228
Testosterone
    avian behavior and, 154–155
    primate research and, 214
TFIIIA, 63–64
Thalassemia, 182, 227
Thalidomide, 215, 239
Thermophiles, 31
Thiobacilli, 22
Thiobacillus ferrooxidans, 32
Ti plasmids, 117, 122–123
Tobacco, embryo development, 112–113
Tobacco mosaic virus
    resistance to, 126–127
    self-assembly in, 105
Toluene, bacterial degradation, 29
Tomato, genetically engineered, 122, 126–128
Toothed whales, 193
Torpedo californica, 137
Totipotency, 105, 110–111, 117
Toxic wastes, transformation, 29–31
Toxoplasma, 78, 80
Toxoplasma gondii, 77
Trans-acting factors, 182–183
Transcription factors, 19–20, 63
Transcription, retroviral, 10
Transferred DNA (T–DNA), 122–123, 128
Transferring gene, 182
Transfer RNA, 18
Transformation, plant, 121–124
Transforming growth factor β, 61
Transgenic animals, 179–189
    gene expression in, 181–183
Transgenic chickens, 180
Transgenic mice, 179–188
    mutations in, 186–188
Transgenic technology, applications, 183–188
Translation, retroviral, 10–11
Translocation breakpoints, 230
Transplacement, 41
Transplantation, 240
    studies with rats, 167–169
Transplantation antigen, 185
Transposon Tn5, 27

Transposon mutagenesis, 94–95
Transposons
    bacterial IS1, 11
    C. elegans, 70
    Drosophila, 95, 101n
    in plants, 117
    yeast Ty, 5–6, 11
    See also Retrotransposons
Transposon tagging, 117, 128
Trifoliin, 29
Trophoblast antigens, 169
Trypanosoma, 78
Trypanosoma brucei gambiense, 81
Trypanosoma cruzi, 78
Trypanosomes
    antigenic variation, 77–78, 81, 86
    gene expression in, 86
Trypanosomiasis, 81
TS13 stem mRNA, 108–109
Tubulin, 42, 52

Ubiquitin, 42
Ultrasonography, 236–238
Ungulates, 193, 196
Urethral obstruction, 238

Vaccines
    against parasitic infection, 82–86
    development, 219–220, 239, 241
    malaria, 83–85
    testing, 238–239
Vectors
    A. tumefaciens, 117, 121–124
    P elements as, 96
    prokaryotic, 181
    retroviruses as, 2, 13–14
    shuttle, 41
    yeast, 40–41
Vibrio, 26
Vibrio cholerae, 20
Vir (virulence) genes, 122
Viruses
    fish neoplasms and, 141
    herpes B, 221
    Marburg, 221
    oncogenic, 220
    type C, 220
    See also Retroviruses
Vision
    primate models, 211, 213
    rabbit model, 211
von Gierke's disease, 238
von Recklinghausen's disease, 184

Wastewater treatment, 32–33

Weed control, genetic engineering and, 125–126
Whey acidic protein (WAP) promoter, 184
Wilms' tumor, 165
Wilson's disease, 227
wingless, 9

Xanthomonas, 27
Xanthomonas campestris, 28
Xenobiotics, biodegradation, 29–31
Xenografts, 168–169

Xenopus laevis, 57–64
Xenopus oocytes, 98
XTC mesoderm-inducing activity, 60–61

Yeasts, research on, 39–44

Zebrafish (*Brachydanio rerio*), 136, 140–141
Zinc finger, 63–64
Zoogloea ramigera, 33
Zymomonas mobilis, 31